Solidarity Road

The story of a trade union in the ending of apartheid

Jan Theron

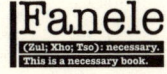

First published by Fanele, an imprint of
Jacana Media (Pty) Ltd, in 2016

10 Orange Street
Sunnyside
Auckland Park 2092
South Africa
+2711 628 3200
www.jacana.co.za

© Jan Theron

All rights reserved.

ISBN 978-1-928232-27-8

Design & layout by Shawn Paikin
Cover photo: National officebearers of FCWU and AFCWU, circa 1977.
   (Photographer unknown, courtesy of UCT Special Collections)
Endpapers photo: Press conference at the end of the Fatti's & Moni's strike.
   (Courtesy of Independent Media and Robben Island Mayibuye Archive)
Set in Sabon 10/13pt
Printed and bound by Creda Communications
Job no. 002830

See a complete list of Jacana titles at www.jacana.co.za

To Sandy

Contents

List of acronyms..ix
Acknowledgements..xi
Preface..xiii

Part 1: Beginnings
1. A birthday of sorts..3
2. Private road..13
3. Corporation Street..26

Part 2: 1976
4. Klein Drakenstein Road..35
5. The road to Mbekweni..43
6. The road to Ashton..53
7. West Coast road..58

Part 3: Reconstruction
8. Waterkant Street..69
9. Side streets..82
10. The N1 at Richmond..98
11. Durban road...108
12. Modderdam Road...128
13. The road from Namaacha..148
14. The road to the crèche..166
15. The Graaff-Reinet road..187

Part 4: Recognition

16. The road to Kidd's Beach ... 201
17. Malta Road ... 212
18. Off Bhunga Avenue ... 227
19. The road to the cemetery ... 239
20. Congella Road ... 259
21. The road to Pretoria West ... 274
22. Hospital way ... 289
23. Albert Road ... 303
24. The Transkei road ... 321

Part 5: From the nineties till now

25. The desolate market ... 339

Endnotes ... 367
Index ... 426

List of acronyms

AFCWU	African Food and Canning Workers Union
AZACTU	Azanian Confederation of Trade Unions
AZAPO	Azanian People's Organisation
BAAB	Bantu Affairs Administration Board
BMWU	Black Municipal Workers' Union BMWU
CAHAC	Cape Areas Housing Action Committee
CCAWUSA	Commercial, Catering and Allied Workers Union of South Africa
CCMA	Commission for Conciliation, Mediation and Arbitration
CIA	Central Intelligence Agency
COPE	Congress of the People
COSATU	Congress of South African Trade Unions
CTMWA	Cape Town Municipal Workers' Association
CUSA	Council of Unions of South Africa
DA	Democratic Alliance
Elfco	Elgin Fruit Packers
FAWU	Food and Allied Workers' Union
FCWU	Food and Canning Works Union
GAWU	General Allied Workers' Union
GDR	German Democratic Republic
GWU	General Workers' Union
HOCC	head office controlling committee
I&J	Irvin & Johnson
ICFTU	International Confederation of Free Trade Unions
ILO	International Labour Organisation
IMSSA	Independent Mediation Service of South Africa
IUF	International Union of Foodworkers
LRA	Labour Relations Act
MACWUSA	Motor Assembly and Component Workers' Union of South Africa
MAWU	Metal and Allied Workers' Union
MK	Umkhonto we Sizwe
MSF	Médecins Sans Frontières
NAAWU	National Automobile and Allied Workers' Union

NACTU	National Council of Trade Unions
NEDLAC	National Economic Development and Labour Council
NUM	National Union of Mineworkers
NUMAROSA	National Union of Motor Assembly and Rubber Workers of South Africa
NUMSA	National Union of Metal Workers of South Africa
NUSAS	National Union of South African Students
NUTW	National Union of Textile Workers
PEBCO	Port Elizabeth Black Civic Organisation
RAWU	Retail and Allied Workers' Union
RFF	Rhodes Fruit Farms
SAAWU	South African Allied Workers Union
SAB	SA Breweries
SABC	South African Broadcasting Corporation
SACCAWU	South African Commercial, Catering and Allied Workers Union
SACCOLA	South African Consultative Committee on Labour Affairs
SACOS	South African Council on Sport
SACP	South African Communist Party
SACTU	South African Congress of Trade Unions
SACTWU	Southern African Clothing and Textile Workers' Union
SAD	SA Dried Fruit
SAPCo	South African Preserving Company
SASO	South African Students' Organisation
SATAWU	SA Transport and Allied Workers' Union
SB	Special Branch
SFAWU	Sweet Food and Allied Workers Union
SRC	Students' Representative Council
SWAPO	South West African People's Organization
TRC	Truth and Reconciliation Commission
TUACC	Trade Union Advisory Coordinating Council
TUCSA	Trade Union Council of South Africa
TWIU	Textile Workers' Industrial Union
UDF	United Democratic Front
UK	United Kingdom
USSR	Soviet Union
UWO	United Women's Organisation
UWUSA	United Workers' Union of South Africa
WCTA	Western Cape Trader's Association
WFTU	World Federation of Trade Unions
WPGWU	Western Province General Workers Union
WPP	Western Province Preserving
ZANU-PF	Zimbabwe African National Union – Patriotic Front

Acknowledgements

I HAVE TRIED IN THIS BOOK to recognise the range of individuals within FCWU (including AFCWU and the Fruit and Vegetable Canning Workers Medical Benefit Fund) who played a role in its revival in the late 1970s and 1980s, as well as all those in other organisations and on the fringes of the trade union movement that played a role in its establishment. Inevitably, however, it has not been possible to recognise everyone. I have also tried to acknowledge the role of individuals from other trade unions in the formation of FAWU. Let me simply emphasise that this was a collective endeavour, in which there were many unsung heroes.

When I took long leave from FAWU to write a book, I received funding from the Canadian Labour Congress and the Church and Work Commission of the South African Catholic Bishops Conference. The book I had intended to write then was not this book. I am nevertheless deeply grateful for their support. It enabled me to formulate my ideas about the kind of book I wanted to write, when the time was ripe to do so. I am not sure I would have persisted with this project without it. I would also like to thank the Academic and Non-fiction Author's Association of South Africa (ANFASA) for awarding me a grant in 2009, in terms of their grant scheme for authors. This enabled me to take time off work to make a start.

I have also received encouragement in writing this book from more people than I am able to mention by name, but special mention must be made of those who read one or more drafts or sections of the book, and gave me their comments. These include (in alphabetical order) Debbie Budlender, Athalie Crawford, Colleen Crawford Cousins, Bill Freund, Rob Gaylard, Doug Hindson, Sandy Liebenberg, Jeff Lever, Erika Oosthuysen, Karen Press, Dennis Rubel and Shauna Westcott. I would also like to acknowledge the helpful comments received from two anonymous reviewers.

Special mention must also be made of the role played by Edward Webster and Karin Pampallis, who co-ordinated a review of the text, and have supported me in various ways in getting this book published. This might never have happened but for a generous grant towards the cost of publication from the Bertha Foundation. My thanks go to them, and to attorneys Chennells Albertyn for administering this grant.

A number of people helped me locate the images used in this book. Clive Kirkwood of Special Collections at UCT was unfailingly helpful, as was Graham Goddard of Robben Island Mayibuye Archive. Independent Media generously agreed to let me reproduce images from its titles. Paddy Donnelly, Chris Ledochowski, Gavin Younge and Mike Gavshon each let me reproduce their excellent photographs. I was particularly encouraged by the support given by the current general secretary of FAWU, Katishi Masemola, both in this regard and more generally in respect of the book.

Athalie Crawford has given her unwavering support for this project over many years, and I greatly appreciate the interest in it my eldest son Leif has always displayed. Lastly, I must thank Sandy and Sam most profoundly for their forbearance, in putting up with my working after hours and on weekends for more years than I care to mention.

Preface

THIS BOOK IS A HISTORY DEALING WITH EVENTS in which I was more often than not a participant. Most of these events are documented, in the minutes of the meetings of FCWU or FAWU, or in annual reports presented to the annual national conference, or in some instances circular letters sent out by the head office to the membership. Wherever it seemed necessary to do so I have referred to these documents. Interested readers may see for themselves what they say in the Food and Canning Workers Union Archive. This is kept by the Special Collections and Archives Department of the University of Cape Town Libraries (referred to herein as Special Collections, UCT). Unpublished documents that are not in this archive will be in the Jan Theron Papers, which are also at Special Collections, UCT.

The meetings in question were, in the case of FCWU, the management committee (MCM) which ordinarily took place monthly, the national executive committee (NEC) which ordinarily met twice a year, and the annual national conference. In 1985, preparatory to a merger between FCWU and other trade unions, the decision-making structures changed, with the introduction of a head office controlling committee (HOCC) and quarterly meetings of the national executive committee. During my tenure as general secretary and during the tenure of my predecessors, a lot of effort was put into ensuring that minutes presented as full an account of issues and debates as possible. This was made easier by the fact that proceedings were translated.

Reference is also made to minutes of meetings with employers and other trade unions, as well as other organisations of civil society, such as those that took place when the formation of the United Democratic Front was being discussed. These minutes have no official status insofar as they were not circulated to the other parties in question, or approved by them. They do however reflect the Union's version of what transpired or was agreed. Again, interested readers can see for themselves what these minutes contain in the abovementioned archives.

Since this is not a sanitised account, I am obliged to recount events which do not reflect well on certain individuals. However I have tried to do so without bad-mouthing anyone or settling personal scores. Where I have been critical of individuals and they are (or were) public figures, I have had no alternative but to name them. Where they are not public figures, I have preferred not to name them explicitly. Instead, I have either abbreviated the name, or used a nickname, or coined a name for the person. In the latter instance, I have marked such names with an asterisk to indicate it is not a real name.

The images in this book are from various sources, and include photographs taken by myself as well by others. Where the photographer is known he or she has been credited as such, but in some instances it has not been possible to identify the photographer. These include photographs commissioned by FCWU when I was general secretary or given to me, at a time when it was not considered necessary to keep a formal record of the donation of an image. These photographs have now been placed at Special Collections at UCT, and are credited as such in the book. In the event that the identity of any such photographer become known upon publication of the book, I will endeavour to ensure that such person is credited accordingly in any later edition of this book and in the records of Special Collections at UCT.

PART I
Beginnings

'It is an easy thing to rejoice in the tents of prosperity:
Thus I could sing & thus rejoice: but it is not so with me.'
– WILLIAM BLAKE, ENION'S COMPLAINT, IN *THE FOUR ZOAS*

1
A birthday of sorts

SOMETIME IN 2006 the newly elected general secretary of the Food and Allied Workers' Union (FAWU) phoned me to ask if I would give an interview for what he termed a commemorative DVD. It was to be the 65th anniversary of the trade union, he said. I immediately agreed. I was once the GS of this trade union (GS is how people in struggle circles generally refer to general secretaries). Ever since I resigned, my policy has been to talk to anyone in the Union who still wants to talk to me.

Even so, his request raised a question to which there was no easy answer. It was actually the anniversary of the founding of the Food and Canning Workers' Union (FCWU). In law, FAWU was the successor of FCWU, and I refer to them interchangeably as the Union in this book. But is it really the same organisation? This depends on how you understand a membership-based organisation like a trade union. People from my class do not have much understanding or experience of membership-based organisations. What I learned in the Union, I had to learn from scratch.

First, I had to learn about its particular model of organisation. What took longer for me to understand was how this model was animated by a particular tradition. This tradition defined many things that could not necessarily be read from its constitution: how leaders should conduct themselves, for example, in relation to the members, and what the relationship was between the organisation at a local level and centrally (or nationally). Above all else, it concerned the values that held the organisation together, such as solidarity.

Not long after the phone call, I got a letter from the GS, telling me more about the DVD. It would be 'the property of the working class' under the custodianship of the Union. It would give 'a historical account of the role that FAWU has played in the struggle for workers' rights', etc. In some ways this was speaking to the tradition I recognised. Solidarity in a trade union does not simply mean standing by your members, or by organised workers. It means solidarity with your class.

The working class was a defining concept for me when I became GS. At the time, in 1976, the working class was fragmented. Working for a trade union was part of a project to unite a fragmented class, and to give it a voice. Without being unduly grandiose about it, this was the historical project to which a number of people from a certain intellectual background were drawn. This would be our contribution to the struggle: what we did to end apartheid. It was a struggle for democracy, but democracy did not just mean everyone getting to vote every so often in national elections. People also had to eat.

The most obvious way in which the working class was then fragmented was in terms of race. The Union put its commitment to solidarity into practice by uniting workers of different races in factories manufacturing food. To do so it had to overcome divisions among workers created by the ways in which government had structured employment, in terms of the law, which the bosses were able to exploit. Nowadays 'bosses' seems like a dated term, yet this is the term workers used to refer to the people for whom they actually worked. It is also no less important today than it was then to differentiate between those who control the factories and mines and those who operate at their behest.

My spell in the Union was the formative experience of my adult life. Membership of the Union was also a formative experience for ordinary workers, especially those who joined after 1976. I know, because many have told me, even if not in so many words. It was their first experience of democracy in that, for a time, they had a leadership that was accountable to them and a real say in decisions affecting their everyday lives. Perhaps it was their only experience of democracy, apart from standing in queues to vote.

Someone who became my trusted informant about what was happening in the factories was John Pici, a worker who was about my age. John sought me out some five years after I had left the Union, to ask my advice. We came to an understanding: I would give him advice, on an ongoing basis, and he would tell me what was going on in his workplace. This would be his story, we both understood, but it was not just a story about himself. It was about what we both had in common once, as comrades in the Union, and had now lost. The understanding endured, and we became friends.

It is of course easier nowadays for South Africans to make friends across racial lines. However, friendship across class lines is a rare thing. John was, by any definition, working class. His father worked on the mines. After he died, John left his Transkei home, with a few years of primary school education, to join his brother in Cape Town. He found a job in the same bakery where his brother worked, and that was where he was still working in 2006. From my own observation, he was in some respects worse off than he had been when he first joined the Union some 20 years before. This was as a consequence of the way in which employment was being restructured, not only in the bakery, but elsewhere.

Law played its part in this restructuring, to be sure.[1] However, it was primarily because South Africa began its transition to democracy at the start of what can be called a neo-liberal era, during which certain policies held sway globally. These neo-liberal policies became increasingly influential within South Africa, following the failure of the first democratically elected government to develop an alternative to them, and of the trade unions to resist them.[2] As a consequence, the working class became once again fragmented: as fragmented as it ever had been, if not more so, although not along the same lines. This goes to show, you might argue, that the historical project to which I and others were committed had failed.

Could this have been avoided? When I left the Union, the PC was just beginning to become an indispensable item for those with education and means. No one foresaw the far-reaching changes in technology which its advent symbolised, or how it would facilitate the restructuring of employment. But John did not lose his job due to technology. He was told, and initially he believed, that he was being 'empowered', at a time when empowerment was becoming a code word for enriching well-connected individuals. This was a peculiar idea of 'empowerment' for a trade union to support. Nevertheless, the Union did so, along with other 'empowerment' schemes, as I will later explain. So did other trade unions, as well as the federation of trade unions which the Union had played no small part in bringing into being, the Congress of South African Trade Unions (COSATU).

The blame for failing to resist neo-liberal policies cannot be placed entirely on the trade unions' embrace of 'empowerment' schemes. However, it says a lot about their inability, already evident by 2006, to overcome the ways in which employment had been restructured. To understand how this came about, you need to go back to the start of the movement which eventually led to the formation of COSATU. This movement played no small part in bringing about the transition to democracy, even if it is no longer fashionable to say so, and the Union was there at its beginnings. But the Union then, as John would not tire of telling me, was 'totally, totally, totally different from what it is today', pausing for emphasis before each 'totally'. Knowing all this, I balked at the idea of writing a history that would be 'the property of the working class'. Which section of a fragmented working class would that be?

Old bones

I did not hear anything further about either the interview or the DVD. Instead, I received an invitation to the 65th anniversary celebration. I had been invited to the odd ceremonial occasion before, to shake hands, and receive some trinket which a union functionary thought 'veterans' would value. What I did not expect was a phone call, days before the event, to ask me to speak.

Contemporaries of mine in the trade union movement were jostling for positions in the new order even before the transition began with F.W. de Klerk's announcement of the unbanning of the African National Congress (ANC), the Pan Africanist Congress (PAC) and the South African Communist Party (SACP). The question was whether these one-time unionists would use the opportunities that were opening up for them for the benefit of the workers who had propelled them there, or for their own material benefit. The outlook was not promising. The clearest indication of this, for me, had been a decision of the Union to 'buy' a complex of buildings from the Premier Group, including the hostel where John was staying at the time.

Self-sufficiency was another important value that defined the tradition of the Union. It began with being financially self-sufficient. Members were schooled to be most particular about where their money came from, and how it was spent. They were also repeatedly warned to beware of the bosses bearing gifts. Premier, as we called the group of companies, played a leading role in persuading other bosses to embrace negotiations with the ANC. It also employed more Union members than any other business. The Premier complex of buildings was a gift, so far as I was concerned, even if I was not in a position to prove it. How had the Union been able to jettison its values so easily?

Events that took place while I was on long leave from the Union, and at the same time as the unbanning of the political organisations, went some way to explain how this came about. There were a series of coups in the Union, in which the national leadership of the Union – the president and other office-bearers – engineered the removal of elected leaders at a local level. What was most shocking for me was that workers had been expelled for challenging the national leadership for having breached the constitution. The expulsion of workers was unprecedented. It was also fundamentally at odds with the values of democracy, as practised in the Union.

Somewhat vainly, my first thought was that the coups were calculated to oust my supporters. Although that may have come into it, I now think they were calculated to oust leaders that were seen as being too independent, at a time when negotiations were taking place clandestinely between government proxies and the ANC.[3] Neither party could allow cracks to show in what they regarded as their constituency. COSATU could also not allow cracks to show in the parallel negotiations then taking place with the bosses.[4] The Premier complex must have been one sweetener among many for deals like this.

So at the very point at which the country was moving to formal democracy, democracy was being dismantled at a local level, in the Union. At this point the option of returning to the Union was still open to me. The powers that be would surely have let me hang around in some sinecure, so long as I was prepared to keep my head down. They did not want to alienate those who knew me, like John, because I had visited them at their factories, and stood by them in their struggles. This was another aspect of the tradition of organisation that the Union represented: paid officials were trusted, and accountable.

At some point, perhaps, I would have been 'kicked upstairs', into Parliament. I surely had as good a claim to be there as some of those whom COSATU sent to Parliament in 1994. Instead, I decided to blow the whistle on what was happening in the Union. I wrote a lengthy article for the *South African Labour Bulletin*, which had once served as a forum for debating labour issues.[5] It was the response of an intellectual. It also represented a point of no return for me.

I did not for a moment think a 65th birthday celebration would be the appropriate occasion to dig up these old bones. Apart from anything else, I did not expect workers I knew to be there. Many had lost their jobs due to the reorganisation of production or the restructuring of the workforce. Others had voted with their feet, by resigning or by breaking away to form new trade unions. Some of these break-away unions were formed in direct response to the coups; others were formed afterwards.[6] I never endorsed a break-away, even though the workers were doing what I had always maintained was their right: to vote with their feet. The Union was still my organisation. Or was it? This was the root of my ambivalence about the occasion to which I was invited. The feeling was clearly mutual. The invitation for me to speak was clearly an after-thought.

Vuyisile Mini
I was still reflecting on what to say when I took the turn-off from the N2 to Gugulethu, on my way to the anniversary celebration. The road takes you past what used to be single-sex hostels, where migrant workers used to stay. Former migrants still stay there, but in the time since I used to be a regular visitor to the townships, shacks and informal businesses had sprung up around and between them.

My destination was the complex of buildings that the Premier Group had donated, where the Union's head office was now located. Perhaps it was apt that the celebration was taking place there: 'the emporium', as dissidents within the Union had labelled it. The official name was the Vuyisile Mini Centre, after an official of the South African Congress of Trade Unions (SACTU) who had joined the armed struggle and been executed.[7]

This particular Saturday afternoon the entire parking area in front of the main building was occupied by a huge white marquee. A clutch of security guards all wearing special commemorative T-shirts steered me in the direction of a metal detector of the kind you go through at airports. The

President would be in attendance, someone told me. Had it not been for all this security, I would have supposed the guards were talking about the president of the Union, or perhaps of COSATU.

The commemorative T-shirts bore the faces of Elizabeth Mafikeng, Liz Abrahams, Oscar Mpetha and Ray Alexander. Mafikeng was the only one whom I did not know in person. She was a charismatic leader in Paarl in the 1950s, but had by all accounts been broken by her years in exile. Liz Abrahams was the only one of the four that was still alive. I found her sitting unobtrusively on a couch in the reception area, with Lizzie Phike beside her.

'The two Lizzes', as we used to refer to them, had both been workers before becoming paid officials of the Union: the one coloured, the other African. You still have to use to these terms to understand post-apartheid identity politics, and perhaps also the rivalry between them.[8] Since leaving the Union, Liz had become a member of Parliament, and Lizzie a member of the provincial legislature. The third person on the couch was known as Aunt Lossie; she had never made it beyond Sea Harvest, the factory where she had worked.

The only person I encountered that afternoon who was still a factory worker was Lizzie Phike's faithful friend Luska, who was another of my points of reference. Luska started at H Jones and Co. in 1967, my final year of high school. She was still there nearly forty years later. There was a rally earlier that day for the Paarl workers. Evidently we are at a more select gathering, to which ordinary workers were not invited. The rally, Luska told me, was a flop.

People started drifting into the marquee, and I was asked to take my place on the platform. On the left of the podium was a table with seven chairs behind it. The fourth chair, several people kept on saying, must be left open for the President, who was to arrive later. I angled for the sixth chair, wondering who would be joining me. Liz Abrahams was one, although she seemed reluctant to come forward. Although we had drifted apart in the years prior to her retirement from the Union, we were still comrades.

The last time I visited her at her Paarl home was after the term of office of the first democratically elected Parliament had expired. Not much had changed, apart from the memorabilia on the walls. I wondered what she made of the emporium – or of the brash master of ceremonies who was conducting proceedings. He had clearly been hired for the event, and was working from a script. This would also have been unthinkable in our day.

Jeremy Cronin, representing the SACP, was another guest of honour. 'The Party' is how it is commonly referred to, even by those who do not believe it has a claim to be *the* party. Ebrahim Rasool, the Premier of the Western Cape, arrived at about the same time, followed closely after by

Zwelinzima Vavi of COSATU. 'Comrade Jan, how are you?' Rasool said, as we embraced. I had a recollection of our being in the same meetings in the 1980s, after the formation of the United Democratic Front (UDF), but do not recall our ever exchanging a personal word. He would likely have regarded me with some mistrust at the time.

Vavi moved along the table, embracing each of us. 'I know you,' he said pointedly, when I introduced myself. He must also have been at meetings I had attended, perhaps during the 1987 mine workers' strike, when the country was on the brink of a conflagration. He played no part in the protracted process that led to formation of COSATU, however. I would describe him as belonging to a second generation of trade union leadership.

It was at this point that Chris Dlamini materialised, and greeted me warmly. At a personal level there had been no animosity between us. I was nevertheless somewhat surprised that he took a seat next to me. Ever since FCWU had become FAWU and he became its president, we had been at loggerheads on a number of issues, which I will later discuss. I cannot do so without referring to Chris by name, because more than any other worker leader he had been the public face of COSATU after it was formed. We had not spoken since the coups.

Rasool was the first to speak, once the formal proceedings commenced. His role was to welcome those from other provinces to the Western Cape, at what he described as 'this very difficult time'. He was alluding to the local government elections that had taken place a few weeks before. Although no one party emerged with a clear majority, the ANC had just lost control of the City of Cape Town to a coalition led by the Democratic Alliance (DA).

FCWU once played a crucial role in overcoming divisions 'between the different communities in the province', Rasool said, meaning the coloured and African communities. Some attributed the ascendance of the DA to blunders by the Africanists who were in charge of the ANC in the province, which had alienated coloured voters. I would say the rot had set in far earlier. As a matter of fact, the 'very difficult time' was about to get worse. The Union had indeed played a crucial role in overcoming racial divisions, but Rasool had only to look at the make-up of the gathering before him to realise the Union was no longer playing this role. Apart from Aunt Lossie, who had retired, I did not see a single coloured worker present.[9]

'Long live the 65-year-old special relationship between the South African Communist Party and the Food and Allied Workers' Union' was Cronin's opening line, with the customary clenched fist in the air. I was curious to learn about this special relationship. Instead, he told us about Johnny Gomas being expelled from the ANC for not toeing the line.[10] This went to show that 'the ANC was not always what it since became'. The point of this digression, I think, was that President Mbeki had recently rebuked Cronin publicly. The subtext was that the ANC under Mbeki was not what it had been when it ushered in the transformation to democracy.

As he skipped over the Union in the 1940s or 1950s, the glory years, I was curious to see how Cronin would explain the evolution of this special relationship over the next few decades. The 1970s was a low point, he said, but what was remarkable was that the Union survived. He did not explain why it survived, however, and his account petered out at about the time I left the Union. So there I was, listening to a story in which I had played a central role, told by someone who had no part in it. As I did so, any space I might have had to say what had happened was being closed. This was the Party being the custodian of the history of the Union.

The fact of the matter was that there was no special relationship with the Party in the late 1970s and 1980s. Although I had never been hostile to the Party, I was also fiercely protective of the autonomy of the Union. This was enough for people like me to be labelled as 'workerists'. As I listened to Cronin speak, I formulated the historical question I would have liked to pose: if everything of consequence in the Union happened in the late 1970s and 1980s, and nothing worth mentioning had happened since, what had been the value of this 'special relationship'? I doubted whether the occasion would ever arise.

Cronin was still on his feet when President Mbeki arrived, with Sydney Mufamadi in tow. The crowd broke into a struggle song, interspersed with ululations, and Cronin wound up his speech with an uncomfortable attempt at humour: the kind of joke a teacher might make when the headmaster walks in, to find out why his class is so rowdy. Vavi's speech was laced with rhetoric about the unflagging commitment to the workers' struggle, for which some speech writer was presumably accountable. It also contained an historical inaccuracy: the suggestion that the Union had had a hand in bringing about an alliance between workers and students engaged in the Soweto uprising. He lost my attention, as I focused instead on what Mbeki was doing with his hands beneath the table. Mbeki was dapperly dressed in a pin-striped shirt and tie. Beneath the table, he was fidgeting furiously with his fingers.

Relations had become increasingly strained between COSATU and the ANC under Mbeki. Perhaps Mbeki had brought this on himself. 'Call me Thatcherite,' he had joked in very bad taste. As pilots of the neo-liberal ship, she and her ally President Reagan gave the bosses a freer hand in all sorts of ways. These included going to some lengths to sink organisations based on values of solidarity, and especially trade unions. Mbeki was blamed for letting the ship into South African waters, although it might be said that it was his predecessor who did so.

The most recent tension between COSATU and Mbeki was over his sacking of Jacob Zuma as Deputy President, because of the charges of corruption Zuma was facing. Firing Zuma, I thought, was one of the best things Mbeki had done. It sent out a message that leadership implicated in corruption could not hold public office. That was also the position taken in a Leadership Code which the Union had adopted. 'Where someone in a leadership position is involved in dishonesty or corruption,' it stated, 'he/she is not fit to be a leader.'[11] But Vavi was a Zuma acolyte at the time. COSATU took the line that Zuma should be regarded as 'innocent until proven guilty'.

'I want to say *baas* when I see this name,' the master of ceremonies said, after calling me to the podium once Vavi had finished speaking.[12] I do, of course, have a very Afrikaans name. While I was coming forward, still wondering what to say, he went on to call Chris to the podium and, after him, Alan Roberts. So this was it, it dawned on me: the façade of unity. A ceremonial presentation of what someone perceived to be the leaders of the three trade unions that had formed FAWU, although there were in fact others, including the union of which Sydney Mufamadi was then GS. Comrade Syd, as he was known, was now Minister of Local Government.

'Archbishop Tutu would have loved this scene,' the master of ceremonies went on, drawing from an apparently inexhaustible store of bad-taste jokes. In this instance, he had a point. In terms of apartheid racial classification, Roberts was coloured, although in most other countries we would have been regarded as indistinguishable, racially speaking. The non-racial solidarity the Union achieved in the late 1970s and 1980s had nothing to do with the cheap spectacle of the 'rainbow nation'. Organising across racial lines was about creating a common understanding among workers that race was not the root of their problems.

It was not clear if we were now expected to say anything, and Chris was inclined not to. I urged him to speak, but he broke into a freedom song in isiZulu instead. The audience picked up the refrain. Despite feeling upstaged – leading a crowd in a freedom song is not something an *umlungu* should attempt – I resolved not to be silenced.[13] There was something that needed to be commemorated, even if it was not possible to do so within the sliver of opportunity that was presented to me. So I said something about rebuilding the tradition of Union, despite the low point at which I found it. I meant to imply by this that even now it was possible to rebuild what there once had been.

It was now the President's turn. 'I do not know which of the three, no four, drafts of speeches before me I should use,' he said, shuffling ostentatiously through the papers in his hands, 'so maybe I will not use any of them, and speak from the heart instead.' There were murmurs of approval. He was addressing the criticism that he was remote from the people and too cerebral. But it soon became evident that this most cerebral of presidents had little insight into what trade unions had been doing while he was in exile. The coup that would remove him from presidential office was only a couple of years away. Comrade Syd, who should have had more insight into these things, would go with him.

I did not know then, and do not know now, what, if anything, of the tradition of the Union remained. The way in which any tradition is kept alive, however, is through stories. The untold story of the Union is about the women who kept its tradition alive, and who are more easily forgotten than the men. It is also about the role unremarkable people played in rebuilding an organisation into something they, at least, saw as remarkable at the time, as I did. Above all, it is about what later went wrong, and why: not just with the Union itself, but with the trade union movement of which it was part.

There was also no one better placed to tell this story than I, and I would like to have done so sooner. But I could not while I was too close to events. It was also not as clear where this neo-liberal ship would land us. Soon, however, it would begin listing. The Union may never again be what it once was. Trade unions may never ever again assume as important a role as they did in ending apartheid. Yet it is equally clear that we will not be able to launch the lifeboats without workers being organised or without organisations committed to solidarity. So while Mbeki, Rasool, Vavi, Cronin and the others applied themselves to cutting an enormous birthday cake – I was not invited to join them – I slipped away, with fresh resolve to write this book.

2
Private road

SINCE THE STORY I HAVE TO TELL is also my story, I need to explain about the tradition I came from, and how it brought me to the Union. It seemed to represent a radical break with all that my parents stood for, at the time: in some ways it was, but as I have got older I have become more interested in the ways in which it wasn't.

I grew up in a well-to-do suburb of Cape Town, the third of four children, living with both parents in a gracious, free-standing Victorian house at the bottom of a gravel road with no name. 'Private road', a sign said, with a handful of the names of residences listed underneath. Another sign said 'Trespassers will be prosecuted', although no one ever was.

My earliest conception of South Africa was of a country that people were always leaving, to make a life elsewhere. 'Leaving for good' we termed it as children, when families we knew emigrated. This could not be because of their material circumstances. I learned from adult conversations that it had to do with the political situation.

I was not yet ten when Robert Sobukwe launched the PAC's anti-pass campaign. Then there was Sharpeville. The images of people burning their pass books, and of the rows of coffins, made their way from the *Cape Times* to *Life* magazine. Soon afterwards Philip Kgosana led a march of 30,000 people into the city. This must have been when our teachers at preparatory school told us to walk home in groups. 'If any of you see strangers approaching, on your way home, especially a group of strangers, you are to go straight to the nearest house and ask to be taken in.' Race would not have been mentioned. Ours was essentially a universe that belonged to people who looked like us, in which there was no need to draw attention to our whiteness; it was something you took for granted.

We knew people who were not white, of course. These were people employed to provide people like us with services: domestic services, municipal services and the like. We valued the services, so we were bound to like and trust them. To this extent they had ceased to be 'other' than ourselves. On the other hand, as schoolchildren, we did not need to be told what kind of strangers to avoid when there was danger. These would be dark strangers. I do not think our outlook in this respect was much different from that which prevailed in the countries people were emigrating to. However, in the park on the way home from school were benches with wooden slats, with 'Europeans only' stencilled across the topmost slat. Evidently our whiteness could not be taken for granted, after all. 'It was all so unnecessary', my mother would say of the 'Europeans only' signs. It caused unnecessary offence. It was she who introduced us to the notion of 'prejudice', and instilled in me the belief that as a family we were free of racial prejudice. Her family was descended from Scottish missionaries who came to Africa believing all were equal before God, and in a society in which merit was rewarded, regardless of race.[1]

The Therons, on the other hand, traced their family line back to the French Huguenots. It was because 'we' had been in South Africa so long, and had nowhere else to go, that 'we' were staying put in South Africa, come what may. That, at least, was what my father always said, and my mother never contradicted him. So, from a young age it was clear that this situation was to be our political legacy, and somehow our responsibility.

'Separate but equal' was the justification for apartheid, which Dr Verwoerd – leader of the National Party, or 'Nats' as we called them – preferred to describe as 'separate development', but a child could see there was no equality in separation. Along the N1, from the back window of my father's Studebaker sedan, were the corrugated iron shacks where 'the natives' lived, standing in pools of water after the winter rains. Someone with a dark sense of humour had named the place Windermere, after the town of that name in the English Lake District. Along Prince George's Drive, the road to the beach, were the sand-blasted housing estates for coloureds.

The squalor and uncertainty and the fear were all the Nats' doing, I thought, and it was because of the Nats that people were leaving. This was of course in large part a reflection of the views of my parents. The great political divide of my parents' generation had been over World War II, and both had joined the armed forces. Many Nats supported the other side, actively so in the case of those like B.J. Vorster, who was Minister of Justice in Verwoerd's cabinet.

Yet what did it say about our relationship to the people of South Africa that the electorate voted for the Nationalists in increasing numbers? For some time I believed that the Nats were not truly representative, not because the electorate was white, but because the constituencies in which people voted were weighted in favour of the *platteland*, where their support was strongest.

The turning point for me was the referendum on whether South Africa should become a Republic. I remember my dismay as I lay next to the radio shortly after my tenth birthday, listening to the results, as the 'no' votes in the urban constituencies were gradually overhauled by the rural 'yes' votes. What kind of people were these, I wondered, a few months later, when jubilant crowds welcomed Verwoerd on his return from London, having just announced South Africa's withdrawal from the Commonwealth. Not long afterwards I was playing with my siblings in the back garden when my mother told us Verwoerd had been shot. Thinking he was dead, and that all he personified would die with him, we cheered with delight. My mother reprimanded us, and, as it turned out, he survived.

Class and prejudice
My mother was concerned about causing unnecessary offence to people of colour, I think, because it offended her notion of class. There was no need to tell those who did not have the money to do so that they should not get into a first-class carriage, any more than that they should not shop at the department store Stuttafords. It was the Nats' folly to seek to deny to people of colour who had clawed their way into the middle classes the benefits of their status. Even then, in the circles she moved in, the received wisdom was that the stability of the political order depended on building an African and coloured middle class.

'There are three classes,' my mother once told me, on her way to the municipal wash-house, probably by way of explanation as to why someone I wanted to play with was not suitable: the upper classes, the middle classes and the lower classes. The middle classes could in turn be divided into the lower and upper middle classes. 'We belong to the lower middle class,' she pronounced, with a slightly regretful tone. For a long time this puzzled me. We were not ostentatiously rich, but I did not understand how she could relegate us to the lower middle classes. The reason, I later realised, was that she had married below her class. Or her father thought so.[2]

I have it from my mother that her father never forgave her for marrying an Afrikaner. One of the unresolved questions of my youth was how to explain this kind of prejudice. Evidently it was class prejudice. In England, class had been a rallying call for the owners of inherited property to safeguard their privileges against the lower classes. In the Eastern Cape countryside, where her father owned a prime farm, the Afrikaners were *bywoners*: tenants living on farms, or, in other words, whites without land, from the lower classes.[3]

The irony was that my father's father, Oupa as we knew him, was a Sap (so called after the South African Party, the party of General Smuts), which stood for reconciliation between Afrikaners and English-speaking whites. That is why my father went to an English-language university, and was comfortable about us growing up in an English-language environment. Occasionally he was seized with the idea of getting his children to speak Afrikaans at home, but persisted with as much vigour as his project to acquaint us with the Bible. We did not get beyond Genesis.

This was not so much because my father had no affection for Afrikaans, as because of his distaste at the Nats' endeavour to impose the language. This was an important part of their nationalist political project, along with resisting South Africa's incorporation into the Anglophone world. Consequently it was *suiwer* Afrikaans you were expected to be able to speak, untainted by English. I came to think of it as Pretoria Afrikaans, as distinct from the expressive language spoken by coloured people, who couldn't care less about purity, nowadays referred to as *Kaaps*.

I was always at a loss as to how to address white officialdom: if in English, then risk being rebuked for not speaking Afrikaans, or if in Afrikaans, risk derision for my poor pronunciation of *U*, the formal mode of address in that language. What would make matters worse was my surname. 'What kind of Therons were we', Uncle Johan would growl at us, in Afrikaans, out of sight of the other adults, 'that we could not speak a word of Afrikaans?' I began, like my mother, to develop a block about the language.

Johan was my father's elder brother, but had become a Nat, it was said, from spending too long in the Transvaal. One Christmas Day, after the mince pies and tea, Uncle Johan got into a political argument with Oupa. When he called Sir De Villiers Graaff, leader of the United Party (the successor of the SAP), a cabbage, my father intervened: 'That is no way to speak to Oupa', he said, in a clipped, almost light-hearted tone, which suggested supreme self-control at the point at which you knew he was about to lose it. He then took a swipe at Johan, breaking his jaw.

My father called his children together the next day to apologise for his conduct, but secretly I was impressed. It had been a blow against the Nats and their political project. It also went to show that political values trumped family values. Unsurprisingly, this was the last of these extended family Christmases. What was not as clear, however, was what my father believed was the alternative to the Nats, or what his position was on the all-important question of the franchise.

My views in this regard were shaped more by my mild-mannered mother. When a group of members of Parliament broke away from the United Party to form the Progressive Party, she joined. The Progs, as they were known, stood for a qualified franchise: educated Africans, and those with property, should be able to vote. More significantly, she joined the Black Sash. As the Nats increasingly consolidated their hold on the white electorate, my opinionated father became less and less politically vocal.

The rule of law
I knew the vote was an all-important issue since I had grown up in the shadow of a constitutional crisis. When the Nats had come to power, some suitably qualified coloureds and some Africans in the Cape had the

vote. There were provisions in the Union of South Africa constitution, the 'entrenched clauses', that were intended to safeguard this right. On my father's telling, the Nats' gerrymandering to get around these provisions so outraged him that he gave up smoking in protest.[4]

My father was an advocate at that time, and what was at issue for him, I think, was the rule of law. This clearly required the courts to uphold the constitution, in both letter and spirit. In doing so, the courts were preserving a political space within which solutions could be found to the problems facing the country, without disturbing the existing order. It was not their role to question the justice of an order in which the majority of people were denied the vote.

It had been determined at a fairly early stage in my upbringing that I would follow in my father's footsteps, and become a lawyer. 'You argue like a sea lawyer,' my father used to tell me, whenever I took issue with him. I did not know what a sea lawyer was, but a dictionary definition is 'an argumentative sailor who questions orders'. This questioning of orders must have related to the rules of our home, 'the law' of our household, which we were expected to internalise.

The ground rule was that the children were treated alike, and the best way to get what you desired was to argue that you were not being treated in a like manner. On the other hand, it was clearly not possible for everyone to get all that they desired, or all sense of order would be corroded. There was a fine line between questioning rules and defiance of authority, as my sister, the eldest, found out to her cost. Like any other system of law, it was underpinned by the capacity of those in authority to enforce it and, at a push, by terror.

Outwardly my father was the most controlled of individuals, scrupulously polite to everyone, even charming. But at home his barely controlled rage at things that disturbed his personal sense of order was terrifying. It could be a pile of leaves in the driveway that had been disturbed, or cupboards against walls (because they left a mark). Perhaps I was only able to adopt a questioning role because my position in the family made it safer to do so. As the third of four children, I was not in the front line.

The promise of the law was justice, and justice necessitated questioning whether like was truly being treated alike. My hypothesis about these mad rages was that they expressed a raging personal conflict between a need to maintain order and a passion for justice. It was his passion for justice that impelled him to push boundaries, and to challenge those in authority whenever he thought power was being abused, with no regard for the embarrassment he was causing us as children. It became much more difficult to contain this conflict once he became a judge.[5]

'We always considered it our duty to accept appointment,' my father would say about being a judge. It was not clear to whom this duty was owed, but it weighed heavily on him. There was the death penalty, to which he was opposed. Then there were the increasingly draconian security laws: first, it was 60 days for which persons could be detained. Then it was 90 days. My father's 'we' included professionals like himself who had earned a position in authority on merit. It excluded 'outsiders', of which political appointments by the Nats were a prime example.

This was another aspect of the Nats' political project that infuriated my father. They operated more like a political movement than a party, in disregard of the tradition of an independent civil service that South Africa had inherited from the British. However, they did not break entirely from the British tradition of appointing judges from among the senior advocates. This, in my view, was because the advocates regarded as the most senior were those that made the most money. Since the most money to be made (then and now) was in working for those who had the most money, they could be trusted not to bite the hand that fed them.

'Bring back the rule of law', the Black Sash placards said, when Vorster, as Minister of Justice, piloted the introduction of what became popularly known as the Terrorism Act. The cartoonist Dawid Marais began portraying Vorster in SS uniform and jackboots; but Vorster, who had himself been an advocate, must surely have understood that the most effective counter to the charge that the government had abandoned the rule of law was that there were judges like my father who were known not to be Nats.[6]

The government certainly knew this when they appointed my father as the head of a delegation of three judges to investigate conditions on Robben Island. This was his opportunity to give expression to his other self, his passion for justice. We knew the story of his confrontation with the head of the prison from his retelling of it, on many occasions. There is a recognisable version in Mandela's autobiography, except that I remember hearing, while sitting around the dining room table, that it was my father that told off the commanding officer for threatening Mandela in his and his fellow judges' presence, not Mandela himself.[7]

Initiation

The day Verwoerd was assassinated was my seventeenth birthday. I lay sick in bed, listening to the endless dirges on the radio, and feeling profoundly depressed at the prospect of becoming an adult in South Africa. The next day my worst fears were realised: 'our authority remains unassailable', the set faces of the white men clustered on the steps of Parliament seemed to say. These were people who would brook no challenge to their authority, and Vorster was their new leader.

I was hoping against hope I would not be called up to do my military service, but some months before I was due to write my matriculation examinations, I received a postcard saying I had been drafted. The one option open to me was to defer military service until after I had completed my university education. Ultimately I did not have to do so, because of a lucky accident. It cost me a loss of sensation in three fingers of my right hand, notably my trigger finger, but got me exempted on medical grounds.

There was no war at the time that I knew of, but there were rumours of war. During my first year at the University of Cape Town I took my first step toward choosing sides in what would become an undeclared war. I joined a march from the mountainside on which the campus was perched down to the administration building, in protest at the university's withdrawal of the appointment of an African lecturer, Archie Mafeje. As with other public buildings at that time, there were no security guards to bar access, and no turnstiles or gates. The administration did not even take the elementary precaution of locking the doors before we got there. This was the start of the sit-in of 1968. We stayed in the corridors and Council chamber of the administration for a week, and at first it seemed as though I was outside all the institutions that had structured my life thus far. This, of course, was an illusion. The university was supposed to be autonomous, and 'governed' by a Council representing its various 'stakeholders'.[8] It was also supposed to be 'open' but was not. The student body was almost exclusively white, as were its staff. Institutional autonomy is always relative, and the Council had jumped when government let it be known it would revoke Mafeje's appointment if it did not do so itself. 'Let the government do its own dirty work', was our demand.

I remember the debates about what we hoped to achieve, once it became clear that the Council would not accede to our demand. I remember even more clearly sitting in the corridors, peering at my feet to avoid being seen by Aunt Catherine, a friend of my parents, as she picked her way between the mattresses and ashtrays on the way to her office, with a look of distaste on her face. I was the child of an establishment which had a base in this institution and others like it. It was an English-speaking establishment, which subscribed to liberal values, such as making appointments on merit and regardless of race; but it was also not going to compromise its base by needlessly provoking the government of the day. Or disturb an order of which it was part.

I felt even more alone on learning that the leader of the Council's delegation in negotiations with student representatives was none other than Uncle Marius, my godfather. I knew him as an urbane and witty family friend. Now I had to listen to a tearful president of the Students' Representative Council telling us how the judge had ridiculed him. This was also my first experience of struggle. As in any struggle, there are *bittereinders*, for whom the struggle has become an end in itself. The test of leadership, I could see, was to know when it must be ended. An ultimatum was conveyed, I can no longer remember how: either we get out of the building the next day, or the Stellenbosch students, who had come to demonstrate their hostility to the sit-in, would remove us by force. This might have been an idle threat. On the other hand, their university was a bastion of the Afrikaner establishment, and they beat us at rugby.

Although we told ourselves we were not going to be cowed by threat of force, it was noticeable how few were present as darkness began to fall that Friday. When there was still no sight or sound of Stellenbosch students, our spirits started lifting. Then we heard what can best be described as baying, and through the bay windows of the Council chamber we looked down at a crowd of more than a thousand massing in front of the building, many of them doubtless drunk. 'Get away from the windows, it is only provoking them,' someone said. We were not going to resist if they got in, was the line; but when a window shattered, I decided not to be a lamb for the slaughter and went with others to investigate. If there was going to be any break-in, it would surely be at the back entrance of the building, out of sight of any passing police patrol. Not that we would have called for police protection.

Out of the rhododendrons at the back of the building, a phalanx of police emerged with dogs. I could see an officer out of the corner of my eye orchestrating the manoeuvre. It was magnificently executed. Thanks to the police, the *betogers* were saved in the nick of time from the wrath of their fellow students.[9] The irony was not lost on me. Not long afterwards, I was approached to stand for the Students' Representative Council, and elected the following year.

This was my first exposure to people who delight in the pursuit of power, and the *schlenters* they resort to: underhand manoeuvres to secure positions, which had more to do with gratifying their egos than with those they were supposed to represent. I decided I was not a pure politician, and had no interest in becoming one.[10] Also, our position as a student leadership was a false one. We were trying to take those whom we ostensibly represented in a direction the majority did not want to go, as if we represented a vanguard. But our freedom to do so, as the police had so eloquently demonstrated, was consequent on our being the beneficiaries of white privilege.

Black students were to make a similar point. My first encounter with black student leaders was through the National Union of South African Students (NUSAS). It was government policy that 'non-white' students attend a university or college for their own racial group, except if there was a course these universities or colleges could not offer, and the South African Students' Organisation (SASO) had recently been established to bring together African, coloured and Indian students into one organisation.

Turfloop, as it was then known, was the only black university still affiliated to NUSAS. Its representative was one Abram Tiro. If NUSAS were to lose its last black affiliate, its aspiration to be a non-racial organisation would be doomed.

I remember the hand-wringing and anguish this prospect provoked among NUSAS congress delegates, as well as my irritation at the fawning attitude some displayed toward Tiro. It was an attitude I was to encounter often enough in later years, among whites desperate to prove they were not racist: so much so that they jettisoned their own self-respect and what was of value in their own tradition. The SASO leadership, by contrast, seemed self-assured and clear. They did not want our help. Our problem, of being politically relevant, was not their problem. A group of us met with them at the Phoenix settlement, outside Durban, one evening. Steve Biko was the person they all deferred to, as we stood outside in the half-light, because the building was probably bugged, and chatted briefly. Someone suggested that we should focus on our own. That, at any event, was the line taken by SASO towards white students.[11]

I met Biko on another occasion, when he visited Cape Town, with Barney Pityana. I was at this point the very young president of the SRC, and must have seemed politically naïve and impressionable, as indeed I was. When Pityana asked me 'What do you think of Lenin's theory of revolution?' he might have been putting me down, in the kindest way. Even if there was an element of posturing in his talk of revolution, Pityana had put his finger on the issue: while white students like me might say they wanted political change, black students were not interested in the kind of incremental reforms we had in mind. Reform and revolution were then seen as mutually exclusive projects, and 'real change' could never come about by operating within the law. Not long afterwards I jettisoned the idea of practising law, although I would complete my degree. A couple of years after doing so, Tiro was blown up in Botswana.[12] Biko would be murdered some three years after that.

As long as the situation in South Africa was conceived of in black and white terms, whites remained politically irrelevant. If it was conceived of in terms of class, rather than colour, other possibilities opened up. Class was not merely a social hierarchy, as I think my mother conceived it. It was the product of the prevailing economic system, capitalism, and your perspectives were determined by your position in this system. Material conditions determined consciousness. Those who embraced capitalism, like the Progs, argued that economic growth would bring about an eventual withering of apartheid. Yet there had been an economic boom throughout the 1960s and into the 1970s, and apartheid was more entrenched than ever.

At about this time, in circles of which I was part, a variety of documents began to be circulated among us: copies of published papers, extracts from dissertations, reading lists, roneoed documents, barely legible copies of handwritten notes. All went to show that some of the worst aspects of apartheid, such as the migrant labour system and pass laws, originated long before the term was coined, as a means of controlling African workers and to facilitate the development of capitalist production. 'Real change' was therefore scarcely conceivable unless African workers, and the class they belonged to, were organised. The idea of an alliance between students and workers had been popularised in May 1968, when the students in Paris called on the workers to join them. The question was whether such an alliance was possible in South Africa.

'Most black South Africans are workers', begins an undated document someone passed on to me. It expresses the kind of assumptions I and others subscribed to at the time. 'We believe, therefore, that to understand the problems facing black South Africans we must begin with the labour situation. It is the situation in which there is the greatest potential for forging new organisations through which blacks can reclaim their human dignity.'[13] It was not clear what kind of 'new organisations' these should be, but 'workers' referred to those employed in production. They were part of the working class, comprising people with no other means of subsistence than to sell their labour. Unlike elsewhere in Africa, traditional farming was no longer a factor in South Africa, and there was no peasantry to speak of.

For the most part, the workers we had in mind were also men. The situation of the women that stayed at home, wherever home was, was not our concern. We were also not concerned with those engaged in subsistence activities. They were in any event not part of the working class, most would have said, even if they depended on the wages earned by the men. Ours was a quite narrow idea of class. It was also not informed by experience. The nearest anyone I knew had got to actually linking up with real workers was the Wages Commission, established under the auspices of NUSAS to investigate workers' wages. No one I knew was involved in the organisation of workers, and the working class, when I left for Europe. Oupa had left each of his grandchildren some money and I had decided to 'blow' mine by going to Europe. It seemed necessary to say I was not leaving for good.

London, 1975

What I was intending to do, on my return, was to work with or for a workers' organisation. Soon after I arrived in London, where I was based for the best part of two years, I heard about the 1973 strikes in Durban. It confirmed the soundness of my plan, but had no influence on it or my views about worker organisation, until my visit to Durban years later as the guest of the South African Allied Workers Union (SAAWU), as I will relate in a later chapter.

My spell in Europe was a way of coming to terms with what working for 'real change' in South Africa would entail, and also a way of coming to terms with my mad family, at a time when RD Laing was articulating the idea that madness was symptomatic of a dysfunctional society. I needed to prove to myself that I could leave for good if I had to. Working with madness was what I would do. So I avoided ex-South Africans, whom I saw as preoccupied with the privations of 'exile', when in fact their 'exile' was self-imposed.

I also tried to erase my traces, by softening my accent to something that said more about my upbringing than my nationality. But national identity is of course not a choice. This was brought home to me rather publicly when a patient at the psychiatric day hospital where I was working arrived in a dapper white suit, with a fluffy white dog on a leash, and launched a tirade against a certain South African whom he declined to name. He had gone to the trouble of painting his face and hands black.

Although I stayed in a house in North London for the best part of a year with people from various parts of Britain, all of whom spoke in the working-class accents of the regions they came from, I had nothing to say to them. It was as if I could not bring myself to become involved in another country's working class. Perhaps it was because the working class in Britain reminded me of the white working class in South Africa, which had a standard of living on a par with the lower middle class.

I only realised how small-minded I was being after the mine workers went on strike in 1973. The issue was industrial legislation that the Conservative Party government under Edward Heath wanted to introduce. Heath retaliated by introducing a three-day work week, supposedly to conserve electricity and break the strike. The three-day week only served to boost the popularity of the mine workers, and the following year Heath resigned and a Labour Party government came to power.

The mine workers had demonstrated the potential power of the working class if they were organised. The fear of trade unions, and the 'hold' they supposedly had on workers, was now one of the principal preoccupations of the ruling class in Britain and elsewhere. It would soon inspire a counter-offensive that would also have a bearing on events in South Africa.[14] What had a more immediate impact, though, was the military coup in Portugal. A significant number of officers in the Portuguese armed forces allied themselves with the Left, including organisations of the working class, and seized power. This precipitated the independence of Mozambique, and the much longer struggle for independence in Angola. I travelled twice to

Portugal during that period, on holiday. The second occasion was during the European summer of 1975, after I had moved to Paris.

'A revolutionary situation exists in Portugal,' said a publication I brought back with me to South Africa, its cover torn off to make it less obvious which Left-wing group had published it. 'Either the crisis will be resolved by the working class or the forces of reaction ... In the last resort, all the political tendencies in the working class movement are to be judged by their willingness and their ability to lead the working class forward to power in the time of crisis – or by their contribution to its defeat. Today, Portugal is the touchstone for organisations claiming to be socialist or communist.'[15]

There were a dozen or so such organisations in Lisbon at the time, including a Maoist party that was not under any conditions prepared to work with the others. Together with political tourists from all over Europe, mostly young, I joined a march through the streets of Lisbon, in a massive show of support for the provisional government. There was a large contingent of Portuguese workers in their overalls. No one was sure how well organised the workers were, or indeed whether the working class was large enough to sustain the provisional government. Nevertheless, there had been a revolutionary break with the old order without loss of life. Or was it merely a reform?

I travelled from Lisbon to Paris, to say goodbye to people I had known there. One was David Cooper, an ex-South African and colleague of RD Laing. 'There is nothing you can do about the situation there,' he had said to me, in dismay, when I told him I was going back. Cooper was an extraordinary-looking man, with a huge pot belly and flowing ginger locks. He also exemplified what I did not want to become: someone who had erased his traces so completely there was no way home. From Paris I went to wind up my affairs in London. If I was going back to South Africa, I could no longer defer choosing sides.

It was 1975, and the only political organisation I could have joined, given where I had come from, was the ANC. Since it was a banned organisation, the time to do so was while I was outside the country, and hopefully outside the Special Branch's field of vision. I went with a friend to what you might say was an exploratory meeting with an ANC representative, in some shadowy South London pub. We were given a name and contact details, if we decided to join. My friend and I did not discuss the meeting at the time, or whether we would or would not join, nor did we refer to the meeting ever again. It was understood this was a decision you had to make alone, and these were not memories you wanted to retain if you were ever interrogated. I still see this friend from time to time, and till today do not know what she decided.

I contacted the person whose name I was given, and was asked to explain why I wanted to join in writing. I typed a motivation on my portable typewriter, on one page of thin blue copy paper. It described my involvement in student politics, and how I had come to the conclusion that protest politics were futile. I had become a revolutionary, I said, committed to the overthrow of the existing order, even though my father could be regarded as part of it (better disclose that he was a judge, in case they found out and held it against me). I did not sign or put my name to it, in case it should end up in the files of the Special Branch.[16] It was embarrassing to describe myself as a revolutionary, and brought to mind a poser in the NUSAS leadership who used to wear a Che Guevara beret and floral shirts to its congresses. It is by your deeds that you are known, and I had certainly done nothing to earn such a title. I was not also sure to what extent I was willing to subscribe to a revolutionary morality, in which the ends justified the means. The argument that this led inevitably to a Stalin seemed to me hard to refute.

Yet I did place myself firmly on the Left of the political spectrum, among those opposed to the capitalist system. It was also safer to describe myself as a 'revolutionary' than as a 'socialist'. The ANC famously defined itself as a broad church that included different classes, notably the black middle classes. It was bound to be suspicious of anyone who called themselves socialist. Probably such a person was not a communist, and unlike members of the Party, would be critical of the 'two stage' theory, in terms of which a transition to socialism, however defined, could only be achieved after what the ANC called the National Democratic Revolution, or NDR.

In response to my application, I was asked to suggest somewhere central to meet, but not too busy. I chose Park Crescent, near the canal along which I used to ride on my bicycle to work. There I would find someone wearing a grey raincoat, at the nearest bus stop. I was not told his name, but it turned out to be Reggie September. My application had been accepted, he told me, and he asked what I was intending to do when I got back. He liked my plan to become involved in organising workers, and suggested someone in Cape Town who might help me, the principal of a prominent coloured school. This did not sound to me like a good idea.

'There is someone else you might speak to,' he said. 'She is a working-class woman who stays in Paarl. Tell her I sent you.' He mentioned the name, but I did not write it down, of course. How, in any event, would I be able to find her? I had forgotten the name by the time of my return, but did later find the person he must have had in mind, without ever having to look for her.[17] 'Stay in contact. All you have to do is write to us, every month or so, telling us about this or that. You can use the same address we have given you, and the name Jackson.'

3

Corporation Street

THE FIRST THING I HAD to do on my return to South Africa was to earn a living. The Western Cape office of the Institute of Race Relations was looking for a regional director, and I got the job. Race Relations, as we knew it, was an institution of the English liberal establishment. Although I had no wish to become part of it for this reason, it gave me the opportunity to find out about the state of worker organisation and make contact with actual workers, notably farm workers through a research project I initiated. At a time when like-minded people were establishing circles and reading groups of all kinds, I began looking for a circle of like-minded others to join.

My plan of becoming involved in organising workers had begun to crystallise. To bring about 'real change', the working class needed autonomous organisations to enable it to find its own voice. It also needed intellectual allies. These would inevitably be drawn from the middle classes. I would be an ally of the working class, and an agent for social change. The life I lived should also reflect my allegiances. I would not become preoccupied by the maid or the bond, or providing for retirement when not yet 30. In fact, retiring was not something I ever thought I would do; and if I did, it would not be in the kind of society there was then. There would be no disjuncture between the personal and the political.

Two circumstances shaped my choice of lifestyle, one personal, the other material. The material one was 'inheriting' a rented cottage on the outskirts of the city from the friend of a friend. What was most unusual about it was that the rent was controlled. Rent control was one of the few remaining perks government had introduced long before to protect the white working class, although what was left of the white working class in 1976 was mainly in skilled and supervisory jobs. It would have been beneath them to live in a place like that.

The personal circumstance was meeting and falling in love with Athalie. Like me, she was from the white middle classes and seeking a life outside white suburbia. Like me, she was also a returnee, although she had left in more dramatic and difficult circumstances. This went to show that others from the same background were making similar choices, for similar reasons.

You could also argue it went to show that the middle classes in South Africa were not reproducing themselves altogether successfully.

In my father's eyes, the personal and political choices I was making proved I was a communist. I have this from my mother. 'Of course he is not,' she had responded. 'You can see he is not [a communist].' I am not sure what my mother understood a communist to be, or what made her so sure, but 'communist' more or less equated with being a traitor, and some would say that by allying myself to the working class I was a traitor to my class. As I saw it, however, I too was doing what 'we' considered our duty, where the law was clearly not capable of fulfilling its promise of justice.

With the benefit of hindsight, you could also argue that my choices were informed by enlightened self-interest: that it was a case of the (upper) middle class reproducing itself intelligently, and with foresight. There was a potential contradiction, however, between allying yourself with the working class, with its own autonomous organisations, and becoming a member of an existing political organisation like the ANC. To whom, in the final analysis, was I accountable? Frankly, I did not give this question much consideration at the time. I was also too busy finding my feet to think of writing to Jackson, and I never did so.

Perhaps a pamphlet posted to me played a part in this. It was a call to arms, but it appealed to the lowest common denominator, by studiously avoiding any mention of workers or class. I had come across the same kind of rhetoric in ANC publications in London. I would not have been willing to distribute a pamphlet like that, I remember thinking. Not long afterwards, someone exploded a bucket near the Grand Parade, as a way of disseminating the same pamphlet or one very like it. There were arrests, and I found out from the trial that followed that one of the accused was Jeremy Cronin, whom I knew slightly. He had been a few years ahead of me at university. Uncle Marius, my godfather, was the presiding judge in his trial, and had no compunction in imposing longer than the minimum sentence.

At the same time as the saga of the pamphlet bomb was unfolding, someone from the circle of like-minded persons I had joined, Jonathan Bloch, came to see me. The general secretary of a registered trade union he knew of wanted to retire, he said. It was looking for someone to replace him. Would I be interested in the job? The person to see was the secretary of the Union's medical fund. I went to meet her at the Union's head office one Saturday morning.

The Grand Parade on Saturday is criss-crossed with trestles laden with rolls of fabric, roots and herbs, and whatever knick-knacks that people will take a train to town on a Saturday morning to buy. Corporation Street leads off it, alongside the old City Hall. To get to the head office, you would walk up it, in the direction of the mountain, beaming down on the city like an overbearing aunt about to embrace a prodigal nephew.

The head office occupied the first floor of a narrow, nondescript building on the corner with Longmarket Street, where the Red Monkey Café used to stand. Taking the stairs, you would exit from a dark passage into an oblong room, decked in shades of green: curtains in the faded green of a used ten-rand note, steel cabinets in a military green, and the floor grey and green linoleum tiles. The general secretary occupied the corner office. Green-upholstered chairs encircled a scratched desk, but the centrepiece of the room was the safe. It was a large safe, with the paint peeling off it, also green. On top of it was mounted a black-and-white photograph of a woman with a severe face, her hair in a bun: not your typical white South African woman, obviously. This was Ray Alexander, who had founded the Union in 1941, and was its first general secretary. I used to think of her as Russian, because you did not differentiate then between Russia and the Soviet Union. She had in fact been born in Latvia. Miss Ray, as the members of the Union referred to her, was in exile. The person I was to meet was the secretary of the Union's medical fund, Miss Yon, and it was she who ushered me in. I was not able to place her that day or for some time afterwards. Her complexion was white, but she did not behave like the whites I was used to. She was dressed in the crimplene and pastels of the Northern Suburbs, and was jolly and large: these were working-class attributes, I later decided.

Miss Yon introduced me to the man I was being asked to replace. Though it was not clear what business she had at the interview of a prospective GS, she did most of the talking. 'The members of the Union were country people,' she said. 'They have nothing else but the Union to look to, apart from the church. The Union was like the church for them. You know what it's like on the platteland.' I knew what she meant. The platteland was the domain of the *boere*, where workers knew their place and *opstokers* of any description could expect short shrift. She was obviously anxious I take the job.

Each year the Union held an annual national conference, and at the previous conference the general secretary had announced he wished to retire. This year's conference was in a month's time. Someone had to be elected to replace to him, but there was no one in the Union who could do the job. They had been looking outside the Union for some time. Adam Small, the coloured poet and academic, had been asked to recommend someone from the University of the Western Cape, where he taught, but it seemed the coloured students were not interested. She had told the Union they needed someone qualified, she said: someone like myself, with a university education.

The salary the Union was offering was R350 a month. It was all the Union could afford. The position was mine for the taking, she gave me to believe. I had only to submit a letter of application, and the conference would appoint me. The general secretary, Johnny M, said nothing to contradict her. This was to be the only time we met. I remember him as a physically large man, but that might only have been in comparison with the coloured men I had to do with on the *platteland*, who were often slight in stature and looked nutritionally deprived. Johnny M was from one of the mission settlements where people had security of tenure and were able to engage with the farmers on more equal terms.

I told him about the conditions of farm workers in the Citrusdal valley, which had been the focus of my research. In response, he told me a story. What was curious about it was that a farmer I interviewed in the course of my research had told me the same story, with minor variation in the detail. This farmer was atypical, in that he wrote letters to the press expressing his liberal views, although there was nothing atypical about the conditions under which his workers worked and lived. There had to be some basis to the story. Truth, on the other hand, resides in the detail.

The locale was familiar to me: as Boy Scouts we had camped nearby. The mountains tower over vineyards and orchards, and the high thatched roofs of the homesteads emulate them, as though to make out that their occupants have been there nearly as long. It was (and is) the heartland of the white farmer. A white man had called a meeting of farm workers there: a Hollander, as Johnny M put it, as though only someone from elsewhere could be as bold. The meeting never took place. The farmers were waiting for him, sjamboks in hand, and beat him within an inch of his life. He took the next plane out.

A new life
'I will think about it and let you know,' I said in response to the job offer, although my mind was mostly made up. Terror, without doubt, underpinned the existing order, and the sjambok was one of its symbols. Johnny M's story, the story of the liberal farmer, was a fable of impotence in the face of the existing order. My personal terror was of a different kind: being stifled comes closest to it. This is what living the life of the white suburbs represented to me.

Miss Yon had given me a booklet to read as I left, which she called a history of the Union, although it was really only the story of its founding. A line drawing on the cover depicted a factory, with what could be an orchard on one side of it, and on the other the sea. Over the sea, the rays of the sun are rising. You could tell it was a rising sun, because over the rays was emblazoned the title *New Life*. It told the story of the Union's early leaders.

I wanted to believe in these people depicted in this booklet: Frank Marquard, the Union's first president, a coloured who grew up on a wine farm and did not drink; Miss Ray, deeply influenced by religion until she was converted to a new faith – 'something that would give her both an explanation of life and a guide to action. She became a socialist.' Botes was the white worker who made up his mind to join the Union because 'he knew that the colour question was not the crucial one. The important conflict lay always between employer and worker, whether white, black or coloured.'[1]

The story of the founding of the Union was the story of a strike, at the fruit canning factory of H Jones and Co. in 1941. The chairperson whom the workers had elected was dismissed the day after they had joined the Union, and his fellow workers had risen to the occasion by walking off the job after him. The scabs the bosses hired to replace them, white women, were not capable of doing the work. The bosses had to learn the hard way that they could not manage without their workers, and agreed to recognise the Union and to negotiate increased wages. It was a story of solidarity, typical of many others I would hear over the years. The strike at Jones, as the factory was known, established the Union as a force to be reckoned with. It also established lasting organisation among the workers in the fruit canning factories of Paarl and beyond. Lasting organisation, I was later to find out, is almost always associated with a founding strike.

'It was a direct and important victory for the Union, a victory that was to have repercussions throughout the food canning industry,' the author wrote. 'For the first time the coloured workers of Paarl had felt their strength. Never again could they be put back to a state of helplessness. They had learned the power of organisation. That was something no one could take away from them. And from the food and canning workers of Paarl others would learn. There was new hope for the coloured workers of the Platteland.'[2]

I was to find out no more about the author of *New Life* than could be gleaned from the inside cover. He had attended an elite school, and written poems for the *Guardian* newspaper. Probably, then, he was a member of the Communist Party or at least a fellow traveller. That would explain his belief that a new world order was around the corner. Socialists, the booklet told us, 'wanted a new world, a world where ... everybody would be free. It was no idle dream, they declared, but an historical necessity. They pointed with pride to the beginnings of their new world in the Soviet Union.'[3]

I could see how the war might have blinded many to what the Soviet Union had become under Stalin. Yet the booklet was published two years after the Nats came to power, and, even before that, the Union had split into a registered union for coloureds only and a 'parallel' union for Africans.[4] It was difficult to see how the writer could have supposed there was a new world around the corner, at least for the workers the Union represented.

R350 a month was a great deal less than I was capable of earning as a lawyer, and also quite a bit less than I was earning at the time. I was 26 years old and the choice I was about to make would determine my career. It was not an economic choice, however. Even though my mind was mostly made up, I wanted to know what others thought about it. Athalie was for me taking the job. So was my friend David Lewis, who also knew something of the history of the Union. David was peripherally involved in a workers' advice office, which at the time went under the name Western Province Workers' Advice Bureau, which I will refer to here as the Advice Bureau.

Everyone else I spoke to thought it would be a bad idea. This was primarily because the Union was registered. Government first recognised the right of 'employees' to join a registered trade union in 1924, in a law known as the Industrial Conciliation Act: but 'natives' (as Africans were then known) were excluded from the definition of 'employee'. The 1956 version of the law which the Nats introduced went further. A registered trade union was compelled to restrict its membership to one race only in terms of its own constitution.[5] This made registered trade unions complicit in a scheme to preserve the position of relative privilege of white or coloured members. This was clearest in the case of trade unions that had adopted a craft model of organisation, which was dominant at the time.[6]

In terms of the craft model, a trade union only organises workers with specific skills. Where such trade unions could control the process whereby workers acquired these skills, and workers could not be easily replaced, they were potentially powerful. Where, however, they utilised their organisation for the exclusive benefit of their own members, they exacerbated divisions in the workplace. This is what had in fact happened. Craft trade unions for 'whites only' were among apartheid's most avid supporters.[7] However, they were not the only ones to give trade unions a bad name.

There was a section of the trade union movement led by the Trade Union Council of South Africa (TUCSA), which claimed to be opposed to racial segregation. Some of its members organised industrially, and had even established 'parallel', unregistered trade unions for African workers, as FCWU had done. However, the consensus on the white Left was that 'parallel' trade unions were a means to control African workers rather than give them a voice. The proof was that despite the Durban strikes, and the fact that the proportion of African workers in the workplace was growing, even in more skilled positions, there were no perceptible signs of organisation in any workplace where TUCSA unions were established.[8]

On the eve of the Soweto uprising, therefore, a critique of trade unions as a form of organisation seemed well founded. It was inevitable, the critique went, that trade unions would focus only on the interests of a section of the workforce, those that were for the time being their members. They would also focus only on wages and conditions at work, and leave politics well alone. Such tendencies were regarded as 'economistic', and were exacerbated when trade unions were drawn into institutional forms of collective bargaining. The system of bargaining at Industrial Councils was a prime example of this.

Some of those concerned with the organisation of African workers preferred to speak of workers' organisations rather than trade unions because they regarded it as an open question whether trade unions were an appropriate model of organisation for workers at all. The Advice Bureau, for instance, was trying to promote a model of organisation based on factory-level committees. People I spoke to were dismissive of my idea of changing the Union from within. It would be useless to try, they said. As soon as the leadership realised that I had a political agenda, I would be isolated and exposed. If I were not dismissed, I would surely be banned.

A banning order meant, among other things, being confined to a magisterial district and prohibited from attending any gatherings. A gathering was defined as meaning any two or more people together. In theory, that would make it impossible to have a social life, although those who had been banned in the broader circles I was part of, like Neville Curtis and his sister Jeanette, managed to do so. It was also not jail. For me, the only question was whether I would have enough time to achieve something worthwhile before I was banned. The most I could hope for, I calculated, was five years. It was probably not coincidental that five years was the usual period of a banning order, as well as the minimum sentence for someone found guilty in terms of the Terrorism Act.

PART 2
1976

'This is the time, and this is the record of the time.'
– Laurie Anderson, 'O Superman'

4
Klein Drakenstein Road

I CAME TO THE IDEA THAT CAPE TOWN was once an island as a child, travelling by train on the suburban line, and noticing how few metres above sea level we were from the signboard at each station. In many ways it still is. The well-to-do gravitate to the high ground, on the slopes of the mountain. The industrial areas and the townships where the working classes live are on what must have been the seabed surrounding it, to the north and the east. The road to Paarl runs across it, in a north-easterly direction, towards what would once have been the nearest land. Well-to-do Afrikaners gravitated there and today, for some reason, so do almost all the people I know who could be called *buppies* – black, urban professionals. Very possibly they are more at home there than they would be among the liberal, English-speaking establishment.

One cold Saturday in July 1976, I was on my way to deliver a letter acceptance to the president of the Union. Although the head office was in Cape Town, the Union did not have a single member there. Its base was in and around Paarl, a heartland of the Afrikaner establishment, who regard it as the birthplace of their language. From the Tygerberg hills, along what is now the old Paarl road, there is a slow decline towards the periphery of the city, where the not-so well-to-do stay. What organisation the Union once had in the city had collapsed following its defeat in a 1956 strike at a nearby meat-processing factory, Spekenam. As you leave the city behind you, you begin to climb again, with the sheer rockface of the Groot Drakenstein Mountains in the distance, on your right, until you descend into the valley of the Berg River.

The rock that gives Paarl its name looms on your left. Most of the factories in Paarl are located on the east side, alongside the railway line. H Jones is first, and next to it KWV, the wine cooperative which dominated the wine industry and the production of brandy; closer to the road there is Monis, another winery. Once it was owned by the same family of Italian origin that gave its name to a flour mill and macaroni factory, about which I will have more to say later. Close by are the silos of Sasko, the only flour mill in Paarl, at that time still a co-operative.

At the Rembrandt tobacco factory there is a bridge over the river that separates the glistening white buildings of the old town and its commercial centre from the east side, which had become a coloured area: the more well-to-do lived lower down, nearer the river, as though they hankered to return to the parts from which they had been evicted in terms of the Group Areas Act. Higher up the hill were the council flats and single-storey houses in which sections of the working class were housed. Across from the bridge, running through the east side, is Klein Drakenstein Road.

It was a wintry Saturday morning when I first drove up it, looking for the Ray Alexander Union Centre, a building the Union had constructed with funds accumulated in the years when it had been stronger. The blackened remains of a barricade of burning tyres lay across the Klein Drakenstein Road, on both sides of the island in the middle. It was about a month after the police had fired on Soweto school students, protesting against the introduction of Afrikaans as a medium of instruction in the schools. The police were unable to suppress the angry response this provoked, and there were riots in different parts of the country. Paarl was one of the places to which the Soweto uprising had spread.

I imagined the Centre would make a bold statement about workers or their organisation. Instead, it was a nondescript commercial building whose name, in plain blue letters, was overshadowed by the neon signage of the hardware store and drycleaners downstairs. These, I later discovered, were tenants of the Union. To get to the Union's offices you had to negotiate your way past the rolls of linoleum belonging to the hardware store and up a concrete flight of steps. There were two offices upstairs, but one was closed. The open one was that of the Dal Josaphat branch. Dal Josaphat was the name of the valley in which much of the east side of Paarl is located.

The members of that branch, 'Dal' for short, worked at a factory which the older workers knew as Moberg's, although it was now part of the Langeberg group. Mrs R was the secretary of Dal branch, and was busy watering a stand of indoor pot plants: the bright variegated leaves of the *Yusuf's kleed* (Joseph's coat). The president of the Union was with her. He was a worker, and a member of the Union, as an office-bearer had to be in terms of the Union's constitution and in terms of the trade union conventions of the time.[1] The workers referred to him as Oom Joe, and that is what I will call him here.

Oom Joe was a lithe man in his forties, with a self-confident manner. He had to be classified coloured, of course, because only coloureds could belong to the FCWU, but he could easily have been taken for an African. I found out later he had grown up in Willowmore, a part of the Karoo that fringes on the Eastern Cape, where the lines separating races that apartheid sought to maintain were easily crossed over.[2] The self-confidence, I found out, went with being in a position of authority in the factory. He was a foreman at Moberg's, and must have been hardworking and ambitious to get there.

The workers will be glad when things were back to normal, Oom Joe said, after reading the letter I had handed him, without comment. He was referring to the burnt tyres and what the Afrikaans press referred as *onluste* – or riots. It was one of those off-hand remarks in which someone declares a position on a defining issue. I took note, but did not respond. Mindful of the warnings I had been given about being exposed as someone with a political agenda, I knew I had to tread with particular care on the matter of the workers' relations to what was going on or to the current regime.

Just as the legitimacy of a national regime is based on its claim to represent the interests of the people as a whole, the legitimacy of a trade union rests on its claim to represent the workers it organises. Nowadays this is set out in a trade union's constitution, with reference to an industry or sector (or occupation, in the case of craft unions) which it seeks to organise. This is invariably defined in wider terms than the workplaces where the union is actually organised. At that time, for a registered trade union, it was defined in terms of its certificate of registration.

According to the Union's certificate of registration, the workplaces where the Union had been organised in Miss Ray's time were in the 'fruit and vegetable canning' and 'fish processing' industries, with a handful of workplaces that did not fit in either of these categories. Practically nothing remained of the membership outside the fruit and vegetable canning industry, however. The main activity of this industry was canning deciduous fruit – mainly peaches, pears and apricots – for export. Vegetables were a sideline.[3]

More than half the Union's members were at two factories: Moberg's was the one, and Jones was the other. Most of the remainder were at Groot Drakenstein, where Rhodes Fruit Farms (RFF) had a factory in the shadow of the mountain of that name, and Wellington, further down the valley of the Berg River. Each of these factories was regarded as a branch, which struck me as strange. However, there were a dozen factories employing at least as many workers outside the area of greater Paarl where the Union had few if any members.

If I had any inkling at the time of what was going on in the Union, I would not have worded my letter of acceptance as I did. An accountant was needed to help with administration, Johnny M reported to the management committee, and the 'head office' had someone in mind, by which he meant Miss Yon, even though he, as GS, was supposed to be in charge. My name was mentioned, although I was not, of course, an accountant. The minutes record that a 'very lengthy discussion' took place. The conclusion was that the meeting did not want a white person.[4]

The management committee (MCM) of the Union met once a month, and was the meeting to which the general secretary was accountable. It was supposed to be a subcommittee of the national executive committee (NEC), which met every six months, but since the Union had not been a national union in anything but name for many years, the management committee meeting was in practice the more important structure. The highest decision-making structure, however, was the annual national conference, and it was here that the national leadership of the Union was elected, including the general secretary.

You could hardly speak of an election, when I was not even present. Had it not been for F, things might have turned out differently. F had been recruited to work as a cleaner at head office by Miss Yon, while still in her teens. This was in accordance with a policy Miss Ray had advocated, of creating opportunities within the organisation for the children of its members. She was the adopted daughter of a shop steward at Langeberg Worcester. If race could be determined from one's complexion and features, she was whiter than I was, but culture determined that she was coloured.

It was F, and not the Union's auditors, who realised that Johnny M and 'the typist' (as the Union's only administrative employee was called) were 'eating the money'. Each had been drawing a double salary for some time. It seems they did not bother to conceal their fraud from her, expecting her to keep her head down and her mouth shut. This she did for some time, until confronted about her poor performance at work. Fearing that her own job was at risk, she blurted out what she had observed. Miss Yon, who did not ordinarily attend Union meetings, presented a report on the extent of the fraud at the next MCM. The lawyers were to recover the money that Johnny M and the typist had misappropriated. What was unknown was how much more had been pocketed, of which there was no record. This was money collected by hand on Johnny M's visits to the branches, which he seems to have believed it was his due to spend, in lieu of overtime. After Miss Yon had left, Johnny M continued with the meeting. In a month's time he was to give the annual report to the conference.[5]

The Joneses

The conference was not 'national' any more than the national leadership was. However, branches which were too far from Paarl to attend MCMs, known as the 'far branches', were expected to be present. Deliberations also took place over two days, based first and foremost on an annual report. Judging from past reports, this was not just about the state of the Union, but about the state of the nation and the world. This seemed to me entirely appropriate. It was already clear by that time that the Soweto uprising was the most significant political event since Sharpeville, and more than 200 people had been killed. But if the security forces were not able to suppress the riots altogether, stones and petrol bombs were also no match for modern weapons. The incidents that took place were confined to the townships, and were sporadic. Outside the townships, life went on much as usual.

The annual report dealt with the uprising in passionate detail. 'Students and workers joined forces in a determined onslaught in the nerve centres of White authority and oppression,' it said.[6] This account, I realised, could not have been written by someone in touch with what was actually happening. Worse still, it was misleading. It was the students and youth who were in the forefront of the action. Many must have been the children of workers, dependent on workers' wages. In some senses what they were protesting against was a schooling system designed to produce a docile working class, and undoubtedly there were workers who supported them. But there were also clear indications of mounting hostilities between workers and students, particularly the migrant workers in the hostels.[7]

Perhaps the struggle rhetoric was intended to reaffirm a tradition the Union had once been part of, a tradition of engagement with the political struggle. Oom Joe's remark in passing was one indication that not much of this tradition survived. Another was the choice of guest speaker: Sonny Leon, the leader of the Labour Party. The Labour Party was a coloured political party that was at the time contesting for support with the Freedom Party. In areas outside Cape Town, like Paarl, this contest was intense. Although the Labour Party preached solidarity with the African majority, the coloured intelligentsia of Cape Town regarded both parties as peas from the same pod, and the representative council to which they sought election as a toy telephone.[8]

So there was a disjuncture between the rhetoric of the report and the political understanding it assumed, on the part of the members. What made this disjuncture all the more glaring was that the Union was clearly not winning victories for the workers or growing. Of the 25 places where the Union had once had branches, listed on the front cover of the report, most were defunct. Years later I heard that the Special Branch used to say of the Union that it was controlled from Lusaka. Like most misinformation emanating from the Special Branch, it had an element of truth in it. Clearly Johnny M had no hand in the report that appeared in his name, for it ended by affirming the need for leadership of integrity, and called on workers to elect honest leaders, leaders 'that are working for the interests of their fellow workers and not for their own interests. Those who work to enrich themselves, are enemies of the workers, and enemies of the union – they must be cleared up [sic].' Miss Ray must have written this.

The minutes recorded that I was 'elected' general secretary, but that is not how I saw it. I had been appointed. Although it was to a position of power, I was by no means in a powerful position. While Oom Joe and others were supportive, I had no base of my own in the organisation, and a power struggle with Johnny M's backers loomed. As if to emphasise my vulnerability, Johnny M attacked my appointment at a meeting of Paarl branch held days after the conference. 'Through all these years this Union was built up by our own coloured people,' he was recorded as saying, 'but today a European had to take over for a monthly wage of R350.' 'The R350 a month I would be getting, I learned, was more than Johnny M was earning at the time, at the end of his working life.

One of the head supervisors at Jones also objected to having a general secretary who was not coloured. The workers called her Grootvoet behind her back, I was later to find out. I am not sure this was because her feet were particularly large, as the name implied, or because all of her was large, especially her rasping voice.[9] Others at Jones objected on other ground that I had not been presented to the workers. I was a 'law student' whose 'father is an advocate'. What I found most difficult to understand was that there were workers who still wanted Johnny M to be general secretary, or said they did. He was not really to blame, it was suggested. Let him carry on working until he has paid off his debt.[10]

Jones's name came from the Australian company that had founded it. Some years after the 1941 strike it had been taken over, and at the time was owned by Picardi, a company which ran a chain of liquor stores, named after the man who controlled it.[11] Jan Pickard was best known for playing rugby for Western Province, but had married the daughter of Eben Dönges, a leading luminary among the Nats. The iron grip which the Afrikaner establishment maintained over a town like Paarl, then and now, derived from old money, especially money made from wine, liquor and cigarettes, and the business opportunities and employment that people like Pickard provided. The workplace was in many ways typical of the manufacturing sector at the time. Whereas formerly there had been owners who actively managed canning factories, now there was a general manager at the top of the workplace hierarchy. Beneath him was a factory manager in overall charge of production and various section managers. This management team was of course white to a man.

The middle layers of the hierarchy comprised an amorphous group of workers (or employees, as management preferred to call them). They ranged from artisans of various kinds to clerical workers to those, like Grootvoet, in supervisory positions. At the bottom of the hierarchy were the ordinary workers, who made up more than 80 per cent of the workforce, a higher proportion than in 'higher wage' sections of manufacturing. The ordinary workers kept production going, and could be characterised as 'unskilled' or 'semi-skilled'. The vast majority of 'unskilled' workers were women, standing at the conveyor belts, trimming or putting fruit into the cans. The 'semi-skilled' typically operated machines or forklifts, or were in jobs of a technical kind, such as quality control. Sometimes they were men, and sometimes women.

This hierarchy corresponded with a chain of command, in which the role of the middle layers in implementing management's decisions was crucial. It included coloured men at the upper end and coloured women at the lower end. So race as well as gender mattered for someone trying to work her way up in the world; and the whiter you were, the higher you could rise. Grootvoet had straight hair and a European nose. Having someone like me in a position one of her own might have occupied would have been galling in itself. What made it even more galling, perhaps, was that I was from a privileged background. It meant I could not credibly be accused of taking the job for my own material advancement. This in turn raised the question of her motives in wanting to preserve the position of general secretary for one of their own: Was it because she believed one of her own could do a better job? Or was it because she and others like her saw the Union as a vehicle for advancing the interests of the section of workers to which they belonged?

It seemed as if the leadership of Union at Jones had been integrated into the chain of command, since all of 'the table', as workers referred to the leadership of the branch, were supervisors. A proposal Grootvoet floated at the meeting reinforced this impression. The factory manager had discussed with them introducing a liaison committee in the factory. Grootvoet thought it was a very good idea. The liaison committee was a workplace structure comprising representatives of both management and workers, which government had been punting since the 1973 strikes as a substitute for trade unions.[12] Although the legislation introducing them applied only to 'Bantu', there was to be a concerted push on the part of government and employers to establish them more generally.

The push to establish a liaison committee clearly came from the top. What the bosses were hoping it would help them achieve, which Grootvoet had failed to mention, was a reduction in the workforce. The resultant savings, according to the factory manager, would be distributed among the workers remaining. This would surely include the supervisory layer of which Grootvoet was part – a clear example of how a trade union could be used to benefit the interests of a section of the workforce. Yet even though the leadership of Jones was made up of supervisors, it transpired that the Union had not been totally integrated into the chain of command. It was Aunt Sabbagh, the other head supervisor, who pointed out what Grootvoet had failed to mention. Most significantly, it was someone who was no longer employed by Jones who sank the proposal. This was Nellie Kilowan, the widow of Abdol, who had been the president before Oom Joe. 'Are you sure the pay of the [workers to lose their jobs] will come into the pay packet of this committee?' It could be your sister or mother or child, she argued, that would lose her job. 'How are we to feel? These people have worked every season ... you must be very careful not to agree to everything the bosses tell you.'[13]

5

The road to Mbekweni

I READ ABOUT WHAT NELLIE KILOWAN had to say during my first week at work, while the clouds of tear gas from across the Grand Parade billowed down Strand Street. Juffrou H, the long-standing secretary of Paarl branch, had faithfully recorded what was said and posted the minutes to the head office, as all branches were supposed to do. When Kilowan said it could be our sister or mother or child who was laid off, she was of course speaking metaphorically, as the bearer of a tradition, about everyone she considered being in the same boat, in that their livelihood depended on a job in the factory. Clearly she had ordinary workers who were coloured in mind. But did this concern extend to ordinary workers who were African?

The policy of the Union, Oom Joe had confirmed, was that the two unions, FCWU and AFCWU, would operate as one. Meetings of the registered union were supposed also to be meetings of the 'parallel' African union. Yet Johnny M told Paarl branch that the Union had been built up by coloured persons, and there was no African present to contradict him. This was the clearest indication that this policy was contested, and not everyone wanted African workers in the same boat. I needed to find out what had really happened when AFCWU was formed. It was an issue about which the Union's history, *New Life*, was conspicuously silent. I also needed to understand why there were two branches of FCWU, Paarl and Dal Josaphat, in the same town. I sensed that these questions were related.

The tear gas marked the first time since 16 June that the riots had taken place in the centre of any of the big cities, and Vorster issued a blunt warning in response: stop the unrest, or else.[1] Instead of heeding the warning, the call went out from no one in particular for workers to show their solidarity with the students, and stay away from work. There was widespread support for the stay-away, but support from workers was uneven: those from Soweto stayed away while those from Alexandra in Johannesburg did not.[2] In Cape Town, press reports said that office and clerical workers stayed away while factory workers, including African contract workers, did not.[3] I put this down to their lack of organisation.

With this in mind, I began poring through all that was written, as you would approach a research project. There was the constitution, past annual reports, and the minutes. The minutes were kept in a set of fat bound books in one of the green steel cabinets. I began working my way through them, starting from the most recent and working backwards. They were typed, and contained a lot of detail, so by the time of my first management committee meeting I had learned quite a lot.

The meeting was usually held on the last Sunday of each month, and shortly before the meeting I found out for the first time that the Union had a motor vehicle, a Volkswagen Kombi. Johnny M did not have custody of it, because he could not drive, but I was entitled to use it for Union business. I drove it to the meeting, and afterwards went straight home. It was the only time in the life of the management committee meeting that I felt free to do so. At the next meeting, and every other meeting afterwards, I gave lifts to people without transport: delegates and sometimes 'paid officials', as they were called.

The term 'paid official' must have been adopted to remind workers of the critical distinction already referred to, between an office-bearer who worked in the factory and who was therefore not paid by the Union, and a functionary. It made no difference that Juffrou H and Mrs R, as well as Annie Adams from Wellington, had each worked in the factory before becoming branch secretaries. They were members no longer. It was not a requirement that a branch secretary be employed full-time, but it was considered ideal if a branch had the money. The fourth paid official was Lilian from AFCWU.

The paid officials were the people I looked to, day by day, to give me the line as to what I should or should not do. A union leader should never see the bosses alone, I was told, lest he or she be accused of accepting bribes or making private deals. At least two of them had to accompany me whenever I visited a factory. The reason, Oom Joe explained, was that if I travelled with only one, especially to the far branches, I might be suspected of contravening the Immorality Act. This was the law that prohibited sexual relations between persons of different races. So two or three times a week, thereafter, we would meet at the Paarl building and set out together.

Relations between us all were formal. They addressed each other as *Mevrou* and *Juffrou*, and referred to me as *Meneer*.[4] This was also how Oom Joe preferred to address me. After the visits, I would also drop each of them off. This was often after hours, in which case I would take each of them home: Juffrou H first, because she stayed closest to the building, and Lilian last. The justification for dropping Lilian last was that I could travel straight through to the N1, and back to Cape Town, but in truth it was that I did not trust Juffrou H. If I had dropped off Lilian first, I would have had to drop Juffrou H last. Relations between the other paid officials and Juffrou H were frostily correct. They also did not have much to say to each other. Even so, no one tried to discuss her behind her back.

Many Africans had lived on the east side, off Klein Drakenstein Road, until the 1950s. They were well integrated with the coloured community there, and many spoke fluent Afrikaans. Prominent among them was Elizabeth Mafikeng, who had been president of AFCWU. Then Mafikeng was banished to a remote region of the old Cape Province, bordering on Lesotho. This was when, despite fierce resistance, the African inhabitants of Paarl East were moved to the newly established township of Mbekweni.

Mbekweni was where Lilian stayed. It was not signposted. Passing through the industrial area of Dal, where the Langeberg factory and Berg River Textiles were located, you took an unpromising road parallel to the railway line, through an unkempt green belt separating Paarl from Wellington. If it were not for the potholes, you would be travelling in a straight line towards a clump of gum trees and a white building that could be mistaken, from a distance, for a farm house. At this building, the road took a sharp, 90-degree turn to the right. To casual visitors, if there were any, the turn would seem bizarre, if not downright dangerous.

The building at the corner was the BAAB office, where the Bantu Affairs Administration Board for Paarl was located. Vehicles going past had to slow down, allowing the apartheid functionaries inside to scrutinise who was going past. A white person was not allowed to enter an African township without a permit, and this was where I would have to apply for one, if I was so minded. Fortunately there was another way out of the township, through the adjacent coloured township of Newtown. So my strategy was to take the sharp turn as quickly as I was able and head back that way, after dropping Lilian. It was on one of these occasions, when we were alone in the kombi, that Lilian made a particular point of engaging me. '*Meneer*,' she began, speaking Afrikaans. '*Meneer* must trust no one ... Not even me,' she pointedly added.

Money matters
The MCM met in the office of Paarl branch, because it was slightly bigger than Dal's. I arrived wearing what I had resolved would be my uniform in the Union, a navy blazer and open-necked khaki shirt. Everyone else was dressed for church. I took my seat at the table, next to Oom Joe, who introduced me to Aletta Amon, the vice-president, an enormous woman from Wellington with an air of great seriousness. The treasurer, a worker from Dal, was also there. Hanging on the pelmet above us was the same photographic portrait of Miss Ray I had already encountered on the head office safe.

The relationship between Dal and Paarl, representing Moberg's and Jones respectively, looked something like the relationship between a government and its opposition, with Dal in power. The Jones delegates sat at the back of the room, in the right-hand corner: Grootvoet was there, sitting with Juffrou H and Nellie Kilowan. The rest of the delegates occupied the rows at the front of the room, and were separated from the group at the back by two or three rows of chairs.[5] There were only two Africans present apart from Lilian herself. One, I suspect, was a friend Lilian had rustled up for the occasion. The other was Moffat Manyosi, who had come with a delegation from Ashton, two hours' drive away. Almost all the other workers present were coloured women, and looked as if they had been around for a long time. I was the youngest person present by far.

I had not set the agenda, and the stay-aways were not on it. The issues discussed were those that were routinely discussed, and I would not have wanted it otherwise: I needed to learn how things were done. Most of the discussion was about money matters, starting with a report by the Union's attorney on the Union's only significant asset, the building in Paarl, and a proposal to put in a new staircase, because the hardware store downstairs was using the existing stairs to store rolls of linoleum. This struck me as ludicrous. The hardware store did not even have permission to do so. The other money matter discussed concerned the subscription, or 'subs'. This was the amount a worker was required to contribute as a member, and it was twenty cents a week, regardless of what he or she earned.[6] For some 30 years the Union had relied exclusively on subs collected by hand, by shop stewards who in turn handed them to the branch secretary.[7] But the only opportunity the shop stewards had to do so was at the end of a long working day. This remained a challenge, especially during the fruit season.

During Johnny M's tenure some bosses, influenced no doubt by what employers elsewhere were doing, agreed to introduce stop order facilities. Provision to do so was then introduced into what simply referred to as *die Ooreenkoms*, or the Agreement, since it was the only collective agreement the Union had negotiated. The Agreement set wages and conditions of work for the fruit and vegetable canning industry, and since Miss Ray's time had been negotiated at a Conciliation Board. This was a meeting convened by the Department of Labour to resolve a dispute, in terms of the dispute resolution procedure of the 1956 legislation.[8]

Where there was a stop order, the employers would simply deduct the money from the workers' wages and deliver a cheque to the Union once a month. Not only was more money collected this way, but there was less opportunity for the shop stewards or branch secretaries to eat it. Delegates resolved that all the canneries should introduce stop orders. However, their depth of feeling on this point, I suspect, had more to do with the corruption at head office than any objective problem with a system of hand collections. There was in any event no way the Union could compel an employer to introduce stop orders, and AFCWU would continue to depend on hand collections, because it was not legally permissible to deduct subs for an unregistered Union from the wages of African workers.[9]

It might be argued that the preoccupation with money matters lent weight to the proposition that trade unions were inevitably economistic, as did the venal attitude some displayed: for instance, at branches where the secretary was working full-time in the factory, and she nevertheless wanted payment from the Union.[10] Yet when Hester Adams, from RFF (formerly Rhodes Fruit Farms), stood up towards the end of the meeting to say that her branch wanted its own cheque book, what I understood her to mean was that her branch wanted control over its own finances. Money mattered to workers because it was the basis of their power in the organisation. The corruption at head office had broken the trust there had been with the branches. One of my primary tasks would be to restore this broken trust.

It was late afternoon, and the women were shuffling through their handbags preparatory to going home, when a frail man from Ashton, Mr Waggenaar, stood up to report that one of the workers at Ashton Canning had been assaulted by a foreman. His report provoked a buzz of outrage, and an almost perceptible sense of a closing of ranks: whatever divisions between delegates there were, assaults or attacks on the dignity of the workers always touched a raw nerve. Preoccupation with money matters was not a true reflection of what the Union signified.

The area of preference
The objections to my appointment on the grounds of my race fizzled out when I suggested we have an open discussion of them. It was not because I was white, the delegates from Paarl branch now said, but because I had not been introduced to the workers, who evidently believed I would be glued to my desk at the head office. To prove that I was not that kind of white, I undertook to visit the factories to meet with the workers as soon as possible.[11] But I also made another point in response to the objection to my appointment. It had been agreed that the first year would be a period of probation. For my part, I was keeping my options open.

The visits to the factories would be an opportunity for me to test whether my race was an issue for ordinary workers. It would also be an opportunity to test how a provision in the Agreement headed 'trade union facilities' was interpreted in practice, specifically regarding 'access' to our members in the workplace. What I found out was that what was said on paper was of far less consequence than what the Union had been able to secure through its organisation. This, in the factories in and around Paarl, was the all-important right of paid officials to hold meetings in the cloakrooms over the lunch hour. But permission was needed, and it remained to be seen whether management would grant it for the new general secretary.

Moberg's granted permission easily enough, so Mrs R, Lilian and I went to meet the workers. I thought I should ask Juffrou H if she wanted to come, but she declined. The cloakrooms were packed with women in green, wearing different coloured *doeks* signifying the sections of the factory they worked in. The sense of solidarity between workers of different races was perceptible. In the front row was Tant Sanna, a woman of light complexion and enormous bulk. In the middle was a tall woman of dark complexion and fine features: this was Aunt Violet, who was to prove to be an orator of note, and whose Afrikaans was peppered with arresting images. Oom Joe, wearing the khaki dustjacket of a foreman, introduced me. There was only a sprinkling of men present.

I was unprepared for the enthusiasm with which I was received, and buoyed by it. I was also received with enthusiasm at Oakglen Canning in Wellington, where Mrs Amon worked, and at RFF. RFF was part of a huge swathe of the Groot Drakenstein valley that had been bought up by Cecil Rhodes, and was now owned by Anglo, as the largest conglomerate in the country was commonly called.[12] I was to experience the same feeling of being buoyed by ordinary workers many times, in many different parts of the country, including places where Miss Ray and the Union had not been before.

My experience at Jones was entirely different. Juffrou H had been holding lunch-hour meetings at Jones for years, and Johnny M had also done so. But the management at Jones were only prepared to allow a meeting at tea break to introduce me. We were let in by the watchman, an African, as watchmen generally were then because no one else was prepared to work their hours.[13] While waiting in the cloakrooms for the workers to arrive for our meeting, we were confronted by an irate security official who wanted to know who had let us in, and ordered us out.

After an appeal to someone higher up in the factory hierarchy, the meeting eventually went ahead, but not a lot of workers were present. Most were still on the production lines, and were only granted a break after we had gone. It was a strategy that was to be used with regularity by different managements at different plants, to frustrate access to the workers. The only workers who were in any way welcoming were the African women, and it was one of them who reported to Lilian that the watchman who let us in had been dismissed by the same official who had ordered us out of the cloakrooms.[14] He had apparently given an instruction that no 'Europeans' were allowed to enter the factory.[15]

There was of course no such thing as a right to a hearing before workers were dismissed, and workers at the lowest level, in particular, would commonly be dismissed by low-level managers for little or no reason. On the other hand, there was a reasonable prospect workers could be reinstated where there was a union to take up their case. It was no skin off the noses of top management to overturn a decision to dismiss about which they had not been consulted. It was also not that important to bolster the authority of management at lower levels, or, if it was, it certainly did not outweigh the importance of preserving good relations with the Union near the start of the season.

The watchman was reinstated the same day, after the officials had taken the matter up. I think Juffrou H and Nellie Kilowan were hoping to impress the new general secretary. The view among Africans at both Jones and Moberg's who were still paying subs was that Lilian was too weak to take up cases on her own. Nominally, Lilian was general secretary of AFCWU as a whole. In effect, since there was no AFCWU to speak of outside Paarl, she was merely the secretary of the branch. This, in the case of AFCWU, covered both Moberg's and Jones, and the bosses at both factories tolerated its existence without formally acknowledging it in any way.

The Western Province, Dr Verwoerd once said, was the area where 'the policy of apartheid can be applied with the greatest ease'.[16] The area he referred to, which was not a province at the time except for rugby-playing purposes, was one in which African people were not yet a majority. The specific apartheid policy devised to demarcate it, and keep it that way, was the 'Coloured Labour Preference Area' policy. You rarely hear mention of it nowadays, probably because it is an inconvenient reminder of how apartheid constituted a racial hierarchy. Yet it gave sharp teeth to every law or policy affecting Africans, including the provision of housing.[17] Its focus was the workplace. An employer could only employ an African if there was no coloured person to do the job.

When the apricots started turning gold at the start of the fruit season, the question foremost in the minds of women of Mbekweni, most of whom must have lived all their lives in Paarl, was how many would be selected, and who would do the selecting. When the flow of peaches and pears abated as the days grew shorter, the question that arose was who would be laid off first. Some foremen and supervisors, including coloureds, did what the Preference Area policy expected them to. Others, like Aunt Sabbagh, were fair-minded and impartial. But you could not always tell who was who, racially speaking, because of Africans who 'tried for coloured'. 'Claasen without an ID' was how Selina joked about her status when asked about working at Jones.[18] Her identity document would have given her away.[19]

So while the Agreement required all workers be paid the same wage for doing the same job, African women did not have as many jobs, or access to the better-paid jobs. Their employment was also not as secure. The position was somewhat different for the men. There were only certain jobs coloured men were willing to do in a canning factory, because of the hours or the pay, or because they had other options. An operator earned only a few rands more than the ordinary worker. So there was not the same competition for the jobs. But there were no African foremen or supervisors at Jones. Indeed, it would be exceptional to find an African in any better-paid job where the Preference Area policy applied.

Lilian took me to meet one of the operators at Jones, a man of 50 years or so in blue overalls, with a firm gaze and a cut on the lobe of his right ear that looked as if it was ceremonial or perhaps decorative: I never felt free to ask. This was John Pendlani, the president of AFCWU, who worked in the jam room: I am from 'PE', he liked to joke, *'die man van die land'*.[20] But *die land* usually referred to the homelands. In fact, Pendlani had a house in Mbekweni.

One day I asked Pendlani about a group of African men at Jones in blue overalls who appeared not to be interested in Union meetings. These were contract workers, and stayed in hostels close to the BAAB office. According to him, they were *'taai'* to organise: tough, in other words. That reflected the conventional wisdom about contract workers, as well as the gulf between them and the permanent residents that I personally would have no part in bridging. It was from the same hostels, I afterwards realised, that Poqo, a militia associated with the PAC, had launched a march which blazed into the imagination of white South Africa, in 1962, by setting off to attack Paarl prison and local police station.[21]

The branch general meeting
Why was the Preference Area policy more harshly applied at Jones than Moberg's, a couple of kilometres down the line? The bosses of Moberg's were no less part of the Afrikaner establishment than Jones, as I will later explain. It can only have been because of the Union and how workers were organised. If there was any truth in the idea of the two unions, FCWU and AFCWU, operating as one, any form of collusion with this policy had to be ended. As a trade union representing a relatively advantaged section of the workforce, FCWU had to make common cause with AFCWU against any attempt to divide workers on racial lines.

The split in FCWU that led to the formation of AFCWU, I learned from the minutes, had been highly controversial, and had been eloquently opposed by Oscar Mpetha, who subsequently became the general secretary of AFCWU.[22] The reason why there were two branches of FCWU in Paarl, I also learned, was the result of another split. To understand the importance of this split, it is necessary to understand the importance of the branch, and the branch general meeting, in terms of the way the way the Union was organised: what I will refer to as its model of organisation.

It is often said that the foundation of a trade union is a well-organised workplace, but that is not entirely true. It is through meeting workers from other workplaces in the same locality and sharing experiences that workers in the first instance develop a sense of being part of something larger. The place where this should happen was the branch, and especially the branch general meeting. Every member was entitled to attend and vote for what he or she believed was in the members' interests. Although there would always be tension between matters that were best decided at branch level and what concerned the Union as a whole, it was a branch matter as to who would lead them, as office-bearers. Similarly, it was a branch matter whether to employ a branch secretary.[23]

The Union's model of organisation was thus a decentralised one, in which branches had considerable autonomy from the head office. This was also in accordance with trade union conventions at the time. When Hester Adams asked for a cheque book for her branch, what she really wanted for the branch was its own bank account, into which it could deposit the subs it collected. She was asking for the autonomy to which it was entitled, as only a percentage of the subs collected was due to the head office in the form of an affiliation fee. Ordinary workers had real power over a paid official in terms of this model: to secure her or his employment, the paid official had to secure her or his relationship with the ordinary workers.

Paarl branch, it transpired, had formerly comprised Moberg's as well as Jones, and Juffrou H had been its branch secretary. But she was from Jones, and there were rumours about the company she kept. She never won the trust of Moberg's workers, and at a branch general meeting in 1965 they were in the majority and elected one of their own to be branch secretary. She responded by mobilising workers at Jones in her support. To avoid a split in the Union, the 'compromise' was to split the branch. As happens with a bad compromise, the wound it was intended to heal festered.

Perhaps Juffrou H's response was understandable. The prospect of being out of a job was of no particular concern to me, because insecurity came with the choices I had made. For someone from a factory, a position in an office represented a move upwards socially, even if it was only a trade union office. Having made the move, she could not easily go back to where she had come from, if indeed the bosses would have her.[24] On the other hand, a 'leader' who was prepared to split the membership to secure her own position put her own interests before those of the organisation. This went with a leadership ethos intended to cultivate dependency rather than values. I did not see Lilian's warning not to trust anyone, in the political climate of that time, as concerning rumours. There are paid officials who are informers, she was telling me as a matter of fact. Was her addition of 'not even me' a ploy to persuade me of her sincerity? Or was it that she had a bad conscience? I was persuaded of her sincerity, at the time.

6

The road to Ashton

THERE WERE 15 FACTORIES canning fruit and vegetables covered by the Agreement. Apart from the core branches in and around Paarl, the two factories nearest the head office were located at the perimeter of the town's eastern sprawl: Deepfreezing and Preserving, or Deepfreezing for short, belonged to a UK multinational and was adjacent to the coloured township of Macassar; Gants Foods was in the industrial area of Somerset West.

I wrote to the manager of Gants to ask permission to meet with the workers, and introduce myself. 'I find it very inconvenient,' the factory manager of Gants had replied.[1] He would always find it inconvenient, I would subsequently discover. There were no workers paying subs at either of these factories, so there was no pressure the Union could bring to bear on either management to accede to our request.

Factories like Gants and Deepfreezing were prepared to negotiate every few years with a trade union that did not represent their workers, because it suited them to have the Agreement. That way, no one among the canners would have a competitive advantage, because of the wages they paid. In fact, an agreement negotiated with a weak, unrepresentative trade union represented the best of all possible worlds, given that such a union would have no influence on how the factory was run. The only downside was having to reply to letters from the Union.

You could understand why the workers had stopped paying their subs at Gants and Deepfreezing. The wages were on a par with what workers in the clothing and textile factories earned. These were the other industries of consequence where women might find work. It was also not obvious the workers were worse off for not having a trade union, either in the clothing and textile factories or at Gants and Deepfreezing. The situation was different, however, beyond the ring of mountains to the north and the east that had been the first barrier to the expansion of the colony.

Prior to 1974 the bosses in the industry had maintained a differential in wages between factories within the orbit of the city, like Paarl and Somerset West, and the more far-flung rural towns. Now all workers on the same grade were paid the same wage, regardless of where they worked. In the *dorp* – a rural town – the point of comparison was always the farm.

Compared with what farm workers were getting, R16 was an attractive wage, especially when there was the prospect of being able to earn overtime.

Communities were also more closely knit in *dorpe* where there were canneries – Grabouw, Tulbagh, Montagu and Ashton – but where there were fractures in the community, they were more visible.[2] Understanding the local dynamics of a community made it easier to identify the persons whose support you needed to secure in order to establish an organisation. My first priority was Tulbagh, where the South African Preserving Company (SAPCo) had agreed to a stop order. Its name was presumably intended to draw attention away from the fact that it was a multinational company based in the United States, although the factory was managed by South Africans.[3] At our first meeting the general manager, Van Schalkwyk*, assured us there would be no problems implementing stop order facilities by the start of the season. I was still at the stage of looking to others to take the lead, and let Juffrou H do the talking.

The next priority was Ashton. Here there were more workers in the canning industry concentrated than any other centre outside Paarl, yet only 20 were paying subs. From Paarl you rejoin the N1 where it begins to climb up the Du Toit's Kloof Pass: in those days there was no tunnel. This was the Great North road of my imagination, connecting the old colony to the economic heartland of the country. At Worcester you branch off in the direction of Montagu, along the R60. Montagu was an old district capital, on the far side of another ring of mountains that marks the start of the Little Karoo. Ashton was on the near side.

The bosses of Jones had taken over a factory in Montagu that still went under the name of Brink Brothers. But Ashton was the more important centre for the canning industry. Langeberg had its flagship factory there, and in the season it was (and still is) the biggest in the industry, employing some 3,000 workers, many of whom were bused in from neighbouring towns: Robertson, Bonnievale, Suurbraak, Swellendam and Montagu itself. This was also the site of Ashton Canning.

The town of Ashton looks across towards where the Breede River flows, not quite visible, on your far right, fed by a stream that at most times of year is a trickle through the middle of the town. The coloured population stayed on the left, as one entered the town. This was a township with the usual two-roomed houses devoid of features except the odd church and a community hall, flanked by gum trees and open veld. There was no middle class visible, bar the houses of pastors, adjacent to their churches.

Most of the white population stayed on higher ground, as is their wont, further down the road. The factories were on the other side of the road. Miles out of town on the Swellendam road is Zolani, where the African population stayed. Some of them worked at Langeberg. Some of them would have worked in Montagu. But there was not a single African worker at Ashton Canning, which was privately owned, because the boss had implemented his own version of the Preference Area policy. It was the only factory that employed just coloured labour.

Codes of conduct

Del Monte Corporation was the parent of SAPCo and, like other prominent American companies invested in South Africa, was a signatory to the Sullivan Code. This was a code of conduct which was intended to demonstrate that you could make a decent profit in South Africa and still do your bit to end apartheid, thus deflecting pressure on companies to disinvest. Among other things, the Sullivan Code required that employers offer the same benefits to its black employees as it did to white employees. This was of course easier (or less costly) to do the fewer blacks (meaning blacks in the generic sense) were employed.

So whereas the distinction between workers employed throughout the year and seasonal workers was fairly fluid at other plants, SAPCo introduced a rigid distinction between 'permanent' workers and the rest. These permanent workers benefited from a medical aid, and pensions and housing schemes for which only they qualified, though they constituted some 5 per cent of the total workforce when the factory was in full production.[4] This was a far smaller percentage of permanent workers than other canneries had, and the percentage of seasonal workers was correspondingly higher. They numbered about 1,500 and were transported in buses from places like Ceres, Wolseley, Gouda and Saron. To get that number to fill in forms at the start of the season was a daunting task, particularly when there was no full-time branch secretary. Having done so, the Union had to ensure that the employers acted upon them. This, then, was the downside of the stop order system. It meant, in effect, relying on the integrity of the bosses.

Juffrou H had addressed Van Schalkwyk in English, clearly also his second language. Among the coloured middle classes, Afrikaans is the language of kinship; English is the language of business and for engaging with the outside world, except of course officialdom or the Afrikaner establishment: *die boere*, as she would refer to them when among her own, and as she would later refer to me. English is also the language of upward mobility, and in this instance her choice of language was consistent with an approach to the bosses that was chatty and reassuring. The subtext of what she was saying was that the Union did not represent any kind of threat, while people like her held office in it: if you make our life easy, the Union can make your life easy.

It established a collusive tone in relations with the bosses that I could not put my finger on at the time. I eventually concluded it was characteristic of relations between the members of the Afrikaner establishment and a certain class of coloured that they regarded as their natural allies, since it was altogether absent when someone more outspoken, like Mrs R, was present. In this instance, Juffrou H's chummy tone probably prompted Van Schalkwyk to tell us about the personnel manager the company was looking to recruit. Having a manager who was specifically responsible for 'personnel' issues was a new development, and the only other factory in the industry that had one was Anglo's RFF. Everywhere else it was the wage offices that dealt with individual workers' issues. It was the factory manager's responsibility to deal with issues of a collective nature.

We are looking for a black personnel manager, Van Schalkwyk said. The problem was that it was extremely difficult to find a black personnel manager with a surname that Americans would be able to pronounce. In the months following this startling admission, it was puzzling to see how, despite submitting hundreds of forms filled in during the season, the amount of money deducted by stop order did not change. It was not even possible to check why not, because the company claimed it was not able to provide a list of workers on stop order. This situation persisted until a coloured man called Baxter* was persuaded to abandon teaching and become their personnel manager. He had a dark complexion to boot.

It is doubtful whether, collectively, the workers of SAPCo were any better off as a result of the Sullivan Code, or that workers employed by the *boere* were necessarily worse off. The bosses of Langeberg were *boere*, literally: it was constituted as a co-operative whose members were farmers, rather than a company, although its management must have had a considerable degree of autonomy. Some years before, the co-operative had taken over the group of which Moberg's was part, Standard Canning. This expansion, Anglo and other bosses complained, was facilitated by government policies that favoured co-operatives.

The Union had had members at Ashton for nearly as long as in Paarl, but you had to wonder if they had ever been well organised. The contrast with the lunch-hour meetings in greater Paarl could not have been starker. Even at Jones, despite the mistrust, workers had paid attention when I was introduced. Here it was a battle to make oneself heard, above the babble of voices and the din of dominoes being slapped onto tables. Clearly their branch secretary did not command the attention of the workers present. Oom At, the chairperson, was nowhere to be seen. Thanks to Mrs R, we managed to gather a small circle around us. She could be a compelling orator, yet her inclination was to preach and *skel*: blaming the workers for their evident lack of appreciation for all the Union did for them. Workers were accustomed to being addressed in this tone by women who were their supervisors, and Mrs R had once been a supervisor herself.

We were able to get from Langeberg to Ashton Canning in time for the lunch hour there. The boss was a farmer rich enough and with enough orchards to set up his own plant rather than cooperate with his neighbours to do so. Mr Waggenaar, who had reported the assault of a worker at the MCM, was pleased to see us. You could not say the same of the workers. They listened out of politeness, but with no hint of enthusiasm at what we had to say. And we all had the same adjective in mind after meeting with Mr Du Preez*, the factory manager, and trying to discuss the assault with him: *skynheilig*, his self-righteousness clearly a cover for his hostility to the Union. Our hands were nevertheless tied, because the worker had not complained to us himself. Doubtless he knew we would not be there to protect him from the consequences of getting his foreman into trouble. The social relations that allowed workers to be assaulted would not change without a strong local organisation, and the Union in Ashton was not that.

It seemed to me somewhat self-serving for a trade union leadership to blame workers for their disorganisation. There had to be a reason why the workers were not paying subs, and this emerged when we began to go through the little blue books in which the branch secretary collected subs. There were pages torn out, and the amount collected was far larger than was handed to the Union. This was the doing of one of the branch secretaries, who claimed she was entitled to payment for her services, and she had been helping herself for some time, in lieu of a salary. The workers probably suspected she was eating the money, and were making a statement by refusing to pay their subs. But that did not explain their boorish behaviour, which was in complete contrast to the sympathetic response of the workers in Montagu, some ten kilometres away. This behaviour remained a feature of lunch-hour meetings at Langeberg in years to come, but gradually dissipated, as more and more workers were organised. Years later, a worker from Robertson told me that they had been put up by the foremen to behave in this way.

It is unlikely the foremen would have done so without a nod from senior management, but manipulating workers was preferable to outright repression. Mr Waggenaar clearly had a passion for justice, and at successive management committee meetings, after the main topics of discussion had been exhausted, he would relay a fresh injustice at Ashton Canning or its surrounds: a family evicted from their home on the farm, factory workers who were put to work in the fields during the off-season, and so on. All too often I would say I would look into the matter and did not. I simply did not have the capacity to do so. This did not seem to deter him. It appeared to serve a therapeutic function to have a monthly meeting where you could express your sense of injustice at the way things were.

7

West Coast road

A FEW MEMBERS AT SALDANHA BAY and Lambert's Bay, on the West Coast, was all that remained of the Union in the fish processing industry, and there was a letter waiting for me upon my return from Ashton, from Hester Goncalves, the secretary of the Lambert's Bay branch. The workers wanted the Union to visit. The bosses wanted to introduce a new piecework system and the workers were up in arms about it. 1 November was the date on which the rock-lobster season opened, and we had to move fast.

I immediately set about arranging a trip, but Lambert's Bay was one of the far branches. We would have to be away overnight, and Juffrou H was the paid official eager to accompany me. Maybe the others had family obligations. I was hesitant to ask about marital status or children. That was personal, and I was anxious to keep my own personal life private. Athalie and I were living together, and maybe they would think this was 'in sin'. Given Oom Joe's concerns about the Immorality Act, I could not refuse Juffrou H's suggestion that Nellie Kilowan come along as well.

From Paarl the road to the West Coast lies through Malmesbury, where you join the N7, the national road to South West Africa, as it was still known. On the far side of the N7 was Westbank, the coloured township, and the graveyard where Oupa lay buried. This was the town to which he relocated after the Boer War, and the contrast between the lush green of the winter wheat and the baked yellows of summer was imprinted in the memories of my childhood.

Outside the town, off the N7, there is the turn-off to Vredenburg, and beyond Vredenburg lie the roads to Saldanha and to St Helena Bay and, at the mouth of the Berg River, to Laaiplek. From Laaiplek there was a gravel road to Lambert's Bay, parallel to the coast. The quicker way was across the rolling hills of the Swartland and over the Piekenierskloof Pass, where the N7 descends into the valley of the Olifants River. There one follows the river until the turn-off to Graafwater, at Clanwilliam, and the sea. But all approaches to the West Coast are similar, as the world flattens into an arid, wind-blown expanse bounded by sea and sky.

There was one tarred road into the town of Lambert's Bay. It led straight in the direction of the harbour, where the 'kreef' boats were being prepared for the season. The breakwater that forms the harbour joins a rocky outcrop in the bay to the shore. On this outcrop is the town's scenic attraction: a densely packed colony of Cape gannets, white birds with yellow and blue markings, surrounded by the more common cormorant. The noise and ostentatious displays of the gannets and the silent to-and-fro of the pitch-black cormorants made a comparison with the segregated town irresistible.

The factory faced the harbour and in the street behind the factory was the Dixie Café. 'The café', where anyone and everyone buys their milk and bread, is something like a national institution, but I had never encountered one like this. There were separate entrances and separate counters for whites and non-whites, as there used to be for liquor stores. It was as though one entered a different social world, where despite the tumult in other parts of the country time had stood still.

The coloured township was to the south of the town, and a short walk from the factory. This had been the flagship of the Oceana group, the largest of the empires built on profits from canned pilchards. However, the bosses had long since fished out the pilchards along the West Coast, and the canneries had been moved to Walvis Bay, where they were in the process of repeating the exercise, by fishing out the resource in what are now Namibian waters. The dregs of an industry that had once been a major employer were left behind.

Among the dregs, rock lobster was the money-spinner. Even though it provided only intermittent employment for some 13 weeks a year, rock-lobster tails fetched outrageous prices in the restaurants of New York. At any time of day or night, during the season, the siren would go off, and the women would don their blue and green overalls and stream to work, clasping orange rubber gloves: these were to break the necks of the freshly landed rock lobster, or to gut them, or to pack the tails in boxes, for export to the United States. *Nekbreekers*, *dermtrekkers* and packers, they were called.

The township itself comprised the usual matchbox-style houses that municipalities used to erect, although in fact they belonged to the company. At one end of the sandy streets were the somewhat more spacious A-frame houses that the more elevated of its employees occupied. This really was a company town. It even owned the only hotel. But at the furthest perimeter of the township there was a string of middle-class houses, where the coloured skippers who owned their own boats stayed.

Voorkamer stories

Visits to branches would usually be truncated affairs. We would arrive just in time for a meeting with the workers, and sometimes meet the committee afterwards; then we would hit the road, to get back to deal with the next crisis. But it was not possible to arrange a meeting in places like this from afar. Telephone communication was difficult if not impossible, and anyway you could not be sure you could rely on the locals to make proper arrangements and ensure that all the workers were informed. So the meeting could only take place that evening. Since the season had not started there was an opportunity to sit down and talk, over a cup of sickly sweet tea made with condensed milk.

Everyone addressed Goncalves as Aunt Hester, and she did most of the talking, prompted every now and then by Juffrou H. Aunt Hester was a great storyteller, a working-class raconteur. The story she had to tell was of the great strikes at Lambert's Bay, and how the workers had stood together and the bosses had eventually had to back down. It was also the story about Annie MacKenzie, the charismatic worker leader who had brought the whole production line to a standstill with a clap of her hands. At the recall of this moment Aunt Hester clapped her hands herself, with an extravagant gesture, as though it was a moment emblazoned in her memory.

It was here, in Aunt Hester's *voorkamer*, that I also heard for the first time about the strike at Red Robin in Wolseley. This was a fruit and vegetable canning factory some hours travel from Lambert's Bay, yet somehow, through the Union, the two industries had been linked, and somehow the same Annie MacKenzie had also worked there and been at the forefront of the Red Robin strike. I later learned from Oscar Mpetha she had been planted there by the Party.

This was also an opportunity to observe Juffrou H's interactions with the workers at close hand. Two explanations came to mind as to why she had been eager to accompany me. One was to spy on me. Her tinted spectacles made it impossible to look her in the eye, and she inspired mistrust. But maybe she also wanted to make it up to me, because of her objection to my becoming general secretary. The *padkos* with which she plied me was lavish in a working-class way: lots of meat, which I did not eat.

What I observed in Aunt Hester's *voorkamer* was that Juffrou H felt among her own people, whose cause she genuinely believed in. But I have no idea what Aunt Hester made of the young white man listening intently to all she had to say. At some point the issue of where I was going to urinate had to be broached. There was no alternative for her but to take me to the shared toilets, with a bucket brimming over. It was my first acquaintance with the bucket system. Then there was the question of where I was going to sleep. The *voorkamer* was almost certainly a bedroom after dark. To avoid further embarrassment, I announced early on that I would be sleeping in the kombi.

Mary, the chairperson of the branch, was present in Tant Hester's *voorkamer* that day, and had listened to her telling of the story without betraying any sense of irony. It was evidently a story she had heard told before, probably on many occasions. A year or so later she told me about a detail concerning the great strike that Aunt Hester had omitted to mention. The bosses had brought in scab labour to try to get production going again. The young Hester Goncalves had been among them.

The discovery that Aunt Hester was a scab did not in any way dilute the power of her story, but gave it a new dimension. I did not doubt the following that Annie MacKenzie had commanded, or that she had brought production to a standstill with the clap of her hands. That signal symbolised the moment of unity, when workers stood together in solidarity with one another. Whatever the circumstance that impel people to scab on others during a strike, they are also workers. The sight of her fellow workers standing together clearly made a lasting impression on her.

Over the years I was to come across people who remembered the Union because of some strike, those at Lambert's Bay and Red Robin in particular, all over the Western Cape. It did not even seem to matter so much what the outcome of the strike had been, as that the Union had stood by the workers.

A gentleman's agreement

The rock-lobster season represented the only chance the women at Lambert's Bay had to earn a wage, apart from domestic work, but rock lobster was not the only things the factory did. Like the canning bosses, the factory at Lambert's Bay belonged to an association representing what it termed the 'inshore fishing industry'. There was still some canning in the industry, mainly of anchovy and pilchards if they could find any, but what provided year-round employment was the fish-meal plants. Fish meal was a key ingredient of animal feeds.[1] As there was no Industrial Council, minimum wages and conditions of work for the industry were set by government in terms of a wage determination.[2] But because it set them so low, the Inshore Association set their own wage scales.

Supposedly these wage scales were agreed with the Union, and technically they were. Each year the Association sent a letter to its members recommending the wages they should pay, and copied the Union, with a request that it accept these wages. This the Union duly did. The Association would then notify the Department of Labour that a 'gentleman's agreement' had been reached, although there had been no semblance of negotiations or consultation with the workers concerned. It was not clear what purpose this notification served, since the Department did not enforce the wages. They were also not especially relevant to the women, who were paid in terms of a piecework system. The object of that system was to get the work done as quickly as possible, and the more each individual worker did, the more she earned.

But the opportunities to earn were also not equal. The packers were able to earn far more than the cleaners (the *dermtrekkers* and *nekbreekers*). What the women were up in arms about was a new scheme that the new factory manager was proposing. At the meeting we had called in the local community hall, there were no men present apart from Oom Hendrick. He was a coloured foreman to whom I had been introduced that morning. At various points in the meeting the workers would defer to Oom Hendrick, who sat impassively throughout the meeting, saying nothing.

No one was sure what the new piecework system entailed that the new factory manager was proposing, but what was clear was that the piecework earnings of all the workers would be pooled and divided equally among them. Although it was only the packers that would be detrimentally affected, all the workers with one voice rejected the scheme. It would be almost impossible to check whether workers were in fact being paid for what they had done, and nobody trusted the bosses not to cheat them.

The meeting with the manager took place the following day. The room adjoining his office was lined with wooden panelling and portraits of the luminaries that had founded the company. The general manager, it seems, had anticipated all that we had to tell him. There would be no change to the way in which the piecework system had operated, he reassured us. It also seemed he had foreknowledge of other complaints that had been raised. Even the names of the committee elected to represent the branch did not come as a surprise. It afterwards dawned on me that his prescience was not so much because he had his ear to the ground, as that he had an informer. I began to wonder about Oom Hendrick.

If the Union had refused to accept the Inshore Association's wage scales, the employers would doubtless have gone ahead and implemented them. No one in the Union even knew precisely how many factories were covered by this 'agreement' or where, apart from Lambert's Bay, they were located. The letters from the Association did not say, and the institutional memory about where the Union had once been organised was lost. So I decided to take a chance and ask the Association for a list of factories to which the agreement applied, not really thinking they would comply.

It turned out that, as well as at Lambert's Bay, the dregs of the industry comprised a string of factories operating along the West Coast, some at places I had never heard of. Up till now I had only visited branches where the Union had some organisation. Immediately to the north was Doringbaai, which I could see from the Association's list also belonged to Oceana. So after meeting the workers of Lambert's Bay, as darkness fell, we took the gravel road northwards, along the Namaqualand coast. Mary took us to a woman she knew. Like Mary, she stayed in a company town, and was employed in the rock-lobster season.

Women and the working class

To claim that a trade union was embedded in the working class might sound like academic claptrap, and it was certainly not true in a place like Ashton, where the Union was (and probably always had been) a flop. Yet even though no one in Doringbaai had seen or heard from the Union for more than a decade, the woman whose house we came to had no hesitation in summoning others to meet us. She was probably a Cloete, as many in those parts were. These workers in turn had no hesitation in welcoming us as long-lost comrades, and undertaking to collect subscriptions from the workers as soon as the season started, which they duly did and without further prompting.[3]

Taking the gravel road back from Lambert's Bay, we stopped at Laaiplek, outside Marine Products, where Oscar Mpetha had once worked. Juffrou H knew the name of a worker there, and asked to see him. He came out of the factory with a colleague, and I was introduced. Here too it was more than a decade since the Union had visited, yet they were excited that the Union was being revived and undertook to arrange a meeting with the workers so that we could re-establish a branch there. This they did.[4] It was the only occasion I can recall when coloured men took the lead in reviving the Union. To the extent that the Union was embedded in the working class, it was owing to the women.

'We are the people from the Union,' we told a woman whom we stopped to ask for directions, on a back road between Lambert's Bay and Saldanha, at what must have been the dwelling of a farm worker, although there was nothing resembling a farm in sight. '*Ons ken vir die Unie*,' she beamed in reply. We know the Union. It stood by us during the strike. Probably she was referring to the strike at Lambert's Bay, and she had moved in the direction of Saldanha after the cannery closed down, lured by the jobs there. This was where Annie MacKenzie had ended up, as I was later to discover. She was elderly but feisty and, like Aunt Violet, had the complexion of an African without particularly African features. She was also fluent in Afrikaans and isiXhosa: surely someone who was the product of an interracial union, who had bridged the racial divide.

The largest employer in town, as on the entire West Coast, was Sea Harvest, built next to the harbour quay where its deep-sea trawlers lay moored. This was where Lossie Rogers, whom I introduced as Aunt Lossie at the start of this book, worked. She met us, wearing the white coat of a supervisor, on the steps leading up to the main entrance of this imposing, modern factory. Over some years she had persuaded a handful of women out of close to 2,000 workers to pay their subs. She apologised for not having posted the last lot of subs she had collected, and hastened to her locker to fetch the money. I cannot imagine what Aunt Lossie told these women to get them to part with their money, other than that '*die Unie is 'n goeie ding*'. This cannot have been because of anything it had done at Sea Harvest, but what it had once been, and also because Aunt Lossie was so obviously a sincere and honest person.

Everywhere I went, it was the women who remembered that past, and the children of those women. It was the women who also understood soonest the necessity to revive the Union on the West Coast, even though many or most were only sporadically employed. I do not agree with those who see the working class as comprised of those who are employed for the time being as workers, and disregard those who are not able to find regular work or any employment at all. Yet it was a class created by processes of industrialisation. Along the West Coast and on the *platteland*, in the case of the canneries, industrialisation had taken place in Annie MacKenzie's lifetime and the lifetime of my parents.

Soon after my return from Lambert's Bay, the news broke: 26 persons involved in the emergent union movement had been banned.[5] Among the 26 banned were friends and acquaintances at the Advice Bureau, and people involved in the organisation of African workers in other centres, several of whom I knew from student politics. The odd person out among them was Miss Yon. The banning order was issued under her married name, and nobody outside the Union knew who she was. She also did not work for a trade union, and even when it became known that the medical fund shared offices with the Union, it was not at all obvious why she had been banned.

It was also not obvious why the government banned the others, but the Minister of Labour would surely have been consulted. He in turn would have consulted his newly appointed adviser, one Nicolaas Wiehahn.[6] If the Soweto riots were a factor, it can only have been out of a fear that an alliance might develop between workers on the one hand and students and youth on the other. But there was no hint of such an alliance. In Cape Town, at least, I continued to hear stories of contract workers being pelted with stones for a stay-away about which no one had consulted them.[7] It was in response to this that I became increasingly opposed to manipulative forms of politics in terms of which workers would be told what to do by people who claimed to know better.

There had also not been any notable advances in the organisation of African workers anywhere in the country. Most probably the bannings were part of a pre-emptive strategy. The people I knew from student politics, and friends and acquaintances from the Advice Office, were organising workers for the same reasons I was, only they had been doing so for longer. The objective of a pre-emptive strategy would have been to nip in the bud an emerging trade union movement that was hostile to the existing order, and preserve a space for the established union movement to supervise the organisation of African workers into parallel unions, as TUCSA was trying to do in its half-baked manner.[8]

'I have no idea why I have been banned,' Miss Yon said when I went to see her. 'It could not be that I still correspond with old friends.' She lived with her family off Voortrekker Road, in a white working-class neighbourhood, and I had to wait until she thought a meeting could be safely arranged. I was not sure whether she feared the surveillance of the Special Branch or of her husband, who worked for the railways. Both bosses and workers contributed to the medical fund, and not all the workers who contributed were members of the Union. It was nevertheless seen as the Union's fund, and Miss Yon had come from the Union, where she had been 'the typist' (as the administrative assistant to the general secretary was known). From her position in the fund she had been the stable influence in the administration of the Union. When it began disintegrating, it fell to her to do something about it. The one friend I was sure Miss Yon was corresponding with all the while was Miss Ray.

The Special Branch must have known that. This was the connection it wanted to sever. It would have served no purpose to ban me and leave that conduit intact. It would also have sent out a wrong message: I was, after all, general secretary of a registered trade union. There was a special meeting of the Union called to discuss the banning of the medical secretary. No minutes were kept, and there were no ringing denunciations of the government. Members reacted with a sense of fatalism, and resolved to collect money at the factories for a parting gift, as was the decent thing to do. I hoped the workers would react more passionately if I was banned one day.

The immediate consequence of Miss Yon's banning was that the bosses tried to take charge of the administration of the fund, concerned no doubt, that the fund was being used to support the Union. They refused to accept a person we put forward as a replacement. Instead they proposed moving the fund to their offices and install their own people to run it. There was nothing in the constitution of the fund to prevent this. When no agreement could be reached about this, they proposed a compromise. No new person would be appointed, and I would become the secretary of the fund instead. In so doing, I think they were hoping to tie me down in administrative work.

What marked the end of the Soweto uprising for me was what happened over Christmas 1976. There was nothing new about the state using terror tactics to put down threats to the established order, but you generally expect it to use its own forces to do so. Instead, it was workers, and the most oppressed section of workers at that. Migrant workers wearing white headbands and carrying arms crossed the line separating the hostels from the townships to attack the residents of Nyanga township in Cape Town. The police stood by watching. The conventional wisdom was that the police were able to use them against their 'own people' because migrant workers were backward, but this begged the question: how was it that the migrants were so easily used? Surely they would not have attacked their 'own people' if that is how they had seen them?

PART 3
Reconstruction

'Traditions, when vital, embody continuities of conflict.'
– Alasdair MacIntyre, *After Virtue*

8

Waterkant Street

THE LEADERSHIP OF THE UNION must have been worried that I would be intimidated as a result of the bannings and shrink from the task at hand. This had apparently happened when Miss Ray was banned, and a young intellectual replaced her.[1] Having lost the person who more than any other held the administration together, the Union could not afford to lose me, so if anything my position in the Union had been strengthened. The replacement who was intimidated, however, was the young woman Miss Yon appointed as typist in the stead of Johnny M's partner in crime. She was from a middle-class coloured background, straight out of school, and could have had no inkling of what she was letting herself in for by accepting a position in an organisation where this kind of thing happened. Now she wanted out.

I was only too happy to let her go. The kind of person I needed as an administrative assistant, I supposed, had to be 'politically aware', with a grasp of what I was hoping to achieve in the Union.

I could not recruit a white for this position without opening myself to the accusation that I was promoting one of my own, even before the racial dust regarding my appointment had settled. I did, however, know of a coloured woman who fitted the bill: someone I had met in my previous job, who was bright and competent, and politically minded. I went to see her, and she was only too keen to resign from her job and join me in the Union. I will refer to her as Vormat, a name that I coined for her, as I sometimes did for persons with whom I had a complicated relationship, and by which I would refer to her privately.

Someone from Cape Town who was 'politically aware' was bound to have been part of the coloured intelligentsia, which had long been under the sway of the Unity Movement. The Unity Movement was known for the purism of its political positions, and its reluctance to get its hands dirty. This was graphically demonstrated to me by a remark that one of its leading lights made about the difficulty of organising in the African townships, because it was unlawful to enter them. I was going into the townships every other day at the time. Vormat, however, was part of a new generation which was willing to take risks. This was also why she was willing to work in a trade union and reason enough to trust her, I supposed, for the long periods during which I would be out of office. The task at hand was to revive the organisation. This, as I initially conceived it, meant re-establishing the structures that had collapsed, starting with the election of worker leaders at the factories and branches where the Union once had a presence. In doing so, we would also demonstrate to both bosses and workers that it was business as usual in the wake of the bannings.

I would drive to Paarl several times a week, and pick up whoever would accompany me, on visits to branches: those where there was still some organisation, and those where there had once been organisation. Where we were able to meet the workers at the factory, we would; where we were not, we would set up meetings after hours, sometimes in the evening after work, sometimes on a Sunday, when there was no management committee meeting scheduled. Saturdays were generally to be avoided because that was when drinkers were most disruptive. I loved being on the road, and was captivated by the conversations about the Union and the places where we had been: it opened up a dimension of reality that would otherwise have been closed to me. When the conversation dried up, we would sing: initially songs borrowed from the church in Afrikaans, and songs like 'We shall not be moved' from the civil rights movement, in English; later, we graduated to musically richer and politically more explicit struggle songs in isiXhosa. '*Senzeni na*' ('What have we done') could be rendered in all three languages because of its simplicity, and became a kind of anthem. There was no radio in the kombi and I also did not want to instal one. Friday was often the only day I had to attend to the head office administration, and after that I could take the road home.

Friday evenings and Saturday mornings were the only time Athalie and I had to ourselves, so I was quite resentful when Oom Joe insisted I meet him at the head office one Saturday morning. There had been an issue about keeping the branch offices in Paarl open on a Saturday morning, because that was the only time workers had a chance to visit. This, I feared, was what Oom Joe wanted the head office to do, even though it had always been closed on a Saturday, because, unlike in Paarl, there were no workers organised in Cape Town who might drop in for a chat. Instead, he brought a delegation from the management committee with him, bearing cakes. Nellie Kilowan and Juffrou H were not part of it.

Once we were all seated with a piece of cake and a hot drink, Mrs Amon heaved herself to her feet, to make a heartfelt speech, thanking me for all I was doing for the Union. What I was trying to do for the Union, I said in reply, on this and other occasions, was to revive what there once had been, and I needed their help to do so. Mrs Amon, and others besides, did in fact respond to my appeal. The fact that she was willing to accompany me after hours to branches, even though she was obviously not in good health, made an impression on me. She seldom had anything to say, either in the meetings or on the way, but she constituted a kind of moral presence, providing a sense of what the Union once had been, and who it was for: ordinary workers like her.

Reviving the Union, I came to see, meant more than electing worker leaders and re-establishing branches that had collapsed. However, the fact that the Union was in a state of collapse was not due to its model of organisation. On the contrary, it had not collapsed altogether, despite a sustained assault by the apartheid state, because the branches were relatively autonomous. In terms of its decentralised model, they were able to carry on even when the head office was in trouble. The damage done by the state was to the values that animated this model, embodied in its tradition. Mrs Amon was a bearer of that tradition, as was Nellie Kilowan, even if they did not understand it in quite the same way. Reconstructing this tradition now became the vision (in the parlance of organisational development) which sustained me on the road. I was no longer able to keep my options open: my probation had ended, so far as I was concerned, and no one else mentioned the matter again.

The road home was short enough to walk: from the head office, across town in the direction of Signal Hill, where each day the noon-day gun spewed a puff of smoke. Skirting the old Malay Quarter, Schotsche Kloof, you cross Strand Street at its crest where there is a disused quarry that once supplied the stonework still to be seen on many old buildings in the City. Mr Ali had a café on the corner with Loader Street, and following that street you came to a cul-de-sac. The view took your breath away: the breakwater sticking its finger into the sea, as if you were level with the clouds.

As you looked down, below, there was an expanse of hill on your left, sometimes green and sometimes yellow, and a path zigzagging down to Waterkant Street. The furthest end of the hill was bounded by a stone wall. Reputedly it was the oldest Muslim cemetery in Cape Town, but the only indication of this was a single grave with a low whitewashed wall around it and, if you went closer, a few broken tablets with Arabic engravings. You could see it from the yard of the cottage opposite where I stayed. Sometimes people from Schotsche Kloof would gather there, to pay their respects. Sometimes there was a *bokmakierie* in the topmost branches of the pomegranate tree that stood beside it.[2]

Reconstructing and reinventing

Athalie and I would drive to Sea Point of a Friday evening and buy a schwarma or go to a movie. Sometimes we would stay at home. Often, at some point in the evening, David Lewis would drop in. Two Advice Bureau workers were among the 26 who had been banned, and David had now moved into a central role there. So our conversations were as much about the goings on in the Advice Bureau as about the Union, and served to contrast the very different model of worker organisation that the Advice Bureau was developing, by establishing worker committees at factory level. David was also much closer than I was to initiatives to organise African workers in Johannesburg and Durban.

We faced many of the same difficulties as these organisations elsewhere, and they would have experienced the same frustrations. Many was the Sunday we would travel to a branch only to find the hall empty, apart from the contact person who was supposed to have arranged the meeting, and many was the weekday when there would be no one in the cloakroom except a handful of familiar faces. There were important differences, however. For one thing, I was not trying to establish an organisation from scratch. This did not make my task easier; only different. The other obvious difference was that what was left of the Union was in workplaces where coloured workers were a majority. Although there was no coloured person who would have said they accepted apartheid, there was a fear among coloureds of what they stood to lose if apartheid were dismantled. The coloured Freedom Party sought to exploit this fear, and also, I suspected, the leadership at Jones. The question was how to undo the damage to the value of non-racial solidarity.

The first thing was to win the confidence of all sections of the workforce. Since I personally cannot remember learning much from sermons, I became increasingly dissatisfied with the format for meetings favoured by Mrs R and others. It would invariably take up at least three-quarters of the available time. '*Is daar enige klagtes?*' one of the paid officials would ask, when there was hardly time to muster the courage to raise a hand. 'Are there any complaints?' It is people without power who complain, expecting others to resolve their complaints for them, so I began to develop an alternative format, which involved initiating a dialogue with workers. 'Who is the Union for?' I would ask, usually in Afrikaans, with Lilian translating; or 'Who does the Union belong to?'

Often the response would be a puzzled silence. For many, the Union was an institution based somewhere else, which was in some way or other concerned with workers' problems: but it was far from clear how or why. I knew that both Nellie Kilowan and Mrs Amon had the same answer to this question, but to get the 'right' answer from workers took a bit of prodding. 'Where does the Union get its money from?' would be the next question, to which the answer would be 'from the subs'. My next question would then be, 'Where do the subs come from?' 'From us, from our wages,' would be the answer. I would then revert to the question with which I had initiated this dialogue, knowing that this time there would be no hesitation: 'Who does the Union belong to?' 'The Union belongs to us,' a chorus would respond.

The affirmation that the Union belongs to 'us', or to 'the workers', brought us to the nub of the issue: 'Who are the workers?' When the workers replied that 'we are', I would confirm that the 'we' included all the workers, men as well as women, Africans as well as coloureds. It was at this point, I suspect, that the traditions represented by Nellie Kilowan and Mrs Amon first began to diverge. It was never unlawful for African workers to belong to a trade union, only unlawful to belong to a registered trade union. It was never unlawful for an employer to negotiate with an unregistered union, although in practice none did.[3] Government would not have wanted them to, and that suited employers as well. Where coloured workers were a majority and organised, the Union could bring pressure to bear on the employers to meet with AFCWU together with the FCWU. Non-racial solidarity entailed reviving this practice.

It was much easier for someone from my background to do so, not only because I was not in the Union for my own material gain, but because no one could credibly claim that by doing so I was really promoting my own community at the expense of another. Yet I had to tread carefully. Reconstructing a non-racial tradition would sooner or later force me into a confrontation with those who sought to exploit the fears of coloured workers. It also raised the political question: why was it not possible for African and coloured workers to belong to the same union? When it came to the political question, however, and the Union's relationship with politics, it seemed that all traces of what it had once stood for had been obliterated.

This was despite (or perhaps because of) the fact that FCWU and AFCWU had been founder members of the South African Congress of Trade Unions (SACTU), and its most important affiliates by far.[4] SACTU's position was that the workers' struggle could not be divorced from the political struggle, and it was part of the Congress alliance, led by the ANC.[5] An alliance seemed to me the correct way for a trade union to take up the political question, and FCWU and AFCWU had never disaffiliated from SACTU. Since the banning of the ANC, however, SACTU had ceased to operate as a trade union federation inside the country. So the question of how the Union should now engage in politics was an open one, on which I was looking for guidance.

Among the documents at head office I found a file of old photographs. There was a group photo, showing Oscar Mpetha, the general secretary of AFCWU: a young man in jacket and tie, sitting cross-legged, with a resolute eye and his arm raised in a clenched-fist salute, with the thumb raised. I copied the image and put it on my wall at home. There was another of the beautiful Mabel Balfour, the Transvaal secretary of AFCWU, holding a placard headed 'SACTU demands ... We must eat more than mealiepap'.

She was banned in 1963.[6] There was another of Liz Abrahams, the general secretary before Johnny M.[7]

These were the elected leaders of the Union. There could be no objection, I reasoned, to my seeking advice from past leaders of the Union. So, as well as still seeing Miss Yon, I began to seek out people who had been involved in the Union previously. Liz Abrahams's banning order had long since expired. I mentioned to the Dal Josaphat circle I would like to meet her, as a previous general secretary, and it was easily arranged. Oom Joe, Mrs R and Lilian accompanied me to her home. This was whom Reggie September had suggested I seek out, when we met in London. At the time I saw it as simply as another step toward reconstructing the tradition of the Union, although it occurred to me years later that I might just as well be said to have been reinventing it.

The paid official

The immediate problem confronting the Union in 1977 was that the Agreement was about to expire, and the organisation in the canneries was in a dire state. Ordinarily, it is trade unions that initiated negotiations with employers, but for as long as anyone could remember the canners had done so, by applying to the Minister of Labour to appoint a Conciliation Board itself. A Conciliation Board was simply a meeting convened to resolve a dispute, held under the auspices of the Department of Labour. To create a dispute, the canners would propose to cut wages by 5 per cent.

'What would happen if the bosses were simply to refuse to negotiate a new agreement?' I asked the management committee. 'What is to stop them doing so?' Negotiations had become institutionalised, in much the same way that they had been institutionalised in Industrial Councils, if not perhaps to the same extent. Some would see that as a good thing, but I did not at the time. The workers needed to wake up, if the negotiations were not to take place entirely on the bosses' terms. Accordingly, when the Agreement did expire, and the canners once again proposed to cut wages by 5 per cent, it was reported to the workers in the context of the question I had raised. The bosses of Jones were alarmed. The divisional inspector of labour, one Van den Bergh, phoned me to complain.[8] It was as though by taking their demand at face value, the Union was not playing the game.

I doubt whether anyone had actually said the bosses were going to cut the wages, but a reduction of wages would occur if increases in the cost of living eroded the buying power of the wages, as had happened after the last negotiations. Although there was provision for an increase each year, costs had increased more rapidly, and at the expiry of the Agreement workers were no better off, or worse off, than when it had started. The point was that if workers were to get a real increase in wages, which raised their standard of living, the Union needed to raise its level of organisation. As it happened, the bosses were to provide a more immediate demonstration of this need. Whether to impress upon the workers who it was that called the shots, or because for some reason they needed to increase output, the bosses at a number of factories decided in various ways to intensify production at about the same time.[9] The scheme that really caught the bosses' imagination was the introduction of eight-hour shifts.[10] The Union of course supported the demand for a shorter working week, but without loss of pay. This was not what the bosses had in mind. They were going to pay for six hours less per week than the minimum negotiated with the Union.

They were not permitted to do this, in terms of the letter of the Agreement, both because the minimum wage was a weekly one, and because to work the shifts they had to do away with the existing tea and lunch breaks.[11] These had been among the few concessions the Union had been able to wring from the bosses, which made its members better off than unorganised workers. These were what we would point to, to prove to doubting Thomases that the Union could make a difference. Eight-hour shifts meant starting or stopping work at awkward hours, late at night or in the early hours of the morning. There was no transport at these hours, and this again was a particular problem for the women.

Predictably, eight-hour shifts were first introduced at factories where the Union was weakest. At Langeberg Ashton I would only find out about it after the event. But where the Union was organised, it was a question of the bosses bending the local committee to their will. That is what Mr Tredoux, the zealous new factory manager at Jones, tried to do. When that did not work, he decided to go over the heads of the committee, to 'explain' to the workers himself the benefits of working three shifts. The factory manager at Moberg's had more success by prevailing on Oom Joe to sell the scheme to the workers.

It was not clear whether Oom Joe did so because he truly believed more workers would be employed on eight-hour shifts or because his job was to ensure that commands were carried out. This was the problem of having a foreman in a position of leadership in the Union. As well as being president of the Union and chairperson of the committee at Moberg's, Oom Joe was somebody of note in the local church. Sometimes he would attend Union meetings in the black suit and white tie of a Protestant elder. There was also a gender component to the authority he commanded: in the canneries, the foremen were in authority over the supervisors, who were women.

The situation that had arisen at Moberg's well demonstrated the importance of having paid officials who were accessible to members in the workplace. The job of the paid official was to inform workers about their rights and to ensure that the Agreement (in this instance) was complied with. She (or he) generally had to rely on bluff and bluster to do so, since the Department of Labour was not likely to back the Union; and resort to lawyers and the courts was in most instances not feasible.[12] Most importantly, a paid official was not part of the chain of command, and had no need to fear displeasing the bosses, because she (or he) was not employed by them. So the paid officials were also able to prevent workers from being bullied, which was why on important issues the Union would always want a paid official present.

The meeting where the factory manager prevailed upon Oom Joe to sell his shift system had been held without Mrs R, the branch secretary, knowing about it. She was outraged. When I telephoned Oom Joe about this, he told me brusquely that a final decision had still to be taken at a factory general meeting. But by the time Mrs R got to it, it was over. The workers had accepted the shifts, Oom Joe said, and that was the end of the matter. This was to be the first of my three confrontations with him.

Organising the unorganised

Pendlani, the president of AFCWU, was the only person to whom I could possibly have complained about Oom Joe's conduct without being seen to challenge an elected leader. But he was junior to Oom Joe in status, even if he was senior in years. When anyone in the Union referred to the president, it was the president of the registered union that was intended. It also did not seem Pendlani was willing to take on Oom Joe. He had boycotted meetings for some time before my arrival, I learned. Now that he started attending meetings again, he would listen intently but say nothing.

The chain of events that would change Pendlani's attitude had already been set in motion when we stopped off at Laaiplek on our way back from Lamberts Bay. True to their word, the workers we spoke to at Laaiplek organised a meeting, and there were more African than coloured workers present. Some among them contacted workers at West Point, a factory on the other side of the Berg River, at St Helena Bay.[13] I received a letter from someone there called Wellington Siphuka, on blue writing paper, in an immaculate, forward-slanting handwriting: the workers wanted to join the Union, and he urged us to visit as soon as possible.[14] This was how the Union was to spread: workers who had joined contacted workers who had not yet joined, who in turn would write to the head office, or telephone, and ask to be organised.

Pendlani was among the delegation that accompanied me one foggy Sunday morning to meet the workers. The Africans of West Point stayed in a grim hostel with blackened walls on the premises of the factory, overlooking St Helena Bay. It was not possible to meet there, so we drove along the coast to a stretch of beach. We were all taken aback by the number of workers who walked to where we had parked, to hear what we had to say.[15] The introductions and explanations were fairly quickly concluded, because it was illegal to hold an open-air gathering, and we were at the point of distributing application forms when a vehicle pulled up behind where our kombi was parked, followed by a police van. A balding white man in a state of agitation came up to us, and accused me of talking to 'his' workers, which of course I had been, even if I did not regard them as 'his'. I assured him we were merely distributing forms.

The response of the workers to the factory manager's tirade (for that was who the balding man turned out to be) was most informative. They were all from the Transkei, on one-year contracts. Most of them had been working at the factory, year in, year out. As with any contract worker, it would be easy for the bosses simply not to renew their contract. But they did not retreat or show any sign that they were cowed by him or, for that matter, the local police station commander hovering behind him. They were more concerned, I think, to see how the Union would respond. The station commander told us to accompany him to his tiny police station, but he was merely doing the factory manager's bidding. He did not seem to be aware it was an offence to hold an open-air gathering, or perhaps he was simply too lazy to phone the local Special Branch at their after-hours number.[16] After a perfunctory exchange in which we maintained we were merely distributing forms, he let us go. We made a point of going back to Wellington and the others to say goodbye before we left.

If the police officer had bothered to check the kombi, he would have found evidence of another offence. One of the sequels of the Soweto uprising was that the government had introduced petrol restrictions. Ostensibly this was to safeguard the stock of fuel, in the face of the threat of economic sanctions. It was also a way of restricting the mobility of troublemakers, particularly on weekends when garages had to close. The trip from Cape Town to St Helena via Paarl and back was beyond the range of the kombi with a full tank. One way to overcome this problem was to travel with a drum of additional fuel in the boot, which was illegal. But on this particular occasion, I think, we had a plastic pipe to siphon petrol and a drum, the intention being to buy petrol in Saldanha after a meeting there that afternoon. This, of course, was also illegal.

It was already dark and a taxi-driver willing to sell us petrol had not yet been found. We were sitting in the kombi joking about our brush with the law when Pendlani turned to me and said, 'Jail is for humans', and proceeded to explain about his own jail experience in the 1960s. He had been detained as a young man along with many others. They all sat gazing at the floor of their cell, despondently, wondering what was to become of them, until Oscar Mpetha joined them. 'What are you staring at the floor for?' Oscar asked them. 'You should be looking at the window.' It had been the beginning of his political education.

A week or so later a registered envelope with my name on it arrived at head office, in Wellington's handwriting. Inside was the duplicate slip from the shop steward book, as it was called, indicating the subs each worker had contributed and his factory number, and the corresponding amount of cash. I wrote out the receipt personally, and every week after that, when the envelope arrived, for as long as I could. I took it that the fact that the envelope had my name on it meant I was trusted, but trust was not something that could be taken for granted, especially given the corruption that had occurred. It had to be re-established, and it would be damaging if money went missing. I would also be implicated. Later other envelopes would arrive, from other factory hostels along the West Coast. The West Point workers contacted the workers at Stompneus Bay, who in turn contacted workers at Sandy Point, another factory along that stretch of coast. With each factory organised, there would be a general meeting to elect committees for both FCWU and AFCWU: and if there were no coloured workers present, the Africans were asked to organise them, which they did, so that by the time we got round to holding meetings for the fourth and fifth factory along the coast, coloured and African workers were present from the outset.

The complexion of the Union had already begun to change by the time of its 1977 conference. Although it had not yet broken out of its rural isolation, there were new faces from branches that had not been organised for 15 or 20 years. But the clearest evidence of change was the presence of delegates from the newly organised factories in the inshore fishing industry. For workers in the canning industry, this was proof that AFCWU was not being neglected. This generated enthusiasm and hope. I was elected unopposed as general secretary of FCWU. Oom Joe and Pendlani were both re-elected.

Pendlani now started becoming more vocal. His leadership style was in marked contrast to that of Oom Joe, whom he would refer to irreverently as '*die president*', using the Afrikaans title even though he was speaking English. The difference between the two was that Pendlani remained an ordinary worker, and had none of the ambivalence of those in higher positions who are concerned that their association with a trade union would jeopardise their career prospects. He was, for me, one of very few men in the Union who could be said to be a bearer of tradition.

We had now been receiving envelopes from the West Coast factories for more than a year, and not once was there less money than there should have been. Why then had it been so difficult to administer the same system at the factories in the canning industry without a stop order? Why was there an expectation of payment, when Wellington expected none? It was when I saw the efficiency with which the supposedly backward and mostly illiterate contract workers administered their finances that I began to realise the extent to which the corruption at the head office had contaminated the branches.

Sooner or later, you might say, someone was bound to pocket the subs, but no African on the West Coast ever did, for as long the system of hand collections was in operation there. Instead, the person who was exposed at the 1977 conference as having pocketed the subs, 'meaning to put it back later', was a FCWU 'stalwart' in Worcester, who we did not even know was collecting them. It was an early lesson about the relation between corruption in the form of eating the money, and corruption in the organisational sense, when there is a leadership that is no longer accountable. These things are inextricably connected, and not just in a workers' organisation.[17]

Cash across the board
In the meantime the Conciliation Board had met. Despite the offensive by the bosses, or perhaps because of it, the decline in membership over the last few years had been arrested, and turned into a steady increase, as workers at factories where organisation had been dormant began to pay their subs. However, there were seven employer groups who belonged to the Association operating 17 factories, and the Union still did not represent a majority of the workers in the industry.[18] There was also an early indication that the danger to which I had tried to alert the workers was not far-fetched. One company to which the Agreement had applied in the past was not party to the application. This was Ceres Fruit Juices. Although a member of the Association, it had decided not to participate in the negotiations. The association was a marketing association more than an employers' association, Mr Glendening, their secretary, told us. There was nothing it could do to compel an employer to participate.

The Conciliation Board was chaired by officials of the Department of Labour, and took place in their offices. This lent it the atmosphere of a court proceeding. The worker delegates came dressed for church, and seldom spoke. When they did, the officials of the Department made little effort to conceal their impatience, and almost everything the delegates said was left out of the official minutes. Despite the way they were treated, it did not occur to them to utter a critical word. This was the only forum they knew of, and their primary concern was what impression their new secretary would make, compared with the last one.

What the Union was demanding was far more than what the workers were getting: it represented an increase of more than 100 per cent. 'How can you justify this kind of increase? No industry could afford it,' was the stock response of the bosses to such demands. 'How can you justify present wages?' was the retort. It was the opening gambit for an argument in favour of a 'living wage' which could only be attained through what in percentage terms were large increases. The argument was buttressed by a study of what was regarded as 'the minimum' a breadwinner needed to support a family, which, even though problematic, came out a lot higher than the minimum wages the Agreement prescribed.[19] The bosses had heard this argument before. The Wages Commission, I was told, had helped draft Johnny M's submission in previous negotiations. Essentially the same argument was put forward at every other wage negotiation I was involved in. If this lent an aspect of ritual to the exchange between the bosses and ourselves, it was nevertheless an important exchange to be having. It concerned the way the bosses were making profits, and how those profits were distributed.

It was all very well to talk about a breadwinner needing so much money, but most of the workers were women, the bosses replied. A woman was not normally the sole breadwinner. It was Glendening, I think, who said they worked for 'pin money'. So what was essentially a moral argument would give way to arguments based on the market, including what workers could expect to earn in the labour market. 'Workers elsewhere are getting the wages we are asking for,' we suggested. It was not a persuasive argument. The wages set by the Industrial Councils, particularly in the clothing and textile industries, were on a par with what the canneries were offering. In any event, the bosses would counter, it does not matter to us what workers in other industries are getting. They are not our competitors.

'There are more than enough unemployed workers who will be only too glad to work for the wages we are paying in this industry. Nevertheless, we are willing', the bosses said, softening their stance somewhat, 'to grant modest increases.' The kind of modest increase the Union held out for was a 'real increase', meaning an increase above the rate at which the cost of living was increasing. Hopefully, this would help raise the standard of living of workers over time. It was as much as could be expected given the low levels of worker organisation. To achieve a 'real increase', however, the Union had to consider how the wage bill was distributed. Workers earn differential wages, and the most problematic differentials were always between workers on the lower and higher grades, and between men and women.

The grades of work recognised in the Agreement more or less corresponded with the grades used in Industrial Council agreements of all kinds and the wage determinations.[20] It had been a stock demand of the Union over the years to reduce the number of grades and eliminate differentials, so the grading system in the Agreement was less complicated system than many. My point of reference here was the agreements of the Industrial Council for the clothing industry, which imposed an elaborate series of qualification periods which workers had to serve before earning the rate for the job. What the bosses proposed, as they always did, was a percentage increase. But a percentage increase translated into more money for the higher grades, and widened the gap between the different grades. This was why the Union's demand was always for a cash increase, across the board, which invariably represents a higher percentage increase for workers on the lower grades. The cash increase we eventually secured amounted to a 25 per cent increase for the lowest paid, well above the increase in the cost of living over the previous year. It was presented at report-back meetings as a victory for the Union, which in a small way it was. The bosses would never have increased the women's wages by as much were it not for the Union, although they would almost certainly have increased the wages of the men by as much or more. Once again, the agreement was for three years. Although there was an annual escalation, the increasing cost of living would surely cancel out some of what had been gained.[21]

9
Side streets

TALK OF 'VICTORY' following the negotiation of a new agreement was too much for Moberg's. Its management had been holding regular meetings with Oom Joe and the committee for some time, unbeknown even to Mrs R. Someone slipped her a copy of the minutes. 'For us the agreement flows out of a collective process, yet the FCWU has a win/lose approach,' the minutes complained. The only indication of who 'us' referred to was that the minutes were headed 'liaison committee/workers' union matters'.[1] It looked as if the Union committee was in the process of morphing into a liaison committee.

This was Oom Joe's doing, and Liz Abrahams's support in confronting him would be crucial. She had been a worker at Moberg's before becoming GS, and was known and respected by the workers, including Oom Joe. I had by this time become a regular visitor at Liz's house. Liz stayed in a side street off the Klein Drakenstein Road, in a section of Paarl East where there were free-standing houses that were privately owned. Across a valley were the flats where the more impoverished sections of the coloured working class stayed, and behind that, in the distance, the mountains of Du Toit's Kloof.

At first I used to enter through the front gate, down a cement path flanked by showy plants: foxgloves and cockscombs and hydrangeas. Meetings were in the lounge. It was the lounge of a member of the well-to-do working class, with doilies on the couches and a display cabinet with crockery and silverware in it. Later, at Liz's suggestion, I would come through the back gate, on my own. You would park in the gravel road round the corner, opposite an open field, and out of sight of prying neighbours who might provide titbits to the Special Branch. In summer you had to duck beneath the overhanging vines, and might stumble into Liz washing clothes by hand at the cement sink outside the kitchen door.

'I am just a housewife now,' Liz liked to say, and our discussions would always take place over a cup of tea. She was someone without pretension, and spoke in a slow, deliberate manner, in the plain language ordinary workers use, because that was the only language she knew. But there was nothing housewifely about what we discussed. As well as the issue at Moberg's, there was the question of how to go about re-establishing branches in Cape Town and Johannesburg.

It seemed there were no traces of organisation left in Cape Town. Crosse and Blackwell, the first factory to be organised by the Union, was now a depot belonging to Nestlé, the Swiss multinational. The core business of the factory called Fatti's & Moni's had once been fruit juices. I was later to find out that it was now a flour mill, an industry that the Union had never organised.[2] The only factory still making what it had made before, so far as I could see, was Spekenam, the meat-processing plant. I knew it for its Vienna sausages. One summer's evening after work I went to Scottsdene, the coloured township where many of the workers stayed, to distribute a pamphlet. Meat processing was another industry covered by a wage determination, and a new minimum wage had just been set. Low as it was, it was worth trying to tell the workers about it and, of course, about the Union.[3] I was glad to see some of them take the pamphlet, but no one contacted the Union in the weeks that followed. This was my first stab at organising a factory, but I could not find the time to follow up.

It was embarrassing to be general secretary of a trade union based in Cape Town without any members there. This was where most of the canneries and the inshore fishing factories had their head offices, and it was the centre to which the workers would naturally gravitate. It was also where there was the greatest potential for AFCWU to expand. The Union needed to break out of FCWU's rural isolation if it was to revive a tradition of non-racialism. It was embarrassing for altogether different reasons to have no members in Johannesburg, but the task of organising a branch some 1,600 kilometres away was of a different magnitude. I needed to find out how it had been done before, and why it had collapsed. Liz could not tell me. The other person I thought of asking was Oscar Mpetha, but Liz knew nothing of his whereabouts.

What Liz did tell me was that she was still in contact with the person who had been the last secretary of FCWU's Johannesburg branch. This was Mary Moodley, who stayed in Actonville on the East Rand. While I was mulling over how to make contact with her, a medical student in his final year at university came to see me. He was Neil Aggett, and he was keen to become involved in the trade unions for much the same reasons I had done. The thing that impressed me most about Neil was a certain quiet determination. So Athalie and I accepted an invitation to supper at the cottage where he stayed with his girlfriend, Liz Floyd: she happened to be the niece of Aunt Catherine, whose gaze I had sought to avoid during the student sit-in years before.

Neil stayed off a remote bend in a wild remnant of Constantia that the developers had still to reach, and we continued our discussion of the political situation and the difference trade unions could make. Toward the end of the evening he reached into the upholstery of a tattered couch to produce a copy of a pamphlet which he suggested I read. It was Lenin's *What Is to Be Done?*[4] I lay down with Athalie on a mattress under the stars, mulling over the significance of this gesture.

It was thanks to the Cillié Commission I was able to locate Oscar Mpetha. Piet Cillié was judge president of the Supreme Court in the Transvaal, and a known Nat, who was appointed to head a judicial inquiry into the causes of the Soweto uprising. The inquiry had been extended to include the events in Nyanga the previous Christmas, and Oscar's evidence on behalf of the Nyanga community was carried prominently in the press. I would later learn that Oscar himself believed his evidence provided incontrovertible proof that the police had orchestrated the attacks on Nyanga residents by migrant workers. A reporter at the *Cape Times* who was covering the Commission told me that Oscar was employed as a security guard at Walls Ice Cream factory. One evening after work I went to look for him.

The factory was in Salt River, in an industrial area that dates back to a time when warehouses were built of brick, at the end of an obscure cul-de-sac parallel to Albert Road, flanked by the railway line. Behind the gates was a security guard in a uniform that didn't quite fit, with a tie that was knotted too tight and thick-lensed glasses. Although he was now an old man, pushing 70, I immediately recognised him from the photograph I had on my wall.

Dispute resolution

'*In ons tyd*' was how Liz used to preface her answer to each of the series of questions I posed to her.[5] She was always careful to preface her advice with an acknowledgment that times had changed, and that she was not acquainted with present circumstances. I had to explain to her what a liaison committee was, and that government was promoting liaison committees as a substitute for trade unions for African workers.

'In our time we always used to say that the workers belong in one camp, and the bosses belong in another,' she said. There were no shades of grey between workers and bosses. The policy was, according to her, that a Union leader, whether a worker leader or paid official, should not accept anything from the bosses, not even a glass of water or a cup of tea. It was not difficult to persuade her why any initiative to institute a liaison committee was problematic.

The underlying issue was that up to this point the Union only ever met with management at the workplace on an ad hoc basis, to discuss a specific dispute or request. Almost invariably these were meetings initiated by the paid official, and the committee members who attended were simply the members of the branch executive committee which the branch had elected. The committee at the factory was not seen as a separate structure, which was why there was not a separate name for it.

Times had indeed changed. Now, for the first time, it was management who wanted to meet with the Union on a regular basis. My reading of this was they were in effect trying to institutionalise a relationship at the workplace in much the same way that the negotiation of the Agreement had been institutionalised. But Oom Joe could see nothing wrong with this. This was the second of my confrontations with him, and after a heated debate on the issue at the management committee meeting, the majority sided with Mrs R and me.[6] This was not so much because everyone shared my apprehension about attempts to institutionalise the relationship at the workplace, as because Oom Joe had agreed to management's proposal without debating it within the Union. Also, ominously, there had been no AFCWU representatives at these meetings.

My apprehensions might seem quaint from a present-day perspective. Workplace relations nowadays are institutionalised to the point that trade union representatives not only meet regularly with management, but are employed by the self-same bosses to be representatives of the trade union, as I shall explain later. Yet the threat to the autonomy of the Union at the workplace was real, not least because it was not clear whether Oom Joe's primary allegiance was to the Union or to Langeberg.

It was also a present threat. The government had appointed a Commission of Enquiry into Labour Legislation some months before, and it was common knowledge that the Wiehahn Commission, as it became known, was contemplating some form of recognition of trade unions for African workers.[7] This was to be government's showpiece reform, in response to the events of 1976. The political calculation underpinning it was obvious. Between them, the bosses and the established trade union movement would easily be able to contain any organisational or political threat that an emergent trade union movement represented. Liaison committees were clearly an important element of this containment strategy.[8] The object was to provide for a form of workplace representation that did not require workers to join a trade union.[9]

This was the first dispute about policy that had arisen at the management committee since I became general secretary, and the way it was resolved was entirely positive. The meetings were at this stage attracting upwards of 40 or 50 people, of whom at least a third were African. Everyone was free to speak, whether officially a delegate or not, and proceedings were translated: despite the fact that it was the conduct of the president that was at issue, there were no personal recriminations, and everyone left the meeting believing the issues had been settled, as indeed they had.

No committee of the Union would meet with management at the workplace without being provided with an agenda, in sufficient time to consult the workers concerned and, if need be, the paid officials. Also, no committee would meet without representatives of AFCWU being present, and the Union was to keep its own minutes of these meetings. These were safeguards to protect the autonomy of the Union committee, as the structure became known. At a later juncture, these safeguards were elaborated and incorporated into the so-called recognition agreements which the Union negotiated.[10]

The minutes were of course the official record of what had been decided, and I tried to write them in the plainest English possible, setting out the issue so that it would be intelligible even to those who had not been present. But although these were circulated to all the delegates, and formally approved at the next meeting, I was never sure how well they were read, or whether I was writing them for posterity. What was clear, however, was that through debate a practice had been established. Some months later, Moberg's management tried to institute a committee for *'opwaartse kommunikasie'* (upward communication) that would have nothing to do with the Union. The workers steadfastly refused to be bullied into electing representatives. 'I told the workers', Oom Joe later reported, 'that even if they are called in one hundred times, they should stand by their decisions.'[11]

Dispute resolution was a matter about which the Wiehahn Commission would have a lot to say. In terms of the official system of the time, the Conciliation Board was the first step in resolving a labour dispute. The next step, in the case of a wage dispute, would have been to go on a legal strike, after 30 days had elapsed. But legal strikes at the time were unheard of, and in any event all strikes in the fruit and vegetable canning industry and fish processing had been made illegal.[12] The alternative, in terms of the official system, was arbitration by a body known as the Industrial Tribunal.

Juices, as the workers referred to it, was a small cannery in Ceres that was still managed by its owner, a Mr Bosman. He was a charismatic character who believed he could bend the workers to his will.[13] If the Agreement was 'reasonable', Bosman told us, Juices would come in line with what we had negotiated with the other canners, as it had done in the past. But Juices declined to do so once the Agreement had been concluded. Instead, Bosman proposed part of the increase the Union had negotiated would be paid as an attendance bonus.[14]

The workers of Juices had been forewarned of this possibility. Indeed, it was with Juices in mind, more than any other employer, that I had earlier asked what would happen if the bosses simply refused to negotiate. The Union could apply for a Conciliation Board itself, to try to compel the bosses to negotiate. First, however, it had to prove it represented a majority of workers in the industry in question, in each magisterial district affected by the dispute.

The Union would never have been able to prove it represented a majority of the workforce in the canning industry as a whole. The standard we were required to meet by the Department of Labour was an extraordinarily high one, particularly for a seasonal industry. A majority of workers had to have paid their subs each week, for a period of three months, so as to be regarded as paid-up members in terms of the Union's constitution. Proof was required, including receipts and bank deposit slips. The Union was nevertheless able to meet this standard at Juices.

It was probably the first time since the 1950s that the Union had successfully applied for a Conciliation Board. Predictably, the Conciliation Board failed to settle the dispute, but we could now go to arbitration.[15] Ordinarily, arbitration of a wage dispute was something no union in its right mind would contemplate, for it meant giving over to an arbitrator the right to make a binding decision. However, in terms of the system of the time, the arbitrator was the Industrial Tribunal, made up of a trade union representative, an employers' representative, and a chairperson appointed by government. It therefore did not purport to operate as a court, applying 'the law'.[16] This, I believed, was entirely appropriate. The issue at hand concerned policy rather than law.

A tripartite tribunal like this was preferable by far to a tribunal masquerading as a court, such as the court which the Wiehahn Commission would later bring into being. It should also not have mattered that it was more cumbersome, since it was only in exceptional circumstances that a trade union ought to resort to arbitration. However, the more immediate problem we had to contend with at the time was the trade union representative on the Tribunal. He was from the established union movement, and I feared he would be hostile to a trade union that had been affiliated to SACTU.[17] On the other hand, the choice with which the Tribunal was confronted was not so much between what was in the workers' interests or their bosses, as between what was in the interests of the Juice's bosses and the rest of the industry. This, at any rate, is what I argued before the Tribunal.

Juices' strongest argument for paying lower wages was one they alluded to vaguely, without providing any specific evidence. Perhaps this was out of class loyalty. Other firms in the area processing fruit, they said, were paying much less than we were demanding for Juices. SA Dried Fruit (SAD) had two factories in the immediate vicinity where, we later discovered, workers were earning as little as R8 a week. This was also what the largest employer in town, Growers, was paying. Although these were not canneries, they were processing the same raw materials supplied by farmers in the area and sold on the same overseas markets. More importantly, they employed ordinary workers, living in the same places as the canning workers. In all the years the Juices' bosses had accepted the Agreement, the workers at these factories had shown no interest in the Union. If Juices had come in line with the other canners, who knows how much longer this situation would have persisted?

What then happened should have been an object lesson for all bosses as to the consequences of taking a hard line. Now that there had been a struggle, news of the Union's victory spread. This time there was no question that the outcome was a victory: Juices were ordered to comply with the Agreement.[18] Juices' workers spoke to other workers and, in a domino effect, one fruit-processing factory after another was organised, starting with SAD in nearby Worcester and Wolseley.[19]

By 1978 the Union was in a position to apply simultaneously for a Conciliation Board for three SAD factories representing the dried fruit industry. This was the highest level of organisational accomplishment the Union had been able to attain since Miss Ray's time.[20] The object lesson for anyone concerned with worker organisation is that it can only be sustained by concrete gains. These also have to be recent gains. If the factories that were organised before my time had once inspired unorganised workers to join the Union, they were not doing so any longer.

Shortcuts
Strictly speaking, the rebuilding of AFCWU was not my responsibility. Lilian had been re-elected general secretary of AFCWU at the conference, albeit with a noticeable lack of enthusiasm, if not puzzlement, on the part of delegates from the West Coast, who were seeing her for the first time. Then there was the matter of the 'cash on hand' in the auditors' report for AFCWU, which had been raised at the conference. 'Cash on hand' was auditor-speak for subs that had been collected and not banked. Lilian was such a nice person that I could not bring myself to suspect her of 'eating' the money, even though, when I asked her about the cash on hand, she said she would pay it back without hesitation.

I thought the problem with repaying the money was that Lilian could not possibly come out on what she earned, which was R18 a week, fractionally more than a worker on the lowest grade. No one could be expected to go out and organise workers at that salary. The other branch officials were earning more than twice that amount, but there were not enough subs coming into the AFCWU account to pay more. Feeling sorry for Lilian, I later concluded, reflected a middle-class misconception of the problem. Lilian was paid so little because the workers were not willing to contribute to the salary of someone who was of no help to them. Probably, even at that stage, they also thought she was eating their money. For me, the penny had not yet dropped.

There was, however, an objective difficulty about my confronting Lilian. I was not general secretary of AFCWU. Nominally, Lilian and I were equals.

As if to resolve this intractable problem, a man and a woman from Sweden arrived unannounced at head office, pretending to be tourists. I kept no record of their visit, and tried to forget the details as soon as possible afterwards, which is why years later I was no longer sure whether they had been from the Norwegian or Swedish LO, as the Scandinavian trade union federations are known. (Fairy godparents coming to the aid of South Africans generally hailed from Scandinavia.)[21] What they were offering was money with no strings attached.

When offered money with no strings attached, your first thought is what you can do with it. One possibility was to supplement Lilian's salary, so she could pay back the cash on hand. The other possibility was to employ an organiser for the Cape Town branch. We would never make progress by relying on me to stand outside the factory gates when I found time. A full-time organiser was needed, and the head office did not have the funds for another salary. Perhaps this was one of the downsides of a decentralised model of organisation. The reason the head office did not have the money was that the affiliation fees the branches paid were just enough to cover the current costs, and the branches also controlled what reserves the Union had. The Paarl branch, in particular, had substantial reserves, and would have to be persuaded to donate them towards organising the unorganised in Cape Town. The alternative would have been to increase the subs, but I was opposed to that. We should rather be pushing up the number of workers paying them.[22]

The fact that branches controlled their own money, on the other hand, forced the head office to engage with them. What a membership-based organisation does with any surplus it generates will always be a political question: convincing Paarl branch that it was in their interests to organise the unorganised African workers was a necessary political task. Money with no strings attached meant that we did not have to bother. But at the time it seemed a good idea to avoid a confrontation over the issue, and also to steal a march on the Special Branch.

State repression creates a climate of heightened anxiety in which it is easy to justify secrecy, to yourself and others. The national office-bearers agreed to accept the money and not report it to the management committee. No minutes were kept. The money arrived. It was not a large sum, but large enough to pose a problem: how was it to be administered? Although there was nothing illegal about receiving funds, the books of both FCWU and AFCWU were audited, and an auditors' report would surely fall into the hands of the Special Branch. So the money was deposited in a savings account that was not reflected in the books of the Union. I now think of it as a slush fund.

The next problem was how to utilise the funds for their intended purpose. I told the management committee I was 'unhappy' about what Lilian was earning, and persuaded them to let the national office-bearers determine an increase for her.[23] I did not tell them where the money was to come from, or consider the implications. The salary of the general secretary was for the conference to decide. I was instituting a practice that I would later renounce, in terms of which the national office-bearers would function as an inner circle, taking decisions that could supposedly not be entrusted to a more open forum like the management committee meeting, for reasons of security.

I also did not tell the meeting where the money to pay an organiser for Cape Town branch was to come from. The organiser we chose was Virginia Engel, a committed young woman from the same circles as Vormat. By this point, I was acquiring a sharper sense of the social distance between the radicalised coloured middle class and the working class in Cape Town. Would someone who was married to a pastor of the Moravian Church be able to bridge it?

About a year after I handed out leaflets to Spekenam workers, one of them contacted the office. She had been dismissed. Now that she had nothing to lose, she felt free to get in touch with us. Spekenam turned out to be the only factory in Cape Town where workers had some recollection of the Union, and it was not a positive one. All the workers had been dismissed following the 1956 strike. A few were later taken back some time afterwards on the bosses' terms. There is nothing the Union can do about your dismissal, I told her, but if you can put us in contact with some of your fellow workers, we may be able to help them. She did that, and we met with a handful of them in a church hall. We asked them to speak to the other workers and come to another meeting, but only a handful arrived.

Organising in Cape Town would be an up-hill battle, and all the more so because there was no other trade union to which we could look for support. The best-known trade unions organising coloured workers were prominent affiliates of TUCSA and bastions of the established union movement. The most important among them was Western Province Garment Workers' Union, because there were more workers in clothing factories than in any other industry in and around Cape Town, and because it was notorious for the top-down way in which it was run. I will refer to it as the Garment Workers' Union.

Virginia needed help, and Athalie did not have a permanent job at the time. My long and frequent trips were becoming a strain on our relationship. There is always some new contact you have to follow up, or some development you need to respond to, when you are trying to win confidence and are on the road. As much as I might try to let her know when I would be back, this was often not possible in the era before cell phones. She would better understand my absences working in the same organisation. We also both believed the personal and the political were intertwined.

Athalie remembers my having opposed a proposal first put to me by Oom Joe and Pendlani in Liz's *voorkamer*, that she join with Virginia in organising in Cape Town. This can only be because I felt obliged to assume the role of devil's advocate: too zealously, it is clear. Although having Athalie in the Union seemed something like a personal necessity, I could not be seen to be creating a job in the Union for my girlfriend, and even less so as a white. It was a similar dynamic that came into play regarding a job in the Union for Neil. I could not be seen to be creating a job for someone who had, in the interim, become a friend.

The salary I was now earning heightened my anxieties about creating a job for my girlfriend. Despite my objections, the conference had insisted on increasing my salary to R400 a month, much more than anyone else in the Union earned, and also much more than the R160 per month it had been decided we would pay Virginia.[24] Athalie and I were able to live quite comfortably on the R400, I thought. So it was agreed that Athalie would work on a voluntary basis. I could see this arrangement gave rise to a gender issue, but for me race trumped gender. The gender issue I was more concerned about at the time was the lack of coloured male members, although in my defence I must say this was also a concern of many women members.

The other problem with this arrangement was that Virginia would be earning a salary while her co-organiser was not. Compounding this problem, to avoid any impression of favouritism to my partner I was at pains to maintain what I saw as a professional distance at work. Virginia, who was a fair-minded person, later criticised me for being harder on Athalie than anyone else in the office. So to solve one problem I was creating another. It was exacerbated by the fact that there was so little perceptible progress in recruiting members. The two of them would wait outside a factory, to try to persuade workers to attend a meeting after work, at a church hall or someone's house. There would be a handful or none at all, and a further meeting would be called, with no better results. It did not even seem to make a difference in cases where workers stayed in the same area, as with Spekenam.

The problem, in respect of African workers, was how a woman from another class and another race approached a man in overalls, or a group of men speaking a language she did not understand, when there was nothing in the social experience of either gender to make such an approach normal. In Cape Town, at the factories the Union was concerned with, African workers were almost entirely male, and the problem with the coloured workers was fear, more than apathy. This was not so much fear of bosses as of their henchmen: workers who had risen to become foremen and supervisors, and were fluent in English. This, for me, was what was different about their situation compared with the *platteland*: on the *platteland*, workers in the middle layer did not seem as close to the middle classes.

It was surely no coincidence that after about a year of trying to secure a foothold in various factories in Cape Town, the first time the Union managed to recruit a significant number of coloured workers was when their foreman took the lead in signing them up and collecting their subscriptions. This was Mr Christian* in the packing department at Fatti's and Moni's.

The importance of security
I had good reason to be concerned how the Special Branch and employers would react to the news that the Union was organising in Cape Town, but even more so that we were planning to organise in Johannesburg. Nevertheless, it had to be attempted: Jones had a factory there, and Langeberg one in Boksburg. The Agreement had been extended to cover both, in terms of the system of extending collective agreements, which is still a feature of labour legislation today, except nowadays it only applies to Bargaining Councils.[25]

The point of extending agreements was then (and is now) to prevent undercutting by 'non-parties'. Non-parties could be workers who were not represented by unions, or employers who were not members of the employers' association and therefore not party to the agreement. Given how controversial the system of extending agreements was to become, it is worth remembering how little controversy there was then among employers. This was because the minimum wages were set so low that none minded.

The extension of collective agreements was an issue for me because African workers could have no say in them. It was one thing to go along with the extension of an agreement at plants where AFCWU was organised and could be consulted. It was quite another to extend it to factories that had not been organised for more than 15 years. It would be a disgrace if we were to do nothing to address this situation. Certainly there could be no basis for us to criticise the established union movement for its indifference to the wages of ordinary workers until we had done so.

The easiest way for the Union to establish a branch in Johannesburg would have been to link up with a trade union there, and the most obvious grouping of unions for us to have tried to establish links with went under the acronym TUACC.[26] It was one of the groupings that made up what was termed the 'independent' trade unions, meaning trade unions not affiliated to TUCSA that were organising African workers. Personally, I did not think 'independent' an apt term for the movement of which we saw ourselves as part, and prefer to refer to it as 'emergent'.[27] TUACC knew about us, because we were both invited to discuss a proposal to form a new trade union federation in 1977. We did not go, for much the same reasons as we decided to go it alone in Johannesburg.

The need to concentrate on building our own organisation was the line the Union took on the proposal for a new federation at the time. If there was to be one, it needed to emerge organically when the time was ripe. Other organisations in the Western Cape took a similar position. The trade unions which would later become the core of this new federation obviously had a different view, but nevertheless they agreed that the emergent trade union movement was weak. It is therefore necessary to debunk the commonly held belief that government recognised trade unions for African workers because of the growth of the emergent trade union movement, as if this constituted an irresistible pressure from below. The converse was true. As I put it to the Union's conference, it was 'certainly ... *not* thanks to a strong trade union movement' (my emphasis today) that the Wiehahn Commission was contemplating the recognition of trade unions for African workers.[28]

Although we could not say so openly, our belief that the time was not ripe for a new federation had a lot to do with mistrust, especially of the trade union driving the initiative, NUMAROSA.[29] It was based in the motor assembly plants of Port Elizabeth, and was a registered trade union for coloured workers. In this respect, it was like FCWU and unlike the TUACC unions. Unlike FCWU, however, it had no history of organising African workers, and had until quite recently been affiliated to TUCSA. Its non-racial credentials were therefore questionable, and its conversion too recent for it to be spearheading the formation of a new federation, so far as I was concerned. I also did not like fellow unionists being referred to as 'brothers' and 'sisters': this was a Cold War mode of address, used by those who wished to avoid calling each other 'comrade' lest they be taken for communists.[30]

My first stab at organising a branch in Johannesburg was when the medical benefit fund flew me to a meeting in Pretoria, to try to persuade the Registrar of Medical Schemes why he should not close it down. The next day I borrowed a Volkswagen Beetle from Annie Smythe. She was part of a white Left circle in Johannesburg, of which I and others would ask favours – a place to stay or, in this instance, transport – as we would with similar circles elsewhere. It was understood that there was no question of payment, and it would probably not be possible to return the favour. This was their unsung contribution towards the struggle. It was also understood that we were not TUACC or FOSATU (the new federation, once it was established), although why we were different was not something we would ever discuss.

I set off in the Beetle toward the East Rand to try to meet Mary Moodley.

Through her I hoped to contact workers who had once been in leadership in the Union. Mary was a *strugglista*: someone whose whole identity was bound up with 'the struggle', and she did not tire of reminding you about it. She spoke in whispered tones about her children in Maputo, because she believed her Actonville house was bugged.[31] When she was still branch secretary, she said, she used to visit Jones and Langeberg regularly, but after she was banned she had lost contact with workers. The following day she would take me to meet someone who could help me organise the Union. We made a careful arrangement to meet at a street corner outside the township in the morning so as to avoid being followed by the Special Branch: 'the SB', as the security police were known.

The next day she arrived with a youth in tow, whom she introduced as Boet. She had said nothing about Boet the day before, and I assumed, when we set off together in the Beetle, that he was a confidant of long standing, even though he did not look the part. The Langeberg factory in Boksburg was close by, near a number of large metal and engineering plants that were to become the stronghold of the Metal and Allied Workers' Union (MAWU), TUACC's most significant affiliate. We stopped outside the gates, and Mary went off to speak to someone. After a long time she returned to say she did not know anyone there any longer. This struck me as odd. If the factory had been organised in the 1960s, there should have been at least someone that remembered the Union, to whom we could have spoken.

Jones in Industria would be a better prospect, Mary suggested, so we drove from East Rand to the West, with Boet directing me. At my insistence, Mary went to enquire at the gate of Jones, and was able to identify an older worker who remembered the Union. Mary introduced me. I explained who I was, and promised that I would be back. I also established from her the whereabouts of Mabel Balfour, the Transvaal secretary of AFCWU. Mabel was now a worker in the factory.

Mary then took me to meet someone she thought might be able to help us. This was Don Mateman, who had been general secretary of the Textile Workers' Industrial Union (TWIU) and had played a leading role in SACTU.[32] We met in the offices of a church hall in the coloured suburb of Eldorado Park, where he had some position. It soon became clear that I was on a wild goose chase and he was not at all interested in reorganising the Union. Maybe this was because the union of which he had been general secretary was now in the TUCSA camp, or maybe he simply did not trust us. After the meeting, we observed coloured women in the church hall filling in application forms for a prospective employer. This turned out to be Armscor, the state-owned arms manufacturer.

If Mateman's indifference had been inspired by mistrust, it would have been for good reason. Mary's concerns about being followed by the SB were needless. I subsequently learned from Liz that Boet, whom I had taken to be a trusted confidant, had knocked on her door looking for a place to stay only a day or so before my arrival. Mary had never set eyes on him before, but it had not occurred to her to suspect him of being a plant. Unsurprisingly, he turned out to be an informer.

Not long after this fiasco, Neil Aggett let me know he had moved to Johannesburg. Shortly after that, he moved into a ramshackle house in Bertrams. He had approached TUACC looking for an opportunity to work in the trade unions, but had been rebuffed. I told him that a delegation from the Union was soon to visit Johannesburg to try to establish a branch there, and when we did so in July 1978, Neil's house became our base.

Giving me the essay in which Lenin famously articulated the need for a vanguard party might have seemed a naïve thing for someone in Neil's position to do. What I think resonated most for Neil about it, however, was the need to engage in both legal and illegal work in a repressive society. This is what Neil was in fact doing by moving to Johannesburg. He was eligible for military service and trying to evade being called up. So I would say that giving me the essay came out of a raw honesty and a seriousness about politics. It was on account of this same quality, by all accounts, that he was rebuffed by the leadership of TUACC.

Our delegation comprised Oom Joe, Lilian, Mrs R as well as Baba Madubula, a young woman from Mbekweni who had recently started attending management committee meetings. On one side of the passage in Neil's house there were two interleading rooms with no furniture whatever. On the other side was the room where Liz and Neil stayed, with not much more in it than a double bed. Mattresses were borrowed from somewhere, and for some reason the others were keen I should sleep in the smaller room, nearest Liz and Neil, while they slept in the other. I did not put up a fight about this.

The day after our arrival, before reporting at the reception of Jones, I established contact with the worker I had met previously with Mary Moodley. 'You remember I said we will be back,' I said to her. 'Let the workers know we are here.' We then introduced ourselves to the management, and informed them we would like to utilise the trade union facilities which the Agreement provided and go into the cloakrooms at lunch hour. True, the trade union facilities were not limited to the factories in the Western Cape. However, the facilities only applied to FCWU, the registered union. There was a flurry of phone calls to Cape Town.

It was an audacious demand. We were trying to utilise a legal space that the Agreement created, and our sea-lawyer argument was that there were coloured workers at the Industria plant, albeit no more than a handful. Our more serious argument was that the registered and unregistered trade union operated as one, as was evident from the composition of our delegation, and we had a moral right to inform the workers about the Agreement to which we were party, and which now applied to them. With the advantage of surprise, by bluff more than as of right, we were let into the cloakrooms, where we were enthusiastically received. Workers who had not seen or heard from the Union agreed to begin paying subs that Friday, and we undertook to collect them.[33] Shop stewards were elected, and virtually the whole factory paid.

I met Mabel Balfour when we came to collect the subs that Friday. She was still the striking women of the photograph, and was glad the Union had returned. She was also still someone the workers looked up to. Yet she expressed something else, as well: a bitterness, which I began to understand when I realised she had been left with no means of support, after she was banned, and felt abandoned. She must have made some promise to the bosses of Jones when they took her back into the factory again. It was because of that promise, I surmise, that she was not prepared to accept a position in the Union.

We had lost the advantage of surprise by the time we got to Langeberg at Boksburg. The bosses were expecting us, but Mrs R and Oom Joe knew the manager, who had previously been at SAPCo in Tulbagh, and he knew the Union. He could hardly refuse when Jones had already agreed to grant us access, so we were allowed to enter the cloakrooms and meet the workers. Although no one knew about the Union, and organising the more than 1,000 workers would be a considerable task, we were able to elect shop stewards.

Bouyed by our success, we sat on mattresses on the floor of the Bertrams house, eating take-outs from KFC one evening. Something about the tone that Oom Joe and the three women adopted in their banter made me aware there was something going on between them to which I was not privy. It was only much later that I realised that Oom Joe must have been conducting an affair with Baba, under our noses as it were.

The need for a follow-up visit to Johannesburg preoccupied me on the long drive back. The Jones workers had wanted membership cards, with which they had apparently been issued before. The Union no longer issued such cards, but it was important to respond to whatever workers wanted, particularly if it would strengthen their sense of identity with the Union. It would also be of great symbolic importance if someone from Johannesburg could attend the conference next month. I had in mind a person who could go to the branch and attend to these things: this was Oscar Mpetha.

At first Oscar did not remember me when I went to see him again at the Walls factory, but he did not take much persuading to take leave the security firm and go to Johannesburg, to return in time for the 1978 conference. He also agreed to be a guest speaker there.³⁴ But the official guest speaker at the conference was the FCWU invitee. It was time to take the Union out of the orbit of Labour Party and coloured politics, I thought, so I proposed someone from the Cape Town Municipal Workers' Association (CTMWA).

A turning point

The conference of 1978 marked a turning point in the fortunes of the Union. The tempo of visits to factories and branches had been steadily increasing, but the objective of these visits had changed.³⁵ I was no longer engaged in a holding operation, to shore up an organisation in a state of collapse. The presence of a new leadership, and especially the presence of a strong delegation from the inshore fishing factories, led by African delegates, generated great excitement. Everyone could see that this was a revitalised organisation.³⁶

The invitation to CTMWA was intended as a statement to this effect to the outside world. It was also to say that the Union was ready to begin building links with other unions. CTMWA was, so far as I could tell, the only registered union with which we might have something in common. However, the invitation actually served to contrast the differences of style between them. CTMWA sent John Ernstzen, its general secretary. He spoke in high-flown English, as though he was addressing a graduation ceremony. I remember Oscar's speech less for what he said than for his transformation from a doddering old man into an assertive, self-confident leader.

10

The N1 at Richmond

I HAD NO FOREWARNING I was going to be nominated as general secretary of AFCWU at the conference of 1978, and when there were no other nominations, I accepted my election reluctantly. I was by no means sure it was correct thing to do, because I was not an African, so I placed on record that I had agreed to fill the post 'temporarily'.[1]

It was much more of an issue having a general secretary who was not an African in AFCWU than having one who was not coloured in FCWU. That was because, in the racial hierarchy established by apartheid, Africans were at the bottom. Even if 'African' was a construct, it was a construct with a pedigree, underpinned by the fact that, like most whites, I could not speak a language spoken by people who did regard themselves as African.

General secretary of AFCWU was Oscar's old position, and it was he who explained to me how the Party had adopted a policy of promoting African leadership. Yet it was a fairly thin line that divided promoting African leadership from African chauvinism, and chauvinism was clearly abhorrent to the worker delegates at the conference. We want one general secretary for both FCWU and AFCWU, they had said. It would be a first, practical step towards becoming one union.

Oscar was one of those who had been most impressed by the turn-out at the conference, and came to see me at the head office shortly afterwards, wearing his security guard's uniform. He had grown up in Mount Fletcher before coming to Cape Town to look for work, and the worker delegates from the West Coast came from the same rural Transkei background as he did.

I am in my seventies and entitled to take retirement, Oscar said, but I still believe I have a contribution to make: my time would be better spent working for the Union than as a security guard. I had no hesitation in accepting his offer, and recommended to the management committee that he be appointed as national organiser. There is no record of the decision being taken, because it was the first time the Union had reappointed someone who had been banned, and I was not sure how government would respond.[2]

In the tradition of the PAC, sympathetic whites had been regarded as African. However, to me they seemed like 'honorary Africans', and what was more important than the PAC's public pronouncements was what they preached in the townships. One Sunday afternoon, along the route I usually took to exit Nyanga township after dropping Oscar, a guardian angel in the form of a young man stepped into the road to redirect me. If he had not, I would have driven straight into a PAC mob on the rampage. Amy Biehl would not be so lucky.[3]

One of Oscar's first assignments was to go to Johannesburg, to consolidate the organisation there. The two of us would go together, to meet with the management and open a branch office. Having done that, the plan was to leave him there. Neil had agreed that Oscar could stay at his place. The challenge would be to preserve the integrity of our model against competing models of organisation, which TUACC and others were punting, although we did not know it at the time.

This was also Athalie's home town, and she wanted to visit her family. We decided to combine the personal with the organisational. Since I did not want there to be any suggestion that I was using the Union's resources for a personal purpose, I decided to take my own vehicle, a Fiat bakkie that I had bought out of the box, with my accumulated savings from my previous job. I had not insured it, because I did not have the money. It would also not have had seat belts in, had my father not insisted on installing them himself. It was not a legal requirement to do so at the time.

Past Worcester, the passage through the Hex River valley and the ascent into the Karoo always seemed like a journey backwards in time. As you leave the comfort of the fynbos behind, and the signs of human habitation become fewer and further apart, the vegetation becomes sparser and the landscape more featureless, until the low, rolling hills give way to the koppies. Then you are reminded that these were once volcanoes and start noticing the rocks that litter the landscape.

On this occasion, on what I always thought of as 'the way up', we took the alternative route through Kimberley. Oscar was a driver, and owned his own car, a beaten-up yellow Valliant, which he used to get around in. He drove extremely fast, he explained, lest he fell asleep at the wheel. It was a quite plausible explanation, because Oscar often fell asleep in the most improbable situations.

I am not sure whether I already knew about his driving habits when I asked him to take the wheel at Victoria West, but I had no choice: Athalie had only recently acquired her driving licence and was not confident on a long trip. I lay on a mattress in the open cab at the back, bouncing around, as my sole material possession veered to and fro across the white line on the way to Kimberley.

After consulting with workers at Langeberg Boksburg, Oscar and I decided to ask the Jones workers to put forward someone whom the Union could employ as a branch secretary for Johannesburg. The Jones workers elected Beatrice*, a worker with connections in the factory but no obvious leadership qualities. Neil helped us find an office in a dilapidated block in Wanderers Street, and I signed the lease.

At least there were only two of us on the way back. Passing through the Free State, I gave Athalie a chance to drive. It was a new stretch of road, with a broad verge, and when she hesitated to overtake, I egged her on: the 'arbiter of reality' was how she described my belief that, objectively speaking, I always knew what was best. The next stretch she drove was outside Richmond, where there was no verge whatever. She started to overtake and I knew what was best, but held my tongue.

What has stayed with me after the event is the slap of metal on earth, as the car overturned the first of many times, and the sensation of watching to see whether we would die. The bakkie landed on its side in the Karoo scrub, on the other side of a fence. The seat belts had saved us. I pulled myself out of the passenger side window, and pushed the bakkie over, to let Athalie out. The bakkie was a write-off, but we were alright, except for a cut over Athalie's eye, which was clotted with blood. For a moment she thought she had been blinded.

Consequences
Something in our lives had shifted after the accident, but there was hardly time to process what it was. On the Monday after my return someone phoned me about a letter to the press that Juffrou H had written. It called for an immediate response, and would undoubtedly bring to a head the tensions at Jones that had been brewing all the while since my appointment. On top of this, I would have to explain to the management committee about the accident. It was as if the slap of metal on earth marked the end of what had, despite the difficulties, been a honeymoon phase in the Union.

I reported to the management committee what had happened on the N1 at Richmond, and Oom Joe then asked me to leave the meeting. It had never happened before that an official was asked to leave while a matter affecting him or her was being discussed, and I was not happy about it. On my return, Oom Joe relayed to me what had been decided in my absence.[4] Mr Theron, he announced in a voice of authority an ordinary worker would never have been able to assume, the Union has come to a decision. It was not a decision about which he would entertain any further discussion or objection on my part. The Union would replace the bakkie with a new one. The delegates present enthusiastically applauded this announcement, and I most probably smiled, as if I was grateful. In fact I had a sinking feeling, and the more I thought about it the more apprehensive I became.

Where would the money come from to buy me a new bakkie? The one contentious issue at the conference had been a resolution that branches that had more money than was needed for their day-to-day expenses should transfer it to head office.[5] Paarl branch told its members that the head office wanted to take its money, yet in fact I had taken no steps to implement this resolution. I realised they had a killer point: the only reason corrupt officials at head office had not eaten even more of the Union's money was that branches controlled their own finances. Although I could not admit as much, I had been won over.

Even so, the suspicion that head office wanted Paarl branch's money persisted, and the fact of the matter was that it was my fault the bakkie was not insured. The last thing I needed was for the workers at Jones to think I was after their money to finance a new vehicle. I was not at all sure the delegates who had so enthusiastically endorsed the decision would defend it as spiritedly if called upon to do so. Despite all the affirmation I was getting, I had every reason to watch my back.

It was not part of Vormat's job as administrator to attend management committee meetings, but she was present at this particular one. I had encouraged her to blur a boundary that I was not sure should exist. In so doing, I had created a stick for my own back. Vormat, I was now sure, was actively undermining me at every opportunity, although this sort of thing is almost impossible to prove. It crossed my mind that she might even have put up Oom Joe to push for the Union to buy me a new bakkie.

Vormat had not been in the office long before she began to complain to me about Penny, who had assumed most of Miss Yon's administrative duties after she was banned. It was not that Penny was a difficult person to get on with: in fact she was sunny and approachable, and Vormat could never come up with anything concrete against her. It was a clash of personalities, I supposed, but I could not but notice that both were from a coloured middle-class background and, in appearance, more white than coloured.

In retrospect, I think the real difference between them was that Penny was apolitical, while Vormat was filled with resentment and anger. Justifiably angry, probably, at all the slights that came the way of a person with a light skin who was not classified white; resentful towards someone like her who would not take her part. It is someone from your own class background that can best see you for what you really are. Towards me, however, Vormat was still solicitous and charming. She was also efficient and 'politically clear'. When it emerged that Penny had bought some pencil crayons and a few other items for her nephews on a work account, Penny was dismissed at my behest, and I felt relieved that the conflict was resolved. It was my first lesson as to why dismissals should not be used to resolve organisational problems.

Vormat then proposed another 'politically clear' person to take Penny's place. I had not hesitated to accept her recommendation. This was Johnny Issel, a politically ambitious person at a time when the politically ambitious were beginning to take note of the Union, and a natural ally for Vormat. It should have been obvious that, for someone like him, an administrative position in the medical fund could only have been a staging post for something else.

A question of branch autonomy
The first I heard about Juffrou H's letter was when Annie Adams phoned me on a Monday morning. Had I read what Juffrou H had said in Sunday's *Ekstra*? she asked. I had not, of course. The *Ekstra* was a supplement of *Rapport*, the Afrikaans-language Sunday newspaper, and I saw it as a government mouthpiece. It had a large coloured readership on the *platteland*, however.

'*Los Kaap vir ons werkers*' was the headline of the front-page story: Keep the Cape for our [coloured] workers. It was based on a letter from Juffrou H, as branch secretary of FCWU, attacking the leader of the Labour Party, David Curry, for saying that the Coloured Labour Preference Area policy should be scrapped. I was not surprised that Juffrou H had such views. What was surprising was that she had said as much in print.

'The fight of the big shots in the Labour Party on behalf of the blacks, or plurals', the article went on to say, 'will, if it is won, not secure the freedom of the coloured but rather the demise of the coloured worker ... I contend that Mr Curry on behalf of the Labour Party is in a subtle fashion handing the productivity of our factories over to foreign ethnic groups.'[6]

I phoned Sabbagh, the chairperson of Paarl branch, and asked her to convene a branch executive committee meeting the next day. All expressed shock at what Juffrou H was reported to have said. Juffrou H's explanation, however, was disarmingly simple: she did not know a thing about the letter. So I went to the offices of *Rapport* the first thing the next morning and asked them for a copy. To my surprise, they complied without question. It was typed, and the signature on it was Juffrou H's.

Maintaining unity in an organisation is never simply a question of preserving the peace or pleasing everyone. It necessitates making hard choices. No one on the branch executive committee was going to agree with the sentiment that Africans were a 'foreign ethnic group', of course, or a call to maintain the Preference Area policy, particularly now that the head office was making an issue of it. That is the difficulty in combating racist attitudes: they manifest on the sly.

The decision we now had to take was whether to go back to the branch executive committee. Strictly speaking, only this committee had the power to discipline Juffrou H in terms of the constitution. Yet there was now documentary proof Juffrou H had lied about knowing nothing of the letter, and all indications were that if we went back to the committee, the workers at Jones would simply close ranks around her, as you would expect with identity politics. For the AFCWU committee, on the other hand, the views in the letter to the press simply confirmed what they already knew about Juffrou H.

Having already held one inconclusive meeting, I was not inclined to risk another. I would also not have entrusted a matter like this to a disciplinary subcommittee, as is customary nowadays, had such a thing been suggested. For one thing, it would have been a gross simplification to characterise the issue confronting the Union as one of discipline. Rather, it was which version of the FCWU tradition would prevail: one tainted with coloured chauvinism or one that affirmed the solidarity between the coloured and African working class. The differences between these competing versions could no longer be papered over.

So the principle of the autonomy of the branch, so far as I was concerned, had to give way to the interests of the Union as a whole. Juffrou H had anticipated that this would be the case. The next week's *Ekstra* carried a follow-up report, headed '*Kom steek my dan*'. Come and stab me: in the back, as it were. The letter about which she had claimed to know nothing, she now declared had been drafted by one Pilcher, an official of the Freedom Party and known friend. This claim was coupled with a threat. '[Juffrou H] was very clear about one thing, and that is that she would take action if she was kicked out of her post. And the possibility is strong [Juffrou H] said, judging by certain peoples attitude at Tuesday's meeting.'[7] Even then, she took no steps to distance herself from the substance of the letter.

Why had she not given the same explanation to the branch executive committee when called upon to do so? Why had she not distanced herself from what was in it? These were the questions I put to Juffrou H at the management committee meeting. Word about the letter had spread, and the meeting was packed. The consensus was that Juffrou H had failed to answer the questions, and Mrs Amon proposed she be dismissed on one week's notice, which was the minimum notice period to which an employee was entitled. No one opposed the motion, but the delegates of Paarl branch were ominously quiet. The meeting was, after all, assuming a power that had been theirs. It could also be said that the Union was attempting to resolve an organisational issue by dismissing an official.

Juffrou H was still working out her week's notice the evening the national leadership arranged to meet the Jones workers to explain the decision. Given that they were already hot under the collar about the head office wanting to take their money, it cannot have taken much to get them even more steamed up about Juffrou H. The FCWU members at Jones had after all been mobilised to support her once before, and that was what had led to the split, in terms of which Jones and Moberg's each constituted a separate branch. For some of the workers it must have been déjà vu.

The offices of the Ray Alexander Union Centre were packed with women in their overalls and *doeke*, and there were also coloured men at the back whom I had never seen at meetings before. Oom Joe opened the meeting by asking me to explain why Juffrou H had been dismissed. Almost simultaneously the workers started shouting: they did not want any explanation, especially not from me. The president stopped the meeting, and Lizzie Phike and her friend Luska formed a cordon of African women around us, as we moved to the door.

'*Hulle is opgepomp*' was Pendlani's comment to me as we climbed into the kombi. The workers were pumped up. Indeed, they had been fit to burst. This was the only occasion I felt physically threatened by workers, though the first of many meetings with the Jones workers to be disrupted.[8] The following week Tredoux, the factory manager at Jones, asked to meet with me. It was the start of the season, he said, and management was concerned about the situation. Also, the committee wanted a meeting. I told him it would be better if the committee contacted head office itself. A few days later a message came proposing a meeting. Oscar and Baba accompanied me.[9]

The purpose of the meeting with the committee, it transpired, was to tell us that the workers intended to break away from the Union. We tried to point out the possible consequences, but the workers who were doing the talking were not prepared to listen. The meeting concluded with me saying that we would prefer to hear this from the workers themselves. So another meeting was arranged, this time at the factory. The entire workforce, some 2,000 in number, assembled in one of the warehouses.

My middle-class training had included public speaking, so I was able to make myself heard above the murmur. When it came to Oscar's turn to translate into isiXhosa, however, the murmur grew to a roar of protest, and I heard shouts to the effect that this was a meeting for 'us coloured workers'. The meeting would have been even more disorderly were it not for the fact that Mr Woods, the general manager, was perched on a pallet behind me, observing proceedings. He came up to me after the meeting. 'You have considered', he said, 'that this workforce must represent about 20 per cent of your membership.' He had done his sums correctly. It would of course be a major setback for the Union to lose as many members, but I was not going to say so to him. I assured him we had considered the consequences, and would not back down.

There was nothing reckless about this stance, as I saw it. The survival of the Union was at stake, to be sure, yet when Woods talked about losing 20 per cent of our membership, he was only talking of the coloured membership. Having actually become a national trade union, with the establishment of Johannesburg branch, we had no choice but to jettison coloured chauvinists, if we were to expand in parts of the country where the coloured working class was a negligible presence.

What Woods also did not realise was that the wind had changed. Elsewhere, the coloured working class would have no truck with chauvinism, and contacts painstakingly cultivated at unorganised workplaces over the past two years were beginning to bear fruit. It was therefore probable that the Union would be able to offset any loss of members at Jones with new members elsewhere, as did indeed happen. Coloured workers organised by the Union would be at the forefront in challenging workplace hierarchy, across the *platteland*.

It is not difficult to understand why events at Jones ran counter to this trend. It was because of the sectional interests the Union had represented there, and because of how its committee was integrated into the workplace hierarchy. 'I work at Jones and I see the workers are in a difficult position, they are scared for their jobs,' Pendlani was later to explain to the management committee. 'Their Union committees [by which he meant the members of the committee] were supervisors, and now the supervisors are forcing the workers away from the Union. They have only one way to organise the workers – to put the workers in fear of losing their jobs.'[10]

It was at this juncture that the wage clerks in what was known as the time office began collecting the names of workers, in order to prepare a list of workers wanting to stop payment of their subs. Once it became clear that the committee were themselves involved in organising for the breakaway group, the management committee decided to suspend the branch executive committee, which it was entitled to do in terms of the Union's constitution, and take over the management of the branch.[11]

Then Norman Daniels telephoned me. Daniels was the general secretary of a number of smaller unions as well as TWIU, the former affiliate of SACTU. Although he typified for me a kind of career trade unionist that I had no wish to become, I think he was phoning out of loyalty to a union that had once been an ally. A delegation of workers from Jones had been to see the vice-chair of TUCSA, he told me, and they were planning to affiliate as soon as they were able to. But first they were going to see Van den Bergh of the Department of Labour about getting registered.[12]

Van den Bergh, I had no doubt, would have given the breakaway whatever help he was able to. If they were ever to succeed in obtaining registration, it would mean the Union would lose its registration for the canning industry in Paarl. This would indeed be a serious blow. But in terms of the legislation of the time, Van den Bergh did not have the power to register a trade union unless it represented a majority of workers in the industry in the magisterial district of Paarl.

So long as organisation at Moberg's and RFF remained solid, the Union would still have a majority in Paarl. It seemed unthinkable that Moberg's would ever try to make common cause with Jones, given the history of enmity between them. Even though RFF was Johnny M's old factory, under the leadership of Hester Adams and others the workers seemed solidly behind the Union. Unless the breakaway group was to be registered, they could surely not survive. I could not of course know that, following the report of the Wiehahn Commission, the legislation would be amended precisely so as to enable such a trade union to get a foot in the door.

An early indication that the wind had changed was at Sea Harvest. This was the factory in Saldanha where Aunt Lossie worked. It turned so-called white fish, primarily hake, into fish-sticks and fillets in various guises. Unlike the smaller boats utilised by the inshore industry, hake was caught deep at sea, by large, ocean-going trawlers. In scale, the factory dwarfed any in the inshore industry, and it was no doubt in order to recover the money which its Spanish and South African bosses had invested that it operated day and night.

'The company had always got on well with my predecessor,' Mr Kramer said, with the hint of a German accent, when I first introduced myself. This was no doubt because the Union had made no demands on the company whatever. Wages were set by the same wage determination as applied to the inshore fishing industry, but, unlike the inshore industry, Sea Harvest paid no more than the minimum. This was a lot less than even workers in the canneries earned.

Mr Kramer let me go into the cloakrooms and address the workers, as Johnny M had done. I was surprised. This was the only factory to grant us access without an agreement, or without our first organising a majority of workers. You realised why once you attempted to address the workers. Much like our experience at Langeberg Ashton, it was impossible to make oneself heard above the din of dominoes.

The domino players were from Hopefield, a small town some three-quarters of an hour's drive from Saldanha, from which the workers were bused to work each day at the company's expense. That made them more expensive than the local women, from Saldanha or nearby Vredenburg. It also made them highly vulnerable, since they could easily be replaced by local labour. Our efforts to organise a majority at Sea Harvest continued to be frustrated by the unremitting hostility of the women from Hopefield until one Sunday, towards the end of 1978, we decided to hold a meeting in Hopefield, and go from house to house to ask workers to attend.[13] Not many workers did, but the fact that the Union had taken the trouble to call a meeting just for them must have made an impression in a tight-knit community. Early the next year a substantial number signed up.[14]

The breakthrough at Table Top was even more unexpected. In terms of the amount of head office time devoted to apparently fruitless house-to-house visits, and waiting outside the factory in the drizzle, our endeavours to organise the factory in George over a period of some two years must have represented some kind of record. Yet we had not even managed to persuade ten workers to come to a meeting. It really seemed workers had resigned themselves to the inevitability of their desperately low wages.

So it was more out of stubbornness than conviction that I insisted we again visit George in early 1979, after a visit to Mossel Bay, and contacted a worker whom I had met before, and who had made an impression on me. She in turn put us in contact with two others, who promised to get back to us, as they had promised before. The next person we heard from was Dominee April, of the Dutch Reformed Mission Church, to say he had been asked by a group of workers to help organise a meeting. We returned the next week to a hall packed with workers. In the space of a single meeting we signed up more than half the factory and arranged to start collecting subs.

11
Durban road

YOU COULD SEE THE WIND had changed from the way in which workers were taking the initiative to organise the unorganised. Sometimes it was simply a question of directing them where to go. Langeberg workers from Boksburg, for example, went to nearby Benoni to organise a depot of SA Dried Fruit (SAD). Sometimes the first we would hear of a factory or depot was once it had been organised. This was the case with the SAD depot at Ceres. A few years later, independently of each other, a worker at each of these depots would become a paid official of the Union. One of them, at the time of writing, still is.[1]

The visits to factories and homes would all be done on a voluntary basis, in workers' own time. The debate at the management committee about paying members for organising work had been quietly dropped. It now went without saying that workers did this because they saw the Union as their own, and because financial incentives for leadership had no part in this organisation.

Nothing the Wiehahn Commission would say could have had any bearing on the direction of the wind. It was some six months before the Commission would table its report, and the relevance of that report for coloured workers would mainly be what it had to say about non-racial unions. Their right to belong to a trade union had long existed on paper. It was the exercise of that right in practice that was the problem. Events at a factory in Worcester with an entirely coloured workforce would show in quite dramatic fashion that this wind was blowing from below.

SAD Worcester was the second biggest factory in town. Once the wage agreement there had been concluded, the news was bound to spread to Rainbow Chicken, the biggest factory in town.[2] Poultry was not listed as part of the food manufacturing industry in terms of the Union's constitution, but, led by its enterprising young secretary, Susan Baxter, workers of SAD had arranged a meeting with the Rainbow workers. It was no time to raise legal objections when Baxter phoned to ask the head office to attend.

So it was that one weekday evening we parked the Union's kombi outside the municipal hall in the coloured township, to wait for the Rainbow workers to turn up. While we were waiting, a car pulled up next to us, and two men got out. One of them introduced himself. Who were we, he wanted to know, and why were we meeting with the workers? This was Hennie Ferrus, a Congress activist who had formerly been imprisoned on Robben Island. Now he was an official of the local Labour Party.

'I would do anything for the struggle,' Ferrus said, by way of explaining why he had joined the Labour Party. 'If they tell me to wear skins [instead of clothes] I will do it.' This was charismatic leader-talk. I was nevertheless impressed that there was a leadership in Worcester attentive enough to notice when workers were being organised. Surely the endorsement of such a leadership could make a difference. Yet the mere fact that a community had credible leaders did not explain how it was possible to recruit the majority of a workforce of close to 1,000 workers so quickly. The gains at SAD must have brought to mind what the Union had once been, before Langeberg had closed the cannery. Something of the tradition of the Union had survived, despite Oom Willem eating the money.

There was also not the dead weight of the TUCSA tradition, as there was in Cape Town.

This is the only way I can explain the contrast between the situation in Worcester and Fatti's & Moni's in Cape Town, where there was not even a memory of organisation among the workers. With Mr Christian's help, we had recruited most of the coloured workforce. Very few, however, were involved in the manufacture of macaroni and pasta, the operations which were listed in the Union's constitution. Pasta, we found out, was a by-product of wheat milling, and the women packed flour into the paper bags in which it was sold in the retail stores. However, the majority of the workforce was engaged in offloading the wheat to be milled, and handling the bags of flour that went to supply the bakeries of Cape Town with bread. This was heavy, physical work. At Fatti's & Moni's and elsewhere, that kind of work was mostly done by African contract workers.

We did not preach about non-racialism simply as a principle: it was first and foremost a practical necessity for a trade union representing the ordinary worker. Fatti's & Moni's illustrated the point well. There was little the Union could do to address the issues that really mattered to the coloured workforce – their wages, and the way they were treated in the workplace – without the support of the African workers. The Africans had a developed sense of collective identity, reinforced by the fact that most stayed in the company hostels in Nyanga. They had even gone on strike before, and won a separate wage increase for themselves. The coloured workforce, by comparison, was a much less homogeneous group. Most must have envied what the African workers had achieved as a collective, and were keen to enlist their support. The difficulty was how to do so. The African workers saw little need to ally themselves with the coloured workers. On top of a generalised mistrust of coloureds which the Coloured Labour Preference Area policy had fostered was the fact almost no one among the coloureds could speak isiXhosa, and few among the Africans could speak English.

Mr Christian was too close to the white management to have been keen about enlisting the support of the African workers, and all of a sudden he seemed to have lost interest in the Union. When Virginia and Athalie tried to retrieve several weeks of subs he had collected from the workers, he was nowhere to be found. Our organisation stood on a knife edge. If we could not recover the subs from Mr Christian, we stood to lose the factory altogether.

I attended what we regarded as a critical meeting with the members in a church hall in Bellville to discuss what to do. One of the workers at the meeting knew where Mr Christian stayed, in a township of Bellville South not far from the hall. So I proposed that a kombi-load of volunteers from the meeting go to his house to tell him that we were not leaving without the money. It was not a strategy which I had ever before attempted, and it could easily have backfired. But Mr Christian must have had a bad conscience and coughed up. The workers were delighted to see a superior get his come-uppance.

The breakthrough with the Africans came in a similar fashion. Oscar learned from speaking to one of the workers outside the factory that a colleague had been seriously injured handling bags and paralysed for life. When his wife came to the factory to claim compensation for him, she had been sent away empty-handed. There and then, without consulting anyone in the Union, or the Union having introduced itself to the bosses, in an act of brazen chutzpah, Oscar took it upon himself to walk into the office of the wage clerk and demand the cheque on behalf of the worker. As luck would have it, it was there, gathering dust, as the cheques of thousands of maimed and disabled workers in offices like this still do, their 'unclaimed' monies from time to time memorialised in the *Government Gazette*.

The wage clerk was sufficiently cowed by Oscar to hand it to him, and Oscar personally delivered the cheque to the worker's family. At the next meeting the African workers were present en masse. It remained to be seen how the bosses would respond. To find out, I wrote a letter to the general manager, stating that a majority of the workers had joined the Union and asking for a meeting to introduce ourselves.[3] This was a more cautious approach than we usually adopted, which was to send a letter of demand. Our usual demands, at the time, were a minimum wage of R40 a week, a 40-hour work week, the recognition of 1 May as a public holiday and a few other things.

Durban Road crosses beneath the N1 at the Tygerberg hills, connecting Durbanville to what was then the commercial hub of Bellville near the station. Across the railway line is the industrial sprawl on either side of Modderdam Road, and between and beyond are the townships where the workers lived. But Fatti's & Moni's is improbably located on this side of the railway line, across from a row of retail stores: the only factory in the commercial district. From the street you walk straight into a reception area adjoining the general manager's wood-panelled office.

It was not often that, from the outset, I took as strong a dislike to a manager as I did with Mr Terblanche. Since I was later to contribute to his acquiring a certain notoriety, I should emphasise that he was not the bull-necked *boer* of the popular imagination, in a safari suit with shorts that could barely contain his thighs. Rather he was an effete man who looked every bit the accountant he had been, and who greeted Virginia, Oscar and me with a limp handshake and in an affected English accent.

The factory conducts different operations, Mr Terblanche told us. The Union was only entitled to represent workers in the macaroni section. There was already a trade union in another section, and the company had no problem in dealing with a responsible union, such as it was. This, it turned out, was one of the trade unions of which Norman Daniels was the secretary. Ostensibly it represented workers in the cones section, which was regarded as a separate operation by virtue of an Industrial Council agreement for the biscuit industry in Cape Town. In terms of this agreement, Daniels's union also had a closed shop. This meant that subs for this union were deducted from workers' wages without their consent. The workers concerned did not even know they were 'members' at the time.

However, the cones section comprised only a handful of workers. The operation where most workers were employed, the vast majority of them African, was flour milling. Mr Terblanche suggested a further meeting, probably to give him time to work out what to do about the fact that the Africans had joined the Union. He also promised to come back to us about certain complaints workers had given us.

Proper organisation
In some ways the breakaway in Paarl had made organising in Cape Town easier. We no longer felt the need to operate clandestinely. To address the mistrust between coloureds and Africans we were able to adopt a more overtly political stand. Indeed, we were compelled to do so because of the breakaway. Even those who were not members at the time needed to understand why the breakaway had happened. In this way organisation had precipitated, even necessitated, an engagement with politics.

The breakaway also meant that I now felt no compunction in disregarding the decision to buy a new bakkie for me. But since I had no savings left, even to cover the deposit for a second-hand vehicle, I was without the one aspect of a middle-class lifestyle I felt I could not do without: having a vehicle to get to Milnerton beach. My solution to the problem was to resort to the slush fund to pay the deposit on a second-hand Ford Cortina, on the understanding that it would be used by Athalie and Virginia to organise in Cape Town, and that I would pay the instalments. The yellow car, as it became known, was to be an indispensable tool for organising the branch, but I was never comfortable about how I had acquired it. I had blurred the line between what was my own and what belonged to the Union.

Oscar and I were now on the road together a lot of time, and I was delighted to have someone with whom I could have a real conversation about the issues of the day. Oscar's observations about the places we visited were also interesting. These were places he had last seen almost two decades before. His overriding impression of both coloured and African townships was how little had changed. What was invaluable for me, however, was to hear about the history of the previous decades from a key participant. Oscar did not have any sacred cows, and loved a listener.

'The ANC was a practically useless organisation until we took it over,' Oscar said more than once in our conversations on the road. He would almost spit out the word 'useless', to emphasise that he had no truck with political organisations that claimed to represent people, and specifically the working class, without ever organising them properly.[4] This was how the ANC had been when it was led by doctors and academics. The primacy of proper organisation was a conviction we both shared: without it, there would never be meaningful political change in the country. But, as with my father, it was not altogether clear to whom Oscar's 'we' referred. I once asked him if he had been a member of the Party. After some thought, he answered obliquely: 'Put it this way, in those days everyone belonged to the Party.'

If one read between the lines, the Party must have mobilised Union members to vote out the middle-class professionals who constituted the old guard, not long before the ANC was banned. Even though you had to make allowances for Oscar's considerable ego, there must have been some basis for his claim of having taken over the ANC. He had, after all, been the last elected president of the ANC in the Cape Province before its banning. The province then included the ANC's heartland in the Eastern Cape. I was not sure what to make of Union members being mobilised in this way. It raised a question about trade union autonomy in relation to a political organisation, although this did not seem an issue in the 1950s.

When all was said and done, however, I was a white man in a position he had once occupied, and Oscar and I were also sizing each other up. Maybe that was why I had not taken him into my confidence about supplementing Lilian's salary from the slush fund. Instead of doing what I had supposed any rational person would do with the extra money, Lilian had carried on as before. Once again, the auditors had reported 'cash on hand' in Paarl AFCWU, and Oscar was tasked with investigating it.

'Auditor Oscar', Lilian described him sarcastically at the AGM in Mbekweni, where it all came to light.[5] It seemed she knew something about Oscar's past that I did not, but she was of course trying to deflect attention from what he had reported: missing shop-steward books and other tell-tale signs that funds had been misappropriated. Also, that Lilian had been receiving payments from an undisclosed source. I was placed on the spot. It would not do to try to shift the blame to the national office-bearers. Even though they had gone along with establishing the slush fund, it had been my suggestion to supplement Lilian's salary. This, if anything, had made matters worse. The AGM decided to dismiss Lilian, and insisted that she pay back what she had taken. Anyone who failed to respect the line between what was his or her own and what belonged to the Union had no place in its leadership. So far as I was concerned, this rule was now regarded as inviolate. But my indulgence towards Lilian had dented my own credibility.

One of the members present at that AGM was Lizzie Phike. At the time she was a worker from the labelling room at Jones, and had begun to assume a prominent role in AFCWU. Because she was fluent in Afrikaans, she was also a key figure in the defence against the breakaway. Lizzie was outraged, not so much at the Union having accepted the donation, as at my failure to take the workers into my confidence about it. I had never before been criticised so bluntly by a worker, and to my face.

The upshot was that I resolved, privately, that the Union would never again accept external funding to cover its running costs or to fund its expansion. This later became a policy that would differentiate the Union's position from that of the other trade unions we would have to deal with. It might be said that this was my way of coming to terms with an embarrassing blunder, which could have been avoided if I had dealt with the money differently. Yet what workers were demonstrating by their willingness to contribute their own time and money to the Union was the value of self-reliance. Reliance on foreign funding was bound to undermine self-reliance.

The painstakingly slow way in which Fatti's & Moni's had been organised had also been about instilling values of self-reliance, as well as establishing trust between workers and paid officials. This was what extracting the subs from Mr Christian signified. It went to show that this was an organisation whose officials would go to some lengths to safeguard workers' money. Instilling values of self-reliance and establishing trust were the foundations of proper organisation. This foundation phase was especially important where the Union had no tradition it could trade on, as was the case in Cape Town.

It was this foundation phase of organisation that people from a middle-class background generally did not understand: whether because they were cynical about solidarity or because they believed that 'the people' were thirsting for the leadership they were better able to provide, and a rousing speech would be sufficient to awaken them. I remember Neil's dismay at how Oscar conducted himself during his first visit to Jones in Johannesburg. Neil had given him a lift in his Beetle, and recounted Oscar's interaction with the workers. Neil saw it as a golden opportunity to make a political address, and raise the workers' consciousness. Instead, Oscar had set about distributing and collecting application forms, and making arrangements for collecting subs that Friday.

Neil was to modify his views, as he was increasingly drawn into the role of a trade union organiser. My rejoinder to him at the time would have been along the following lines: workers do not attain consciousness as a collective through listening to speeches by charismatic leaders, but by being properly organised. Signing up members, or the institution of an effective system for collecting dues, was not simply a bureaucratic requirement. It constituted the material basis for an organisation that served their interests. It was what made a trade union autonomous.

Not that autonomy on its own was sufficient. The trade union of which workers in the cones section at Fatti's & Moni's were 'members' might be said to be autonomous, but it was certainly not democratic to take workers' money without ever properly recruiting them or even holding a meeting. This was, however, commonplace in the established union movement, where there were 'closed shop' agreements, in terms of which workers' subs were deducted automatically. It illustrated what I mean by the dead weight of the TUCSA tradition. The 'responsible trade unionism' it espoused was a tradition without values.

You could not say the same about the breakaway in Paarl, even though subs were deducted at Jones without a stop order authorisation. Workers had once been properly organised, and trusted their leadership because it was accountable. This is why the breakaway represented such a formidable threat. What was at issue was a leadership that nurtured dependence, and a sense that it was indispensable. The coloured workers at Jones were not yet prepared to break with this leadership ethos.

I used to visit Jones whenever I could, utilising the provision of the Agreement that gave union officials the right to enter the cloakrooms over lunch hour, and speak to whoever would listen to me. The coloured women would generally look the other way, although the African women were always welcoming. Having all the years had second-class status at both work and in the Union, the breakaway had given them something to defend. Then, some three months after Juffrou H's dismissal, Jones management told

TOP: Canning workers, circa 1977: The leadership of Ceres Fruit Juices. (Courtesy of UCT Special Collections)

BOTTOM: 'Kreef' workers, circa 1976: Mrs Cloete, the chairperson of Port Nolloth branch (on far right), together with her fellow workers employed to process rock lobster ('kreef'). (Courtesy of UCT Special Collections)

RIGHT: John Pendlani, President of AFCWU, 1978. (Courtesy of UCT Special Collections)

BOTTOM: This is a photograph of a general meeting of striking Fatti's & Moni's workers in the cinema in Bellville South. They are being addressed by a delegation of township women. (Courtesy of UCT Special Collections)

OPPOSITE TOP: Delegates to the 1977 annual national conference gathered in front of the Ray Alexander Union Centre. At this stage the Union is still overwhelmingly a coloured organisation. Sabbah Francke is on the far left in a red dress. To the right of her, in a brown suit is Manie van Graan, later to become the president of FCWU. To the right of him is Hester Adams and Aletta Amon, who was the vice president at the time. (Courtesy of UCT Special Collections)

OPPOSITE BOTTOM: By 1978 the racial make-up of the Union had begun to change. Here we see delegates to a meeting of the national executive council, including African contract workers from inshore fishing factories, waiting for the meeting to begin. (Courtesy of UCT Special Collections)

Fatti's and Moni's workers' meeting during strike, April to November 1979

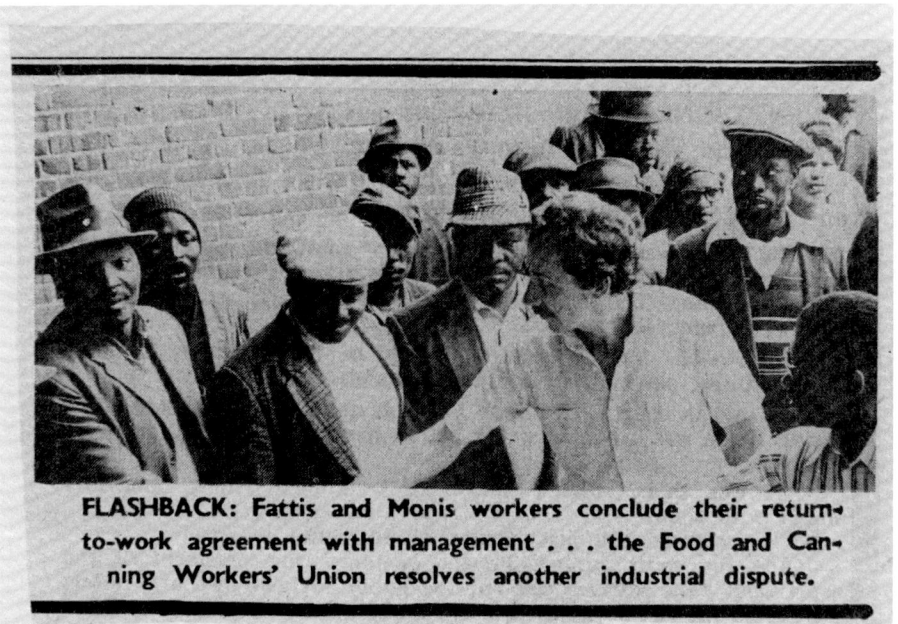

FLASHBACK: Fattis and Monis workers conclude their return-to-work agreement with management . . . the Food and Canning Workers' Union resolves another industrial dispute.

TOP: Fatti's & Moni's workers, 1979: Spasie Saaiman, one of five workers whose dismissal triggered the strike, is on the right. (Courtesy of Robben Island Mayibuye Archive)

BOTTOM: Mr Terreblanche, the general manager of Fatti's & Moni's, shakes hands with the strikers on their return to work seven months after firing them. *The Argus* ran this picture on its front page when Fatti's & Moni's workers went back to the factory. (Courtesy of Robben Island Mayibuye Archive)

RIGHT: This was one of the posters produced in support of the strike and boycott at Fatti's and Moni's. (Courtesy of Robben Island Mayibuye Archive)

BOTTOM: Oscar Mpetha's grandchildren, Oscar and Prince, on holiday with Athalie and Jan. (Courtesy of Athalie Crawford)

RIGHT: Liz Abrahams addressing a meeting of the United Women's Organisation. (Courtesy of Athalie Crawford)

BOTTOM: Langa conference: The speaker is 'Rev' Marawu, an organiser of the WP General Workers Union. Virginia Engel, who was at the time an organiser of the Union, is to his immediate left. Freddie Qayiya, a member of AFCWU, is on his right, and next to him is Athalie Crawford and Spasie Saaiman. (Courtesy of UCT Special Collections)

TOP: The national anthem: Neil Aggett with Harry Magomane and Israel Mokgoatlhe (to his left) singing Nkosi Sikelel' iAfrika together with other delegates of the 1981 annual national conference. Alongside them, Cliff Bestall films 'Passing the message'. (Photograph by Michael Gavshon)

BOTTOM: The *table* at the 1981 conference: Alfred Noko, John Pendlani and Manie van Graan are standing, with Lizzie Phike (on the left) translating. (Courtesy of UCT Special Collections)

TOP: Oscar in chains: This photograph taken during an inspection 'in loco' during his trial well conveys Oscar Mpetha's indomitable spirit. He is accompanied by Ian Farlam, his advocate. (Courtesy of Independent Media)

TOP: When Neil Aggett died, the Union could find no image of him to put on a poster. This one was retrieved by the makers of 'Passing the message', from footage of the 1981 conference. A camera handle partially obscures his head. (Photograph by Michael Gavshon/Cliff Bestall)

us they would stop deducting subs, and would also withhold the cheques for subs already deducted. We demanded through our attorneys that Jones reinstate the stop order, and pay over what was due.[6] Jones reinstated the stop order but did not pay over the cheques, and soon afterwards the letters of resignation arrived. It seemed the Union would be powerless to prevent the resignation of almost the entire coloured workforce en bloc.

The only glimmer of hope was one lunch time when the leadership of the breakaway were not around, and a group of coloured women beckoned to me. What they wanted to tell me was that they had no alternative but to go with the tide. Despite the tumultuous events following Juffrou H's dismissal, it seemed that some of what I said had been heard. It also gave me some sense of how powerless ordinary workers could be when those in authority in the workplace hierarchy were also in leadership in their organisation. In reality, workers were no longer the decision-makers.

Establishing that the ordinary workers were the decision-makers was perhaps the most difficult stage in properly organising them. It meant establishing the importance of general meetings, and securing their attendance. At these meeting, all developments having a bearing on organisation would be discussed. Each and every member should feel free to voice his or her opinion. This was how to establish democratic values. So after meeting with Terblanche, there was a report-back on what had transpired. And when Terblanche failed to come back to us about a further meeting or return our calls, there was another meeting to discuss that. But this was very much easier to do in the case of Fatti's & Moni's than it was for Rainbow, where there was no paid official at hand to ensure that workers knew what was going on.

The Rainbow strike

Not only was it not in accord with the constitution to organise Rainbow Chickens, it was also not in accord with the industrial strategy which I was in the process of articulating. This entailed methodically organising all the operations involved in the manufacture of a particular product, and in so doing uncovering the connections between them, as we had already done in the case of the canning and drying of deciduous fruit, and would shortly do in respect of the packing of fresh fruit, the production of fruit juices and the canning of pineapples.

Rainbow was not part of an industry in which the Union had any leverage. It was in fact an example of a vertically integrated operation: it hatched and raised chickens in its own batteries, which were then dispatched to a state-of-the-art processing plant where the birds were slaughtered. It would be some years before our organisation had advanced to the point of acquiring leverage over the one critical component of the process over which the company had no control, which was what it fed the birds. This was in fact what connected poultry processing to other industries that we organised.

For this reasons I was apprehensive when in February 1979 we decided to announce ourselves to the bosses, by sending them our usual set of demands.[7] The Rainbow bosses were by this point well aware of the Union's existence. It later transpired that they were divided among themselves as to whether a trade union was entitled to organise workers at all, given that the workers were processing 'agricultural produce' from its own farms. Farm workers of any description were at that point not permitted to belong to unions, and even though the processing plant was located in an industrial area, the case law supported an interpretation that this was a farming operation.

A formal approach would force the bosses to adopt a position toward the Union, and it is at this juncture that they were most likely to want to remind workers who the boss really was, and to test how well the Union had done its work. The test that every worker feared was that someone would be dismissed for being a leader of the Union. Victimisation was an offence in terms of the Industrial Conciliation Act, but since there was nothing that required the bosses to state the reason for a dismissal, it was virtually impossible to prove.[8]

The ease with which workers could be summarily dismissed, more than anything else, was what crystallised worker dissatisfaction at the way they were treated in the workplace. That is why workers eagerly embraced the slogan 'An injury to one is an injury to all'. It meant, as we explained to workers at Rainbow as well as Fatti's & Moni's, that if any worker was dismissed for belonging to the Union, all the other workers should stop work as well. It was also necessary to emphasise in the case of Fatti's & Moni's that it did not matter who the dismissed worker was, or what was his or her race. The fact that Rainbow did not employ any African workers must have been part of the deal struck with local government when it decided to set up the factory there. Although next door to Zwelethemba, the African township, the coloured workforce was bused in from across town.

Apart from the low wages and arbitrary dismissals, the main grievance of the Rainbow workers concerned the attendance bonus. This bonus was a popular strategy among bosses paying low wages: if for any reason a worker was not at work, she (most of the workers were women) forfeited not only a day's pay, but the bonus as well. It was also utilised to discipline workers at Rainbow. So when in March of that year a union member failed to carry out an instruction to do someone else's job (he was busy at the time) his foreman simply told him he would forfeit his attendance bonus.[9]

'Before the workers had a union, that would have been the end of the matter,' I recorded in a circular letter to the members after the event. Instead, the next day the workers decided to send two of their number together with the member concerned to complain. As a result, all three were summarily dismissed.[10] This smacked of deliberate provocation. Friday afternoon was a good time to do this, because the workers had not yet been paid. As it turned out, the bosses provoked something bigger than they had bargained for.

Another section was asked to provide substitutes for the three who had been dismissed. Unlike most sections, which were headed by white supervisors, this one was headed by a coloured supervisor by the name of Jan Olivier.[11] When the workers in his section refused to substitute for the dismissed workers, one of the managers clocked the whole section out, and the workers walked off the job. Others followed them and went to sit in the canteen. Another manager then reportedly said to them, *'En julle hotnots wat nog so sit – hoekom trek julle nie uit en laat loop nie?'*[12] This taunt provoked a walkout that brought production to a standstill.

I described what had transpired as a 'lockout' in a circular letter to our members, because of the manager who had clocked the workers out, and the racist taunt in the canteen. Mainly, I did not want to be on record as acknowledging it was a strike. Strikes were illegal and practically unheard of at the time, especially a full-blown strike such as this was to evolve into. Some years later, labour law would distinguish between a work stoppage and a strike, but labour law as we know it today did not exist at that time.

Rainbow management must have realised it had blundered as soon as production ground to a halt, and changed tactics. They would take back all the workers, including the three that had originally been dismissed; but they would not take back the supervisor who had made common cause with the workers, Jan Olivier. The workers, for their part, resolved they would not go to work without him. This was their demand. An injury to one is an injury to all.

The following Monday morning the buses that were used to transport the workers to the factory arrived nearly empty, only to find the workers already there, gathered outside the white walls on the immaculately maintained lawn. They had walked all the way from the coloured township. It was at this point that Mrs R and I arrived on the scene, and after speaking to the workers, we went in to try to negotiate with management. Johan Kets, the factory manager, received us wearing a starched white coat.

Management would not under any circumstances take Jan Olivier back, he told us. It was my first strike and I responded badly. A strike is obviously a weapon of last resort. If a strike had been imminent, it might have been opportune to respond with aggression to management's obduracy, but it was a tactical blunder to do so once the workers were already on strike, as

Mrs R pointed out to me after our meeting. It could only serve to antagonise the bosses, as in fact happened.

An official of the Department of Labour then arrived, and Rainbow summoned the workers outside the gates to a meeting in front of the factory. Mrs R and I looked on from afar. Kets spoke, and then the Department official. He apparently assured the workers that management was within its rights to dismiss Jan Olivier. When the workers asked what the reason was for Olivier's dismissal, Kets would not answer. Although there were indications that some wanted to return to work, the workers managed to maintain a common front and left the premises together. We organised a meeting at the hall where we usually met, at which it was decided that everyone would again gather outside the factory and maintain the same stand.

In fact, not everyone was outside the factory on the Tuesday. Some who had been on strike the previous day reported for work. Fearing that the strike was about to crumble, Mrs R and I went to the front of the factory and asked to speak to Kets. An irate Kets appeared on the steps leading to the reception, and ordered us off the premises. Not long afterwards a squad car pulled up outside the factory gates, where Mrs R and I were standing with the workers on the lawn. Plainclothes policemen, presumably Special Branch, came straight up to me and said I was under arrest. I was later charged for holding an open-air gathering, in contravention of the Riotous Assemblies Act, and with trespass. The trespass related to my entering the premises in an attempt to speak to the bosses. Several workers were taken for questioning, to try to get them to incriminate me.

The effect of my arrest, as well as the attempts to get workers to incriminate me, was to infuriate even those workers inside the factory who had not thus far been involved in the strike. On Wednesday all the workers came out. The company had a major order to export to Saudi Arabia. The final straw, we heard, was when the Muslim slaughterers joined the strike, and the chickens could no longer be certified as halaal. But I knew nothing of what was happening outside the walls of the Worcester police cells until permitted to receive a visit from my 'wife'.

'The garden is blooming,' Athalie told me with a beaming smile, after we had embraced, with a police officer looking on. Although she was a keen gardener, we had only a cement yard. I understood the strike had been won. Mr Methven, the managing director, had come from Natal, and it was his decision, apparently, to take all the workers back, including Jan Olivier.

A 'great victory' was how I described the outcome of the strike. Indeed, I would say it was a watershed event, although it was my arrest, rather than the strike itself, that was publicised. There had not been a strike involving the Union for such a sustained period since the 1950s. Bosses in the Western

Cape and elsewhere certainly took note of what had happened. The strike had also exposed the problem brewing in the head office. The bosses at Jones had allowed Juffrou H back onto the premises of that factory, as the official of a trade union in the process of being established.[13] In a desperate attempt to counter her influence, I went along with a decision to allow Vormat to put aside her administrative work three times a week, and go to the factory over lunch hours. There was not much Vormat could do to poison workers against me that Juffrou H had not already attempted, I remember thinking.

When the news broke that I had been arrested during the strike, Vormat contacted Johnny Issel, and the two of them decided their moment had come. Without consulting anyone else or being invited to do so, they decided to get involved in the strike. For Johnny it was personal. Worcester was his home town, and he was determined to prevent Ferrus from claiming any credit for the Labour Party. That brought him into conflict with Oscar, who had gone to Worcester after my arrest and was collaborating with Ferrus to get Rainbow to negotiate. It was Ferrus whom Methven phoned to say the workers could return to work with Jan Olivier. At a post-mortem meeting of head office staff to discuss what had happened, Johnny acknowledged his error in getting involved in the strike. However, my relationship with Vormat went from bad to worse, not because of anything she said but because she opted to clam up.

I was in jail for the full 48 hours that the law permitted an accused to be held before being formally charged in the magistrate's court of Worcester. The Union's attorneys sent a clerk to represent me, and the prosecution demanded stiff bail conditions: R750, a substantial sum at that time for an offence for which R50 bail might have been appropriate, the surrender of my passport and reporting to the police twice a week. On the advice of the attorneys, I accepted the conditions, and the bail was paid.[14] My father told me I should have stayed in jail and appealed.

My wedding

Like others from our background, Athalie and I had been critical of marriage as an institution: 'We don't need no paper from the City Hall', as the song goes.[15] Yet we had what we both regarded, in today's terminology, a life partnership. After my arrest and the visit from my 'wife' in the cells, a 'paper from the City Hall' suddenly seemed important. The fact that I was an accused and awaiting trial brought home to me the need to secure our relationship against external threats.

One reason for being critical of marriage as an institution had to do with the roles tradition assigned to men and women in the middle-class marriage: the husband as the breadwinner and the wife as the homekeeper. So there was some irony in the fact that I was the breadwinner in the Union, and even when Virginia went on maternity leave, Athalie continued to work on a voluntary basis. However, she was not the only one. Liz Abrahams assisted Athalie on a voluntary basis while Virginia was on maternity leave. Neil was by this time working almost full-time for the Union in Johannesburg and was not drawing a salary. He relied instead on his earnings from working as a locum at Baragwanath Hospital.

The reason they had to work voluntarily was that the branches concerned had no money, and the slush fund had been exhausted. But it was also a useful counter to the smear that I anticipated, that I was promoting my own in the Union: jobs for pals, and a job for the girlfriend. It was by this time clear that Vormat's hostility to me had to do with my being white. This was also clear to Oscar, who proposed, by way of a solution, to employ a white who was a former Congress activist (although he did not put it like that). This was Amy Thornton.[16] The idea was the Amy would assist with the head office administration, now that Vormat was spending some of her time in Paarl.

So the proportion of whites in the Union was an issue, although I would have preferred to cast it as an issue concerning the proportion of middle-class intellectuals in a working-class organisation. Without our having planned it that way, our wedding turned into an occasion for overcoming the social distance that being white created, and acknowledging our partnership in life and in the Union. Friends, as well as a large contingent from the Union, including Vormat and Johnny Issel, and branches as far as Worcester and Saldanha, crammed into our Waterkant Street cottage for a party.

Oom Joe made a speech about Moses having led the chosen people out of Egypt, despite the murmurings against him that he was not an Israelite. But the weekend was hardly over before our moment of personal delirium was overtaken by events. A breakdown in relations with Fatti's & Moni's was now imminent. On the one hand, Terblanche had taken to ignoring our letters. On the other, the workers wanted something done about their wages. The women were getting R17-odd a week and the men R19. One of the African workers with 32 years' service was earning only R31: that was less than R1 for each year he had worked for the company, quipped the workers.[17] The Union had submitted its demand, including a demand for R40 a week, but had no way of compelling the bosses to negotiate with us.

Ultimately, the strategy we opted for was to apply for a Conciliation Board, but we could do so on behalf of coloured employees. The majority, who were African, were not employees in law or able to be represented there. To overcome this obstacle, I hit on the expedient of getting African workers to sign a power of attorney authorising FCWU to negotiate on their behalf.[18] In essence, this represented an attempt to provide a cloak of legality for something the law did not allow but was nevertheless just.

I was interested to see how the Department would deal with what must have been a unique application, but we did not have to wait and see. In the week of my wedding, the management at Fatti's & Moni's called together the workers of each section of the factory and told them they would have to elect representatives to serve on a liaison committee. The Africans refused to do so point-blank. In the macaroni and cones sections the coloureds put forward names, but those in the packing department refused to do so. The bosses could now see where to aim their next blow and appointed Mr Christian, among others, as the liaison committee representative over the heads of the workers.

The following week Terblanche himself called together the coloured workers in the packing department. They would have to choose between the Union and the liaison committee, he said, but if they chose the Union there would be 'difficult times' ahead. Terblanche wanted an answer there and then, but the workers said nothing. He then said they should give him their answer the next morning. That evening we met the workers in the crèche attached to a nearby church, where it was decided I should phone Terblanche the next day.

Terblanche put the phone down on me. I had already established that the company was privately owned, and the head office was in Johannesburg. The members of two families of Italian descent, the Fattis and the Monis, owned most of the shares. The managing director was Peter Moni. I was somewhat surprised he took my call, and listened to what I had to say. The management of the company would never threaten the workers, was his reply. Probably I had been misinformed.

The next day five coloured workers from the mill were called into Terblanche's office after work: three women and two men. They had apparently been identified by Mr Christian and other members of the liaison committee as the trouble-makers. Ursula Andrews was a shop steward and Jean van Kerwel a wage clerk from a well-to-do home in Stellenbosch. Daniel Meloi, Gertie Hendricks and Spasie Saaiman were just ordinary workers. None of them could even be described as outspoken. What they had in common was that they were regularly at meetings, and had perhaps been observed getting workers to sign the power of attorney. Their pay was already made up. No reasons for their dismissal were given.

May Day, 1979

It is commonly claimed to be a historical fact that by relying on legal remedies and resorting to the courts, trade unions secured important rights for workers. The dismissal of the five helps put this claim in perspective. There could be no doubt that the five were being victimised, so far as we were concerned, and victimisation was an offence. The counterfactual question is therefore why the Union did not institute legal proceedings against the company. The short answer is that we would have had to organise the workers quite differently for this question even to have arisen. We would have had to instil in them a faith in legal processes rather than their own judgement. Above all, we would have to teach them that patience is a virtue.[19]

Law was of course important. It constituted a space in which trade unions were able to operate above ground. Where we were able to use it, we did so. Our point of departure, however, was that ordinary workers, generally speaking, were only willing and able to exercise rights in the workplace where they were organised. Victimisation was calculated to destroy their capacity to organise. Patience may be a virtue in some circumstances, but not when you are under attack. In that eventuality, you were better advised to defend yourself than look to the courts. The courts were in any event most unlikely to come to your rescue. To understand why this was so, it is necessary to understand the specific calculation Fatti's & Moni's must have made.

This was that the coloured workers, having seen what had happened to the five, would put aside whatever nonsense the Union had planted in their heads about human solidarity and the like, and keep their heads down. In the unlikely event they did not, the African workers would surely not risk breaking their contracts and being sent back to the homelands by taking a common stand with them. Without understanding this calculation, you could not see the exercise of power involved. While all this was self-evident to us, however, it was quite another matter to prove it in a court of law. It would always be a tall order to prove the motive was victimisation, unless an employer was so stupid as to say so.[20]

Moreover, what was self-evident to us was rooted in our understanding of an economic and political order that the Union was challenging. Although the challenge was not overt, it was nonetheless radical. We were organising workers at the bottom of the workplace hierarchy, across both the gender and racial divide. In the process, workers were discovering they had real power, the power to bring production to a standstill. It was not simply far-fetched to suppose the courts would assist the Union in furthering such a radical challenge. It fundamentally misconstrued the role of the courts. In respect of employment, this was to constitute a space in which the bosses were able to structure relations in ways that suited them and maximised their profits.

Terblanche understood as well as any manager would that the more time that elapsed after the dismissals, the more the workers' fear at losing their jobs would eat away at their sense of outrage. That would be why the five were dismissed after work. As a consequence, their comrades only found out what had happened in the course of the next day, and different sections of the workforce found out at different points. When some workers stopped work and demanded an explanation, others were still working or did not know what was going on. At lunch break, African and coloured workers gathered in their separate cloakrooms, because they did not have anywhere where they could meet together.

When the workers who had stopped work asked him to address them, Terblanche refused, and called in the Labour Department instead. The inspectors of the Labour Department, whom the workers took to be *speurders* (detectives) because they were in plainclothes rather than police uniform, explained 'the law' to the workers: the law was that if the workers did not straightaway go back to work, they would be put in jail and fined.[21] Terblanche then issued an ultimatum: return to work or be dismissed. The workers started trickling back, until there were only another five coloured workers left still demanding an explanation for the dismissals. They were dismissed on the spot, bringing the total number dismissed to ten.

At some point in the afternoon I phoned Terblanche. This time he took the call. My aim was to set up a meeting, to try to get him to back down. Somewhat to my surprise, he agreed to meet with us. I went with Mrs R, whose powers of persuasion were considerable, I thought, as well as Baba, to have an African present. The dismissed workers were breadwinners, Mrs R said. Think what this will mean for their dependants. She called this strategy pleading with the bosses. It was what you had to do in a situation in which you had no actual power. Terblanche saw through it. This is something the Union should have thought of in the first place, was his response.[22]

His exercise of power was meant to inspire dread: a modern-day head on a stake. But if pleading had no effect, there was still a chance we could shift his stance if he was not able to get production going. We met with the workers that evening. It was also their first opportunity to take stock of the day and reflect how disorganised they had been. The next day was different. The workers clocked in and all sat down together. Again, Terblanche refused to speak to them, and called in the Labour Department. The first thing the inspectors tried to do was to separate coloured and African workers. One of them was on the point of physically pulling one of the women to one side, when he noticed a stirring among the contract workers and thought better of it. When the workers failed to heed further attempts to separate them, Terblanche appeared. They were all dismissed, he said, and should get off the premises. They could come and collect their pay on Friday. There were 88 workers in all.

I was in George that day, and had phoned in from a call box to find out what was going on. Events at Table Top seemed to be taking a parallel course. The Union had submitted its demands, but had received no response. However, the company was in the process of establishing what it termed a 'works committee' with the 'main object', according to a management hand-out, 'of improving communication between workers and factory management'.[23] Oscar was in Johannesburg to attend a general meeting of the branch – it was the first time an official had flown at Union expense.

'The workers want to know where they must go' was the question I was asked over the phone. The magnitude of the situation began to dawn on me. Let them go to the head office, I advised. It was important to keep them together until we had worked out what to do. Then I tried phoning Peter Moni, hoping to persuade the company to meet with us. The company had adopted an official policy not to deal with the Union, he said, and there was nothing further to discuss. It also transpired that the company had straightaway begun to recruit replacement labour.

That would ordinarily have been the end of that. Unlike in the canning factories, the company was not working with a perishable product, which had always been the Union's trump card. With a perishable product like fruit or fish, the employer stood to lose its raw materials in a strike as well as the hours of production involved. But Fatti's & Moni's had got its timing wrong. Everyone was waiting expectantly for the Wiehahn Commission report to be tabled. Precisely because the workers had made a stand, their case was now before the court of public opinion. It could not have been at a more topical time.

This was just the time that newspapers discovered they needed reporters who understood about labour. The company had no wish to talk, but I always regarded it as important to put out the Union's story. So it was our story that dominated the front page of the *Cape Times*. It captured the public imagination. After decades of the Coloured Preference Area policy, African and coloured workers had taken a common stand. African contract workers were prepared to put their jobs on the line for coloured workers. The fact that Parliament was in session gave it national significance.

The workers never went to collect their pay that Friday. That would mean, as they saw it, that they accepted they had been validly dismissed. They were on strike, and they had only one demand: that all the workers, including the original five, be reinstated. On Sunday a delegation of workers travelled with me to Paarl to ask for the management committee meeting's support. By this point there were two worker leaders among them who were outstanding. The one was a man with a beard of biblical proportions and an infectious smile: Friday Mabikwe, a Mfengu from the Peddie district. The other was a shy young coloured woman called Spasie, who had been one of the five. Of the two, only Friday could be described as charismatic or a natural leader.

My entire experience in the Union led me to distrust the concept of a 'natural leader', however. Friday and Spasie were outstanding because they did not put themselves first, but the collective of which they were part, and they did not act on their own. This had been proven in the painstakingly slow process by which the workers had been organised, and the tests Terblanche had set them. In the case of Friday, this quality was combined with a strong moral sense. He was a bishop in a township church, without the self-importance and piety that the title suggests. In the case of Spasie, it was combined with a keen intelligence, despite her lack of formal education.

'The Union would stand by them,' I recorded Oom Joe telling the delegation. 'Although it could not pay what they got in the factory, they should be content with what they got, because nothing was won without suffering. The Israelites had left the fat land, Egypt, to go into the desert to reach their freedom, and they had been provided for. So too the workers on strike would be provided for.'[24] It was the first time that the delegates from the rural branches had met the new recruits from the city.

The workers were indeed provided for in the week that followed. From Monday to Friday, they met in a cinema in Bellville South, owned by a local coloured businessman sympathetic to the strike. Each day there would be a roll call, to establish who was there. If any worker was absent without explanation, the committee would have to investigate. Then there would be reports (often unreliable) of what was going on in the factory, and messages of support would be read: anything that would keep the workers spirits up. Lunch was provided, and at the end of the week each worker was paid R15, for which they signed an acknowledgement of receipt.

The money for the strike pay was drawn from the trust that the Union's attorneys controlled. It was thus money the Union had contributed, from the Paarl building, but it would not be reflected in the books of the Union as strike pay. Although we called it strike pay among ourselves, on paper it was described as a relief fund. If this was a strike, as the Union maintained, it was certainly an illegal strike. The Labour Department had quite extensive powers to investigate the affairs of a registered trade union, and I did not want to give them cause to do so.

We regarded it as self-evident that it would not be possible to maintain a strike for longer than a week without paying the workers something. The fact that the strikers numbered less than a hundred, as opposed to several hundred or several thousand, made it much easier to do so. The Union could afford to pay R15 a week, and maintain that payment for perhaps three weeks. Three weeks was about the time we estimated the bosses would need to get production back to normal. If there was no indication within that time period that the resolve of the bosses was weakening, the Union would not be able to keep up the payments any longer. There would then be no alternative but to concede defeat.

May Day had in the meantime come and gone, in the first week of the strike. In matters ideological, the South African government usually followed the United States, which had shifted its 'labour day' holiday to August to dissociate it from celebrations on 1 May. So it was a curious day for the Minister of Labour to choose to table the Wiehahn Commission report.[25] My hunch is that he was hoping to impress the International Labour Organisation (ILO), which was the only United Nations agency from which South Africa had been expelled. The first step toward being readmitted to the ILO had to be a commitment to freedom of association.[26] The report gave this commitment: one of its key recommendations was that 'both trade unions and individuals should be afforded full freedom of association'.[27] The burning question was how they proposed this should be done in practice.

The other key recommendation the report made was to establish an Industrial Court. The problem with the Industrial Tribunal, it said, was that it was not a 'judicial body'. What the report advocated was a court of law that would determine 'disputes of right'.[28] We did not believe it could be in the interests of workers to have a court with such wide-ranging powers. Yet unlike the burning question regarding freedom of association, nothing we said or did was likely to have affected the implementation of this recommendation.

In fact, the Commission was divided over how to implement its recommendation regarding freedom of association. Was it proposing that government recognise racially separate unions for Africans? And was it proposing contract workers be allowed to join trade unions on the same basis as other African workers? The Fatti's & Moni's strike provided our answers. It highlighted the absurdity of a coloured and African workforce being compelled to belong to separate unions. It also highlighted the absurdity of proposing that only permanent African workers have trade union rights, when most African workers in the Western Cape were on contract. The notion that these workers were not South African nationals was the folly of grand apartheid.

However, the report was certainly not intended to make the task of organising lesser skilled workers easier. The African workers it hoped would benefit from being able to exercise their freedom of association were skilled, because skilled workers were needed for the economy to grow, and there were not enough whites and coloureds to go around.[29] Also, 'blacks [meaning Africans] are no longer, from the point of view of their unionisation, "mainly unskilled"'.[30] There is also no basis whatever for a historical narrative that suggests that the report came in response to pressure from the 'independent trade unions'.

The converse was true. It was because the 'independent trade unions' – which I prefer to call the 'emergent trade unions' – had so little to show for themselves, particularly in industries which relied on workers with lesser skills, that the reforms were opportune.[31] For this reason, both the bosses and government had good reason to believe that TUCSA, whose general secretary sat on the Commission, would be able to incorporate African recruits without any loss of control, and carry on its tradition of responsible unionism as before.[32] Doubtless these new recruits, in time, would begin to utilise trade unions to advance their own sectional interests in much the same way as the craft trade unions had done.

The strike had also highlighted how much of an outlier FCWU was among registered trade unions. As if to underscore this, a most bizarre invitation arrived at head office, addressed to me in my capacity as general secretary of FCWU. It was a gilt-printed card, inviting me to what the Minister called a *conversazione* at some Cape Town penthouse. The Minister, it seemed, wished to be seen as a globe-trotting sophisticate. Obviously the invitation had gone to the secretaries of all the registered unions, without anyone making the connection with the strike that was then attracting such unfortunate publicity.

12

Modderdam Road

THE SAME WEEK THE FATTI'S & MONI'S WORKERS went on strike, Rainbow workers decided, without informing the Union, to confront their bosses about their wages. A delegation went to see Kets, who said he would get back to them. He did, that Friday, with an offer to increase wages by one cent an hour for the woman and two cents for men. The workers were not happy with that, and proposed the attendance bonus be incorporated with their wage. When Kets refused to agree, a large group walked off the job.

The first the head office knew about this was at the management committee meeting that Sunday.[1] The only advice the meeting could give was for the workers to report to work that Monday, which they did. Only those workers who had not walked off the job were taken back. Some 400 workers lost their jobs. No doubt their victory in the first strike had made them over-confident. Perhaps they had been organised too quickly. There were outstanding individuals among them, but the workers had not yet had the opportunity to learn which of them they could trust. I also suspect there were agents provocateurs in the factory, planted by the Special Branch to pour petrol onto a smouldering blaze.

Whatever the case might be, the Union had lost its majority and there was not much it could do about it: it was already committed at Fatti's & Moni's and we could not offer the same support, particularly with my case pending. I tried phoning Methven (Rainbow's MD) in Hammarsdale to plead to take the workers back. He suggested I write them a letter. The only response was an invitation some months later to meet a lawyer who, it turned out, was a director of Rainbow as well as a partner in an old firm of attorneys in the city. But he proved more concerned with dissociating the company from the trial then taking place than with the workers who had lost their jobs. Pleading with the bosses was never going to reverse a decision as fundamental as that.

In the meantime, the public response to the Fatti's & Moni's strike was immediate and overwhelming. Individuals and organisations, well known and unknown, issued statements to the press supporting the workers, and the labour reporters of the *Cape Times*, *Argus*, *Burger*, *Cape Herald* and *Muslim News* ensured that they covered them. At a later juncture, newspapers in other centres took up the story, which was kept alive by a succession of attempts to break the resolve of the strikers, such as when the Security Police detained one of the workers for questioning. As a result, hardly a day passed without some coverage of the strike.

He 'did not believe Mr Theron represented ... workers at my factory or any factory in the Western Cape,' Terblanche told the *Cape Times*.[2] This suggested not only a level of antagonism towards me personally, but that he had discussed the dispute with other bosses and believed he was acting in their collective best interests. But apart from claiming the Union did not have a majority of the workforce, probably because he included management and sales staff in his calculation, Fatti's & Moni's opted to keep mum and wait for the story to blow over.

There were a number of trade unions among the organisations who came out in support of the strike: some were unions we were already on good terms with, and some were unions we had had until then nothing to do with. There was also a variety of educational institutions, including teacher and student organisations, faith-based, sports, civic and women's organisations, and organisations representing traders and shopkeepers. The longer the strike went on, the more organisations surfaced. Some of them were established in the course of the strike, and perhaps even because of the strike.[3]

Support took the form of some kind of declaration and a donation: usually of small amounts of less than R500, a cheque from a trade union here, a collection from a group of students there. Every so often there was a larger donation, yet as the first week gave way to the second, and the second to the third, we were still far from having accumulated enough to cover one week's strike pay. Even though the messages of support raised the workers' spirits, the Union could not draw on its own resources to support the strike much longer.

One of the organisations that had made a statement in support of the strike was the Western Cape Traders' Association (WCTA), which claimed to represent 2,500 shopkeepers and traders in the Western Cape. It was the WCTA that first threatened a boycott of Fatti's & Moni's products. Others followed. By the third week of the strike it was clear that production at the mill must be back to normal. This meant that any prospects there had been of the company taking workers back was fast fading. The question was whether a boycott could turn that situation around. It would be an empty threat unless it was something the bosses could actually feel.

It was about this stage that the strikers were asked to move out of the cinema where they had been gathering, into the nearby Belmont Hotel. Athalie and Virginia, on her return from maternity leave, were in daily attendance, and on most days Liz came through by train. Other officials made guest appearances: Oscar on his return from Johannesburg, Mrs R and Baba from Paarl, and Annie Mentoor from Somerset West.[4] My role was to liaise with the media, and look after public relations in general. But each day I would drive along the N1 to Bellville, to report to the workers what was going on. If for any reason I could not do so, I would convey my apologies.

The Belmont Hotel was located on Modderdam Road, and has recently been renamed after Robert Sobukwe. You would come to it after crossing over the railway line that separates the commercial heart of Bellville, where the mill was, from the coloured residential areas. Modderdam Road connected the industrial area of Bellville South with the townships of the Flats, including Nyanga, where the contract workers stayed. The landmarks of the industrial area were visible on the skyline, above the gum trees. These included the silos of Good Hope Bakery, although I did not take note of them at the time.

The hotel was a newish, face-brick building, flanked by a tarmac parking area with no trace of greenery. Inside there was a lounge and next to it a dining area, with tables and chairs. So far as one could see, no one ever stayed in the hotel or took meals there. The proprietor agreed to let the Union meet with the workers during the day, as long as it needed to, at no charge, but by the afternoon the workers had to be out. In the evening it was a drinking hole.

The most boycottable product of Fatti's & Moni's, as we discussed it, was its pasta. This was a product the middle classes consumed, and the middle classes were at that time in the process of transferring their allegiances from the local grocer to the supermarkets that were opening up all over the country, spearheaded by a new retail group called Pick n Pay. Maybe if the workers of Pick n Pay had been organised, there would have been other options open to us. As it was, it was difficult to see how such support as there was for the strike among the middle classes could translate into an effective boycott.

Even if it did, it would take too long for the company to feel it. The workers would never be able to hold out that long. Legal liability was another problem. If the Union called for a boycott, it would surely be sued. We did not at that stage attach much importance to the flour Fatti's & Moni's produced, because the Union had not yet uncovered the connection between the milling of wheat to produce flour and the bakeries.

A lot of capital was needed to set up a wheat mill. To ensure a market for their flour, the mills had acquired the bakeries which produced the standard loaf of bread sold throughout the country. This was the connection. Bread was regarded as a staple food, and the price of the standard loaf was strictly regulated. The government loaf, as it was also called, was one of the perks the working class had under the Nats. The workers must have known that Fatti's & Moni's owned Good Hope Bakery, but it had never come up in any meeting with them, including meetings at which we had discussed the question of a boycott. I first learned this from the WCTA.

The members of the WCTA were the owners of corner cafés and small stores. They supplied the middle classes with what they had forgotten to get from the supermarket, and the working classes who could not get to a supermarket because there were none near them. In racial terms, they were both coloured and Indian, and Christian and Muslim. Mr Allie, the secretary and spokesperson, had phoned me at head office and invited me to meet with their executive. A voluble and excitable man, he was descended, I would guess, from immigrants from Gujarat, as many Indian traders were.[5] Mr Allie introduced me to his colleagues, Mr Ross and Mr Khan.

Mr Ross, it turned out, was a man with a plan that he wanted to present, and the venue was also his choice: a café off Greenmarket Square where people of colour were able to sit down at a table and place an order. This was Cape Town, then the most liberal of South Africa's urban centres, a reputation it has since squandered. But before listening to Mr Ross's plan, I had to hear out Mr Allie: about the managers of Fatti's & Moni's with whom he had dealt over a far longer period of time than the Union had, and their high-handedness. This was the resentment of the small retailer towards the manufacturer on whom he depended for the supply of product, compounded by white arrogance. The WCTA had been to see Terblanche about the Fatti's & Moni's workers, and had obviously not been persuaded by what he had to say. In fact, they would dearly love to bring him down a peg.

Mr Ross's plan was to institute a boycott of the standard loaf that Good Hope baked. The problem was that the members of WCTA did not sell Good Hope bread. This was because the five or six big bakeries that supplied Cape Town with standard loaves had an agreement between them, in terms of which the city was divided up into different zones: each bakery was allocated specific zones which they had the exclusive right to supply. Nowadays an agreement like that would not be allowed, in terms of policies to regulate competition in a supposedly free market. But the political leader who would play a leading role in promoting free-market policies, Mrs Thatcher, had only just come to power in the UK. In a regulated industry it made good sense for companies to co-operate, and the consumer benefited from a price that was controlled.

The zones allocated to Good Hope Bakery in terms of this agreement included parts of False Bay, which were not areas in which WCTA had members, and the townships of Nyanga and Gugulethu. The bulk of the city's African population resided in these townships. Khayelitsha was still a twinkle in the eyes of apartheid's urban planners. There was a sprinkling of African traders in Nyanga and Gugulethu, but they did not belong to the WCTA. They had their own organisation, the Western Province African Chamber of Commerce. Clearly there were issues of trust between African and coloured traders. Indeed, virtually all the organisations that came out in support of the strike, with the qualified exception of the faith-based organisations, were racially constituted, and this underscored what made the strike extraordinary.

'Our members have discussed the strike, but have not yet arrived at a position,' the vice-chairperson of the African Chamber, Mr Mandla, told me the next day. I had phoned him at Mr Ross's suggestion, and he said he would contact me once they had 'arrived at a position'. His attitude was business-like, and he was not inclined to give anything away: I had no way of knowing how seriously the traders were taking the situation.

The particular conjuncture
It would be safe to say that everyone who came out in support of the strike did so because they saw it as politically significant. First and foremost, it was significant because a trade union had united workers, including contract workers, across the racial divide. In the other famous strike of 1979, which took place more than six months after Fatti's & Moni's, the workers at Ford took a stand despite their trade union.[6] It was also significant because men and women were in it together, and because Terblanche's attempts to divide workers were so transparent: it exposed to the public gaze how much the bosses were part and parcel of the existing order. This was certainly not the kind of outcome the Nats had intended with the reforms they were proposing.

There would have to be some 'buy in' from emergent trade unions for these reforms, of course. Government would have difficulty in persuading the outside world that they were indeed reforms if trade unions rejected them for credible reasons. There were indeed credible reasons for trade unions to do so. Despite what the report had promised, government was still not committed to freedom of association. This explains the particular conjuncture: what the emergent trade union movement said and did in response to the proposed legislation mattered. Although organisationally weak, the unions were politically consequential.

The Minister of Labour, Fanie Botha, of course knew what the Commission would recommend. His response to the report must have been part of a predetermined plan to test the water, and see how far government could go without actually forcing its proposals down trade union throats. An essential part of this plan was that he would have room to manoeuvre. This became clear once government tabled its amendments to the Industrial Conciliation Act. The minimum government could do to show a commitment to freedom of association was to drop the notorious definition of an employee. That it now did by way of the amendments.[7] 'Blacks' (which had now become the Nats' preferred term for Africans) would henceforth be regarded as employees.

The simple fact was that this did not mean an end to trade union apartheid, however. 'Mixing' across racial lines was only allowed if the numbers of one 'population group' were too small to form 'an effective separate union', or if in the Minister's opinion it would be 'expedient' to do so.[8] It was only in cloud cuckoo land that the Minister would exercise his discretion in our favour. In any event, African workers could only join a registered trade union if the Minister, in terms of powers vested in him by the amendments, declared the 'group or class of persons' of which they were part to be employees.[9] There would, in other words, only be freedom of association in so far as the Minister allowed it. This, of course, was no freedom at all.[10]

Utilising the room to manoeuvre the amendments gave him, Botha declared that contract workers would not be allowed to join trade unions.[11] Not even the TUCSA trade unions could accept that. Botha's advisers must have told him this would be so, but at this juncture he was more concerned to prove a point to the whites-only unions and opponents in his own party.[12] The test would be how trade unions would respond when he changed tack, without relinquishing his power to determine whom trade unions could organise. 'Contact with other unions' had now become a standard item on the agendas of Union meetings.

The Minister had no clothes on, and it was safe to say so, because he could not afford to be seen to be shutting us up. Instead, there would be an increase in covert 'attacks', as we termed them. An early example of what we could expect was at a general meeting in Lamberts Bay one evening, although, compared to what the Special Branch later got up to, it seems trivial: blocks of wood with nails sticking out of them, placed behind the wheels of the kombi.[13] There was certainly nothing trivial about what was happening in Paarl, where the breakaway group had decided to adopt the title 'association' rather than 'trade union' on the advice of their attorney, a zealous young Afrikaner who helped them in matters large and small.[14]

You had to wonder how they could afford his fees. If the Special Branch were not funding them, someone from the Afrikaner establishment must have been. For the powers that be in Paarl, the breakaway must have seemed like a godsend. One day the factory manager at Jones was overheard referring to the association as a *'vakbond'*, this being a pukka Afrikaans term for trade union, and Vakbond is how the breakaway group would henceforth be known in the Union. To counter it, the immediate objective was to recruit ten coloured workers. Ten were needed, because that was the minimum number that could form a branch in terms of the constitution. Eventually we recruited 12.

The most prominent among the recruits was Aunt Sabbagh, the former chairperson of the branch. It had been a mistake to resign, Aunt Sabbagh told me. She had been a member since she first started working at Jones, at the age of 12. The others were mostly were from the labelling department, and recruited by Lizzie, the worker who had criticised me over the slush fund, together with her friend Luska. But the reason African women were listened to in the racialised atmosphere the Vakbond had created was also owing to Lorraine*, the supervisor, and the fact that the labelling department operated all year round. No one had to fear being laid off at the end of the season.

Lorraine volunteered to accompany us one winter's evening to conduct house visits. It was a brave thing to do, because Lorraine's husband was known to be on the other side. He was also an abuser of women, I later found out. He beat her severely when she got home that night. For months after that we made no real progress. Aunt Sabbagh, it was said, was too soft-spoken to take on Grootvoet and the Vakbond leadership. The person we needed to win over more than anyone else was Polly Solomons. She was an ordinary worker who knew her own mind and would not be intimidated. She also commanded considerable support among the other workers. One evening we visited her at her home.

'The breakaway is going nowhere,' I told her. She listened politely as I explained why a trade union representing only one factory could achieve nothing for the workers, when wages and other conditions were decided at an industry level. This of course assumed the Vakbond would not be able to recruit more factories to its cause. 'The breakaway is going nowhere,' was her rejoinder 'and I am going nowhere with it.'

I could not blame Vormat for our failure to make further headway at Jones, but it went without saying that she should never have let her issues with me spill out in front of the workers. The Vakbond was surely a common enemy. On the first indication that she was instigating workers against me, Vormat was confined to head office. The line between the roles of an administrator and an official was not crossed again, but she did not appreciate being reined in, and relations at head office took a further turn for the worse.

Although Vormat had no cause to discuss work with Johnny Issel, who was employed by the medical fund, I would often find them in a huddle. They would then lower their voices, with no pretence at trying to include me in whatever they were talking about. They must both have belonged to the same underground structure, and Johnny was soon to play a leading role in the internal revival of the Congress movement. Although not overtly hostile toward me, he did nothing to counter banter with staff from which I would pointedly be excluded, or private jokes of which I was the butt: about the deferential way I was treated in the rural areas, for example, and about coloured workers who insisted on addressing me as *meneer* (mister) rather than by my name, or as 'comrade', as I later became. The subtext was that I was a white boss, rather than an elected leader, and Athalie was the boss's moll.

This was more than just office politics. The Union was now fighting on two fronts, and for all the publicity and support we were attracting at Fatti's & Moni's, the main front was at Jones. We were not proud to proclaim that we had lost the factory where the founding strike of FCWU had taken place, least of all to coloured chauvinism. Few outside the Union knew this. We could not take for granted that the Vakbond would not recruit more factories to its cause. So there was no question of putting unity within the organisation before personal ambition: far from it. A third front was being opened up, in which my leadership was being challenged from within.

The fight on this third front raged for the duration of the Fatti's & Moni's strike. It was all the more debilitating because it was covert and could easily be denied. For members who were not aware of what was going on, Vormat was still just the typist. For those who were aware, it was obvious she was bidding for power. You could see how someone who saw herself as black rather than coloured might have justified this to herself. The strike at Fatti's & Moni's would have been portrayed as one in which coloured and African workers were realising their common identity. Having a white as the public face of this most publicised dispute must have been galling; particularly one with such an Afrikaans name. Yet if this was a form of Black Consciousness, it had more to do with racial resentment than identity. Resentment was also something she had in common with the Vakbond, who still called me, somewhat improbably, *die boer*.

The incidence of whites in the emergent trade unions was surely a factor fuelling the fight on the third front. The irony about this was that it was Biko and his SASO comrades – who did not seem to me to be filled with resentment towards whites – that had told white students to change white attitudes. We had stumbled on a way to do so, by way of example: the example of whites refusing to behave as whites were expected to, in alliance with blacks not behaving as workers were expected to behave. And although I doubt whether this was true of all white officials in the emergent unions, I had admitted into the organisation the black activists now making my life difficult, precisely because I wanted to avoid any semblance of white domination.

What compounded the irony was that, having done all that I could to improve relations at head office, and established that Vormat had no intention of doing what she was employed to do, except perfunctorily, I could not behave like a white boss in the circumstances, or even a reasonable manager. Even though there was a valid reason to dismiss, my word would not be enough to persuade the management committee to do so in the poisonous atmosphere that now prevailed. The only person who could have witnessed for me was Amy, and she was white.

So at a time when the Fatti's & Moni's strike was demonstrating how non-racialism was more than a slogan, and was crossing the line between trade union organisation and politics in doing so, the staff of the Union were polarising along racial lines. They were also polarising over the issue of clandestine versus above-ground organisation. What brought this to a head was when Amy came across Vormat and Johnny Issel printing documents on the head office roneo machine. Roneod documents could be traced by the Special Branch. If the documents were illicit, as they surely were, the Union could be placed at risk. What was even more serious was that when Amy raised her concerns in a staff meeting, Vormat denied any such thing had occurred. Either Amy was a liar or they were.

What finally prompted the management committee to act against Vormat was when it emerged she had tried to lobby Fatti's & Moni's workers behind the backs of the officials, and tried to get them to lie about it, in an attempt to cover her tracks.[15] Someone who was not white blew her cover. Even now, I hesitate to mention her name. Is it because racial solidarity remains an issue till this day, even though it was the right thing to do? The management committee instructed Vormat to apologise to me for lying, which she did. However, she could barely conceal a glow of triumph at having frustrated my bid to have her dismissed. It was the blackest moment in my trade union career. For the first time I contemplated resigning.

If Vormat and Johnny were part of some underground ANC structure, as I think they must have been, the reason they would have felt no need to justify their actions to the Union, or lied, must have been that they regarded themselves as accountable only to their comrades in that structure. It was justified to use the Union's resources to advance the struggle, because the Union was subservient to the struggle. It is an outlook I might also have adopted, had I put my energies into clandestine organisation. Yet it had no regard for the independent political role the Union had to play, and was in fact playing.

There was also a subtext of condescension toward workers in the jokes made behind my back. Evidently it was of little consequence that I had been elected and enjoyed the support of the workers. This support was not blind. It was based on ideas about the direction the Union should be taking, and the values to which it should subscribe, that I had been articulating with increasing confidence. I would be betraying the trust of the workers if I left them in the hands of people intent on using the Union for their own ends and bent on the pursuit of power at all costs. This was the belief that kept me going.

The boycott begins
As long as there were donations coming in, the Union could carry on supporting the workers; but by the fourth week it was clear we were spending money faster than it was coming in. By the fifth week, there was only enough for that Friday. All the while, the bosses were refusing to meet with us or negotiate. I could not justify going back to the Union to draw further on reserves, when there was no prospect of negotiations and nothing more we could do to compel the bosses to negotiate than hope for a boycott to materialise. The Union could not afford to support a lost cause, and I had no idea where we would find the money to pay workers on the Friday of the sixth week.

I was receiving phone calls, messages of support and visits from well-wishers at head office on a daily basis. Whenever people asked what they could do, I would tell them to phone the company to demand they meet with us. What Fatti's & Moni's would then say was that they could not meet the Union because it had applied for a Conciliation Board. It was a ludicrous thing to say, but it created enough confusion among those who had contacted the company for them to relay it back to me.

Some well-wishers went straight to the hotel at which the strikers were meeting, where they would be introduced to the workers and could address them if they wished. These visits would also break the monotony of the daily report-backs and reviews of the position of the strike. This was looking increasingly bleak. The African traders were keeping mum. When I phoned Mr Mandla to ask if his members had arrived at a position yet, he told me they had still to meet; when I phoned him again, a meeting that was supposed to have taken place had not happened. Either they had no sense how precarious the workers' situation was or they were prevaricating.

In the seventh week I personally put the workers on notice: we had managed somehow to scrape together enough strike pay for the sixth week, but there was no prospect of money for the Friday to come. Unless the African traders came out in support of the boycott, we would have to call the strike off. It would be important to acknowledge that the strike had failed to achieve its objective and end it, while the workers were still united, rather than to pretend it was continuing and allow the group to disintegrate, as it undoubtedly would if we could not continue to provide material support. The fact that the group had kept together for as long as this was a significant achievement in itself, but already there were a handful of workers whose commitment was wavering and who were not regularly in attendance at the hotel.

At the start of that week, 4 June, a letter arrived from the Department of Labour asking the Union whether it wished to continue with its Conciliation Board application. It was difficult to know what to make of this letter. It was now close to two months since the application had been made, with not so much as an acknowledgement of receipt from them up until then. Given the role the Department had played on the day of the strike, it was hard to believe they had any intention of facilitating negotiations with the company.

So we formally withdrew the application, and issued a press statement. 'Fatti's & Moni's has consistently used our application for a conciliation board as an excuse for not meeting with the Union. The management no longer has any excuse for not dealing with our Union. The workers have for seven weeks been asking all to be given their jobs back, and they are as firm in that demand as ever.'[16] Our demand was now for direct negotiations between the employers and the Union. This would increasingly be the approach adopted in other negotiations as well.

The Department, it turned out a few weeks later, must have had some misgivings about its role in the strike. Someone had sent them the minutes of the management committee meeting at which the threats made by its inspectors were recorded. Most probably it had been the Special Branch. I received a phone call from a new official at the Department, Mr Marais, complaining about the 'lies' in our minutes. A delegation from the Union – including Oscar – then met with Marais and Van den Bergh, and defended what was said in the minutes as correct.[17] The fact that the Department was prepared to meet with an official of an unregistered union was significant. What was even more extraordinary was that the Department did afterwards conduct some kind of investigation, probably at the instance of Marais, who was clearly a new broom, and then apologised for accusing us of lying. Not long afterwards, Marais replaced Van den Bergh.

On the Wednesday afternoon of the seventh week I again phoned Mr Mandla, expecting to be fobbed off again. He informed me in a matter-of-fact way that his executive had decided to boycott Fatti's & Moni's products, and that the boycott would begin the following day. We were all delighted, but had no idea how they proposed to implement it or if it was just talk. So workers were appointed to monitor what happened at different stores in the township. There was still a question about where Friday's strike pay would come from, but I did not want to say anything to dampen the workers' spirits.

As though by divine intervention, a young dominee from the Dutch Reformed Sendingkerk pitched up at the hotel. This was Allan Boesak, someone I had not previously heard of or met. We must have arranged to meet there, but he would of course not have indicated over the telephone what he wanted to talk about. I also have no clear recollection of what he said to the workers: only, as he was leaving, his handing me an envelope in the parking lot of the Belmont Hotel. It had a cheque in it, for a substantial sum: enough to keep the strike going for a few weeks.

The workers appointed to monitor what had happened reported excitedly their observations earlier that Thursday morning. At each shop, the truck from Good Hope had arrived and the van assistants had offloaded the bread, as they routinely did. Once they had finished offloading, the trader would tell the driver he did not want Fatti's & Moni's bread, and that the van assistants should be instructed to load up again. All the traders in Nyanga and Gugulethu supported the boycott that day, and the next. If there were any that did not, we did not hear of them. This was in contrast to the WCTA, some of whose members were reported to have continued to stock Fatti's & Moni's products. The manner in which the African traders implemented the boycott gave me a new perspective on what it meant to be properly organised.

The effect on Fatti's & Moni's was almost instantaneous. On the Monday, Terblanche drove to Nyanga, together with Frank Lighton of the Cape Employers' Association, and went to Mr Mandla's house. They asked to discuss the matter with him, but Mr Mandla said he would only talk in the Union's presence, and asked them to arrange a meeting between themselves, the WCTA and the Union. The meeting was scheduled for the following week, 14 June, at the Holiday Inn in Bellville, in the afternoon. In the meantime the boycott continued.[18]

Bread was one of the lines on which the livelihood of the traders depended, but they were able to secure an alternative supply from one or more of the other bakeries. This had to be in breach of the zoning agreement between the bakeries, but the Union was not going to blow the whistle on any back-stabbing among the bosses, particularly when it might involve bosses that did not support the Fatti's & Moni's stance anyway.

The Union's representatives at the Holiday Inn meeting included a delegation of workers, elected after much deliberation. It included Friday, Spasie and a genial worker from the King William's Town district called Stanford Booi.[19] Nothing can have prepared them for the spread of animal protein laid out before them, on toothpicks and skewers, and wrapped in pastries. The contrast with the vegetable soup at the Belmont Hotel could not have been starker. This, and the gloom of a venue without natural light, lent the proceedings a surreal character.

Terblanche was there, as well as the distribution manager of Fatti's & Moni's, whom we had not met. The distribution manager made some joke about the shirt Mr Allie was wearing. It seemed he had given it to him as a present, in return for business favours, no doubt. The WCTA and

African traders each had four or five of their executive committee present. It was the first time the Union had met the African traders. Terblanche opened proceedings by addressing himself to Mr Mandla. Mr Mandla's response was to invite me to set out the Union's position, and I could see the traders enjoying Terblanche's discomfort, as I did so. When it came to his opportunity to respond, their enjoyment turned to irritation. The company had no answer to the allegation that it had victimised the workers or sought to destroy the Union. If the traders had been a jury, the guilty verdict would have been unanimous.

When it came to the question whether the company was prepared to take the dismissed workers back, it was difficult to know what Terblanche was really saying. At one point Mr Khan could no longer contain himself, and started shouting at the company from across the table: only what he was going on about was the shirt he was wearing. It had been a cheap gift the distribution manager had given him, and not at all comparable to the largesse the company displayed towards its white clients. After asking for an adjournment, Terblanche tried to get the meeting back on track. Mr Moni himself was flying to Cape Town to attend the meeting, he said, and proposed we adjourn until tomorrow. But Mr Moni was not present the next day, although he was apparently in Cape Town. He had decided it would not be appropriate or tactical to lower himself to attend the meeting. This, predictably, irritated the traders no end.

Terblanche then put forward a proposal Moni must have mandated: to employ the workers at other subsidiaries of Fatti's & Moni's, conditional on the company obtaining government approval for it to employ the contract workers. However, the workers' demand was for reinstatement. Reinstatement, as the workers understood it, meant returning to the same jobs they had done before, in the same factory. The traders backed this position, and the negotiations ended inconclusively.

Perhaps an offer of re-employment two months after workers had embarked on an illegal strike might seem reasonable from a contemporary perspective. It was what the Union was still trying to secure for the dismissed Rainbow workers. In the same week as we met with Fatti's & Moni's, I had another court appearance in Worcester. Rainbow had now formally requested the case to be withdrawn, but the prosecutor was not prepared to do so.[20] Rainbow, for its part, was not prepared to re-employ.

The fear of both employers was no doubt that if workers were re-employed, the Union would have an opportunity to reorganise. But Fatti's & Moni's was now engaged in a war of attrition, and the fact that it had made any kind of offer was a sign of weakness. What was paramount, from the Union's perspective, was to preserve its position of strength and for the workers to remain united.

In the days that followed, the workers exhaustively debated the company's offer. Re-employment in other subsidiaries meant in bakeries like Good Hope and elsewhere, where the Union had no presence. With the group dispersed and individual workers isolated, they would easily be dismissed, one by one. There was in any event no guarantee of employment for the contract workers. The workers had left the factory together. It was decided they should return in the same way they had left.

The workers had now been out for two months, and from this point on the strike would be measured in months rather than weeks. There were two further meetings with the company that month, in an attempt to negotiate a solution. At the last of these, brokered by the Cape Chamber of Commerce, the company made its most concrete offer yet: the immediate re-employment of 23 coloured workers at Good Hope Bakery, but without any firm undertaking to employ the African workers.[21] A stalemate had been reached.

A war of attrition

There were two ways in which the war of attrition was now being conducted: on the one hand there was no letting up on the boycott, and on the other there was what the bosses referred to as 'public relations'. The Union had no difficulty in explaining why it rejected an offer that discriminated between coloured and African workers to the general public. After this was done, it was necessary to strengthen the boycott. So the Union invited 13 of the organisations that had expressed solidarity with the workers to a meeting where the Community Action Committee was constituted. The idea was that this committee would assume responsibility for the boycott.

The committee was supposed to be separate from the Union, even if not altogether at arms length, since workers and officials did in fact attend all meetings. This was both to protect the Union against legal claims against it by Fatti's & Moni's, and to preserve a distinction between their functional roles. The Union, or at least its officials, could not play an active role in implementing the boycott, although the workers were encouraged to monitor it.[22] This committee was later regarded as some kind of prototype for building an alliance between trade unions and the community, and fulfilled an important function in disseminating information about the strike.

What was sobering, however, was how few organisations on the committee had a membership base they could mobilise, and were capable of implementing decisions in the manner in which the African traders had done. Some of the most effective initiatives were ones of which the Union had no prior knowledge. Shoppers in a Saturday morning supermarket would 'discover' to their horror Fatti's & Moni's product in their trolleys at the pay-point. They would then volubly insist on replacing the tainted product with something else, as the queues grew longer.

In some ways the attempts by Fatti's & Moni's to bolster its own position only made matters worse for them. It had thus far held off doing anything about the contract workers who continued to live in its hostels in Nyanga, most of whom were staying in Cape Town illegally, by virtue of the provisions of the Bantu (Urban Areas) Act. This was the key law underpinning the system of influx control, and the hostels were the obvious place to initiate a clampdown. Instead, a contingent of about 20 police arrived at the Belmont Hotel while a meeting was in progress, and began checking the workers' passes and identity documents.

When a contract worker was dismissed, the employer was supposed to endorse his pass to this effect. Fortuitously, however, because the workers had not collected their pay, the passes of the African workers had not been endorsed. This meant there was nothing to show they were in Cape Town illegally, unless the term of the worker's contract had expired. This was the case with Mzamo Mxhanto, who had worked for the company for 13 years. He was arrested for being in an urban area illegally.[23]

This show of force by the police presented the Union with yet another opportunity on the public relations front, and the Union's lawyer secured the release of Mzamo on bail. But it also raised questions as to whether the Union really could defend contract workers against the might of the state. Some must have had their doubts. A worker called Skilpad suggested that a group of them go and see a representative of the Ciskei government. Ostensibly their mission was to seek his protection against being arrested. At about the same time there were indications of growing discord among the strikers.

The curious thing was that, although arrest for being in an urban area was an everyday occurrence, the illegality of contract workers in Mzamo's situation had never been challenged. The wording of the Bantu (Urban Areas) Act was open to a different interpretation from that which the police placed on it. If a worker had been 'continuously employed' for more than ten years for the same employer, the section seemed to suggest, he (there were no women on contract) could qualify for so-called section 10 rights. This was the sought-after right to reside permanently in an urban area.[24]

Mzamo had been continuously employed for more than ten years, if one regarded his yearly return to the Ciskei as annual leave, rather than a break between one contract and the next. Mzamo's case was fought and won in the magistrate's court on this basis. But a decision of the magistrate's court did not have the status as a decision of the Supreme Court, so in consultation with the Legal Resources Centre it was decided to bring a test case on behalf of all workers who had been continuously employed for more than ten years. Ideally, this had to be a worker who did not have any breaks in his employment of longer than two weeks, which was the minimum annual leave to which a worker was entitled at that time. The worker who seemed to have the best prospect of success was Stanford Booi. Legal proceedings had been initiated when the strike took a disastrous turn.

You could say the Ciskei representative was to blame for what happened, as we did, because shortly after being approached by the workers, he met with Terblanche. What they discussed was never divulged, but afterwards he approached workers individually, to offer them jobs at Good Hope Bakery.[25] This was surely what Terblanche asked him to do, and he would not have done so for no reward. The representatives of homeland governments were not known for their zeal on behalf of their 'citizens'. Yet Skilpad and the others would not have been susceptible to such an approach if they did not have doubts about the future of the strike.

They did not voice their doubts in the general meeting. They did not discuss their decision to approach the Ciskei representative, which was in breach of an agreement that each and every initiative that workers took during the strike be discussed beforehand in the general meeting. This was what those left behind said about Skilpad and his group in an attempt to come to terms with their betrayal. But they were not judgemental. It was as if they were speaking of an alcoholic who had disregarded the precepts of keeping clean, out of concern that others would succumb to their addiction.

A solution to the dispute was in sight, Terblanche announced gleefully to the press, with the help of the Ciskei government. Workers were given until 27 July to apply at the bakery for jobs, and 21 did. All but two of them were contract workers, and Stanford was one. Skilpad was another. When their first pay day came, the company forgot to pay them. Some were reported to have begun talking about rejoining the strike, but none in fact did.[26] Their sense of shame at having abandoned the strike was to prove a greater obstacle to integrating them into the Union again than was the case with the scabs the company employed at the start of the strike.

It is not clear what part Lennie played in these events, but the role of translator is a critical one in any group whose members cannot communicate with one another. Without a translator from Afrikaans to isiXhosa, the only way the Africans could communicate with the coloured workers was through the medium of English. But it was a minority that spoke English as a second language. When there was an official who could translate, he or she did; often there was not, and Lennie assumed the role of translator.

There was often a problem of bad translation in Union meetings. What was being said was misrepresented, because the translator did not understand the point or was at a loss for words to express it. But the translator is also in a position of power, and misrepresentation can further a particular agenda. For that reason, and because the translator was sometimes privy to discussions among the leadership, he or she made an ideal recruit for the Special Branch.

Lennie was always a bad translator. But during this critical period, when the future of the strike was in balance, his translation of what was being said in the general meeting was not just bad: it amounted to a misrepresentation that could only be deliberate. There were enough people present to realise this, including Athalie, whose command of isiXhosa had improved markedly during the strike. Moreover, Lennie's objective appeared to be to sow discord between coloured and African workers.[27]

When he was confronted about this after the meeting, he eventually admitted to being employed by the Special Branch. I was not present at the time, but that evening I received a call from Virginia to go urgently to a house in Lansdowne, where I found a terrified Lennie being interrogated by three 'comrades'. One was an agitated young man who, I was told, had himself been detained and interrogated by the Special Branch, and was plainly unhinged. He was waving a dart (such as is used in the game) in Lennie's face, threatening to do something to him with it. Predictably, this prompted Lennie to say whatever he thought his interrogator wished to hear, however contradictory or improbable. I tried to suggest this was not helpful, and put a few polite questions to Lennie myself. He had come from a township outside Johannesburg to work in Fatti's & Moni's a few months before the strike began, he told me, and after the Union had begun organising there.

I think this was the truth. I was not surprised the Special Branch had an informer among the strikers. What did surprise me was that an informer would be planted at Fatti's & Moni's at the point when we were beginning to organise there. It was not obviously a factory of strategic importance, as Rainbow was. It reinforced my suspicion that there had been agents provocateurs there, planted in the same way as Lennie had been. The others agreed to let Lennie go, and I later heard he had been put on a train back to where he came from. To me he was more victim than villain. Not only was the manner in which the 'comrade' interrogated him disturbing, but it was so obviously a case of 'us' adopting the same methods as 'them'.

The re-employment of the 21 was a body blow to the strike, and represented the high-water mark for Fatti's & Moni's on the public relations front. We had to make it clear the strike was not over and the boycott would continue. The only positive aspect was that fewer workers – there were now 66 – meant that the money we had for strike pay could go further. The strike was now entering its fourth month. Although it was a hand-to-mouth situation, we were still managing to pay the workers entirely from donations we were receiving.

At about the same time Joe's three-year-old daughter fell ill. He was one of the few African workers on strike with permanent residence. Joe did not have the money to take her to the doctor, and she died.[28] That he had not been prepared to betray the strike by collecting the pay that was due to him at the factory, was how the workers interpreted his loss. It was literally a life-and-death struggle they were engaged in. A large contingent attended the funeral in Gugulethu.

Then the Union was invited to send two delegates to address students at the University of the Witwatersrand in Johannesburg. This was also an occasion to try and extend the boycott to the Transvaal. We had already established that Fatti's & Moni's had a mill at Isando. Although it was not yet clear whether it also had a bakery there (it did not), it would obviously strengthen our hand if the boycott was seen to be national, as well as boosting the morale of the workers. The general meeting decided Friday and I should go.

Barbara Hogan invited various organisations in Johannesburg to a meeting to form a Fatti's & Moni's support committee. The same kind of organisations was represented there as supported the Community Action Committee in Cape Town, but they also included representatives of the Soweto Committee of Ten and Inkatha.[29] Our contacts in the Catholic Church relayed to us how Peter Moni, the company's managing director, had been devastated to encounter a placard demonstration one Sunday on coming out of the Rosebank church his family attended.

Perhaps because the Union was a relatively weak presence in Johannesburg, some in the committee seemed to have difficulty in understanding the boundary between support and interference. Neil Aggett wrote to tell me how members of the Fatti's & Moni's support committee wished to meet with the workers themselves. 'Personally, I feel that it is an attempt by some students etc. to gain worker credibility, and is opportunistic, as they [the members of the support committee] are primarily not a workers' organization, but an alliance of various political groups with varying aims. We also hear that they tried to get one of their friends employed at F and M in order to "organize" the workers.' Fortunately this unsolicited initiative came to nothing.[30]

At all the meetings we attended in Johannesburg, and also on other occasions at which it was necessary to explain about the strike, we would follow the same protocol: Friday would explain in his own simple but complete fashion, step by step, in clearly articulated isiXhosa, how the strike had happened. I would follow, trying to place the strike in a broader context. The point was that the strike was about the workers' struggle, and theirs was the voice that needed to be heard. Most people we addressed got the point. Friday was a man of enormous presence. In his own community he was, after all, a bishop. So I was interested to see how a bishop of the church to which my mother belonged would respond to him. This was of course Desmond Tutu, who was the secretary general of the South African Council of Churches at the time. We met in their Johannesburg building, and when he asked us to explain about the strike, I as usual deferred to Friday. Friday had hardly begun his account of the strike when Tutu interrupted him, with barely concealed impatience, in isiXhosa. It was me he wanted to hear from, and in English.

The extension of the boycott to Johannesburg, and the continued public interest, helped bolster the morale of the workers, and there were no more defections by the time of the conference. Liz Abrahams was the guest speaker. This represented her 'coming out', and her formal return to the Union.

'What will happen to dismissed workers is still in the balance,' I reported to the conference. 'If they do not get their jobs back, the coloureds may be able to get other jobs but the Africans will be sent back to the Ciskei, where there are no jobs at all. Even if this happens, the Union will not have lost. The workers involved have learnt lessons that will stay with them for the rest of their lives.'[31] I still think this was true, but it was also putting on a brave face, in the event we had to concede defeat on this front. The elections of national office-bearers would be significant for the struggle on the other two fronts.

I was again unopposed as general secretary of FCWU, but the position of general secretary of AFCWU was contested. Oscar was nominated by Lizzie and seconded by Luska of Jones, more because they wanted an African in the position, as I saw it, than because of my handling of the Lucy issue. I won the vote by a small margin.[32] A worker from SAPCo, Alfred Noko, became vice-president. But the election that was to have the most profound consequences was for president of FCWU. The president in a trade union is in a sense the embodiment of the workers that it organises. More than anyone else, she or he is also the bearer of the tradition of that organisation. Oom Joe was for the first time opposed, by Manie van Graan of RFF, a diminutive man who had always impressed me as being sensible and level-headed. Although he also wore the khaki overcoat of a foreman at RFF, Van Graan did not conduct himself in an overbearing or oppressive manner, and was clearly popular with the workers. He won the vote conclusively.[33]

I had no inkling Oom Joe was to be ousted as president, and neither, it appeared, had he. After losing the vote, he was nominated for the position of vice-president, but declined. He was clearly put out. Afterwards, I wondered whether there had been some sort of clandestine plot to oust him, but dismissed the possibility. Workers were capable of drawing their own conclusions, and no longer wanted someone they feared rather than respected. The Union was a school for democracy, and the outcome of the election went to show how far the members had progressed. Perhaps they also did not want someone who was tainted by corruption. Long before I knew about it, the Paarl workers must have known that Baba, Oom Joe's lover, was corrupt. That would explain a certain scepticism they had displayed towards her, whereas I had all along been taken in by her charming disposition. So much so that I thought there must be some mistake when the auditors' report showed an unexplained sum of money missing from Paarl AFCWU.

'Paarl again,' Oscar reported to the conference. 'Last year it was Lucy, this year it is Baba. Strong action is required in this respect.'[34] Baba decided not to put up a fight. She resigned and signed an acknowledgement of debt, although I doubt whether she ever intended honouring it.[35]

13

The road from Namaacha

WEEKS BEFORE OOM JOE WAS DEPOSED, I appeared for the last time in the Worcester court. I had entered the premises of Rainbow Chickens in order to meet with the management, I testified. There could not be trespass if my intention was lawful, my advocate, Ian Farlam, argued.[1] But, according to the prosecutor, I was guilty of trespass whatever my intention, if I did not have the landowner's permission.

I should have been on weaker ground on the other charge, of contravening the Riotous Assemblies Act by holding a meeting out of doors. I had simply been waiting outside the premises along with the workers, I had testified. That was a half-truth, but there had been no cause to lie. The state had not succeeded in getting any of the workers who had been outside with me to testify.

The hearing had taken two days, and each day the courtroom had been packed with dismissed Rainbow workers. There were cheers in court when I was acquitted, and the magistrate, who was not much older than me, smiled.[2] Afterwards, over his lamb chops, Farlam suggested I had spent long enough on what he clearly regarded as a diversion from the tradition to which I really belonged, and gave me a tie from the law faculty from which we both had graduated. I still have the tie, but have never worn it.

My acquittal meant I could get my passport back and go to Swaziland. Athalie and I had for some time been planning a holiday there, where we could stay with her sister Colleen and Ben. Ben had fled South Africa years before to avoid a jail term for a political offence, so he was persona non grata as far as the police were concerned. But neither he nor Colleen was politically active in Swaziland, so far as I knew. The police also knew of my plans to go to Swaziland: I had told them when we unsuccessfully tried to persuade them to give my passport back

We did really want to see Colleen and Ben, but I also needed a pretext for visiting a neighbouring country in order to meet Miss Ray. If I had been asked, under interrogation, why I wanted to meet Miss Ray, I would have said that she had been the first general secretary of the Union, and there were things about the Union she knew that no one else could tell me. That was the truth, but hardly likely to satisfy the Special Branch. Miss Ray was on the Central Committee of the Party, and also in a leadership position in SACTU. This was the enemy, so far as the South African government was concerned.

There were those who maintained you refuse to say anything under interrogation, but no one I spoke to who had actually been detained believed this was a position one could sustain. It was certainly not a position I could sustain as an elected official of a registered organisation. I could also not plead ignorance of the fact that Miss Ray was a member of the ANC. So it was a situation I would have to deal with as best I could, if it ever arose. If it did, I was fairly certain the workers would have no problem with my trying to see Miss Ray. The same photograph on the head office safe was in the Paarl office and gazed down at the delegates at management meetings, like an icon in church.

Even so, this was to be a clandestine operation, which I could not discuss with anyone, even people I trusted. You did not want to burden anyone with things they did not need to know. Oscar knew about my plans, and of course Athalie. This was in fact my second attempt to meet Miss Ray in Swaziland, but something had gone wrong the time before. The intermediary then had been Miss Yon, but she had since asked me to stop visiting. The intermediary now was Henning, an intense young German who was working in Zambia and had come to Cape Town 'on holiday'.[3]

Since this was to be a clandestine operation, I tried to consciously suppress details about it that could implicate other people, and of course kept no record of it. The interesting thing was that having consciously repressed so much about the trip, and not having spoken about it afterwards, I had no wish to talk about it even years after I had left the Union. It was as if, once pigeonholed in my mind as 'confidential', I was no longer able to access it.

Spring had not yet arrived, and it was already dark when a man arrived at Colleen and Ben's house unannounced and asked for me. His instructions were to convey me to the border with Mozambique, where someone would guide me over. Miss Ray would meet me in Maputo. This I had not expected. Unlike Swaziland, which I had thought was safe enough for both of us, Mozambique was a front-line state, in a state of undeclared war with South Africa. It was certainly not safe for me to be there, and getting there was even less so. I had to decide there and then what to do, but after all the difficulty in setting up this meeting it would have felt like cowardice not to go.

So I said goodbye to Athalie and my hosts, and went off in a nondescript vehicle, with a man whose name I did not know. We stopped off at a block of flats where someone else joined us: he was to be the driver. Then we stopped again, maybe in Manzini, to pick up three youths, one of whom was to be our guide. If the youths had wondered what a white person was doing in the car, they did not say. I did not need to ask why youths of school-going age were crossing illegally into Mozambique.

There was hardly any moon and a moment of panic arose, when the dogs started barking and there were voices. Our guide whispered to me to run for it, and I did, squeezing through a barbed wire fence not much more formidable than an average farm fence. Nearby was the outskirts of a town, and I could see the characteristic painted blue tiles of the Portuguese buildings. This was Namaacha. I was told to lie down in a patch of long grass on my own. I lay there a long time.

Close to midnight a South African couple fetched me and drove me to Maputo. They did not introduce themselves, but I had an idea who the woman was. It was understood that we were not going to speak to one another, but I gathered from their conversation that things had not happened the way they were supposed to. I was not meant to have crossed over in the company of the youths. I was also not meant to have been left lying in the long grass. Obviously this was a known spot.

Yet the same spot was my point of departure for the trip back two nights later. Our guide was the same young man who had brought me over, but my companions on the return trip were different: two men whom I had observed in Maputo, but to whom I had not been introduced. They looked like workers, which I later established they were. One of them had distinctive, pendulous earlobes that I associated with people from northern Natal.

Again, things did not work out as they were supposed to. The crossing was uneventful, but there was no one to pick us up on the other side. After waiting a long time, our guide said we would have to walk. It was a straight gravel road from the border, over even terrain, with no roads leading off it, and no vegetation on either side that could provide the slightest cover from the headlights of oncoming vehicles. After more than an hour of walking, you might have seen the shape of a dwelling of some kind in the distance, but otherwise there was no sign of human habitation that could have provided an innocent explanation as to what you were doing there in the middle of the night.

In one hand I was clutching a Makondo carving which had been a wedding gift from Ray, and in the other a copy of a paperback for Miss Yon. It was *Lord Hornblower* by CS Forester: you could buy a copy in just about any bookshop. On my wrist I was wearing a watch from the German Democratic Republic, another present. All the while we were walking, I was mulling over what I had learned and what I had observed.

I had spent almost all the time indoors, in a flat in a modern block in a central location. The flat belonged to a woman who was an academic at the university there. During the first day I was alone, except for Ray's visit. I was on my own again on the second day when someone knocked on the door. I was reluctant to see who was there, but the knocking persisted, and eventually I opened it. It was Indres Naidoo, to whom Jeanette Curtis had introduced me at his family home before I knew I would be working for a union.[4] He did not remember me, and I did not remind him that we had met. Now he was in exile, and hungry for news about what was happening in the trade unions. Later that day the couple that had fetched me from the border took me to meet Ray in another building. When they heard of Indres's visit they were more than a little put out. My visit to Maputo was supposed to be a closely guarded secret.

At the time Ray was engaged in what seemed to be a training session, with the same two workers who were now walking beside me. There was nothing to be lost by talking to them, and I was curious to know where they were from, and what they had been taught. They had been taught about workmen's compensation, the one said. It had been very useful. The other one agreed. Having observed their session with Ray, I was inclined to believe that this was what they had crossed the border to learn about, though maybe it was a cover story.

We encountered no vehicle along the road, and after what seemed like the best part of the night, we came to a T-junction and a tarred road. Our guide lent me a balaclava, to try to minimise the amount of white face that was visible, and flagged down a passing vehicle. It was fortunate that it was a bakkie with an open cab. I clambered into the back, and stared backwards, fixedly. It was getting light when I clambered out, outside some Swazi village. I was on my own again, and the watch from the GDR had already stopped, never to start again. In the glow of morning I caught a bus to Mbabane.

Those inside, and those outside

Ray had arrived at the flat bearing the wedding gift. I do not think it even crossed my mind to complain about being expected to cross the border. There would have been no point. I had embarked on an adventure, and would have to wait and see how it unfolded.

Wilma had told me Ray had reservations about my becoming general secretary, because my father was a judge. So I felt obliged to explain why I had wanted to work for a workers' organisation, and why I had been approached, not omitting that I had joined the ANC in London, and my belief that by rebuilding the Union I would be contributing to the same objectives as the ANC stood for. Ray did not comment on my narrative. She also gave no indication that she had known I had joined the ANC. My impression was she had not. For my part, I was glad to put a body to the face I knew so well. In the course of our discussions, the Miss Ray of the workers' imagination was to become just Ray.

On the question of membership, my considered opinion is that although there is a difference between membership of a political organisation and a trade union, the ANC (as distinct from the party) always had a weak notion of membership. What mattered more than any application process was whom you knew and associated with, and what you did. The notion of the ANC as a movement contributed to this.

We did not, however, discuss the ANC or how it operated. We did not even talk about SACTU. This might seem strange given that I had been spirited across the border, but I understood that she was respecting the line between politics and trade unionism, even though her own history suggested it was not possible to do so.

She wanted to know what was going on in the Union, and I told her. Her questions were many and varied, and included questions about the Fatti's & Moni's strike, which she had followed keenly in the press.[5] Liz was the only other mother figure in whom I had been able to confide, but Liz was Ray's protégé. Meeting Ray represented engagement at a different level, with an intellectual who also had broad organisational experience. This inspired me to raise my own level of analysis.

I asked Ray how she had dealt with the issue of corruption, at which she launched enthusiastically into the story of Eva Arendse, the first branch secretary of Paarl, who had been discovered pocketing the subs. The nub of the story was that the branch executive committee had refused to take action against Eva. Ray's response, on her telling of it, was that she had suspended the committee and called a general meeting. At the general meeting, the workers forced the committee to apologise to her. A resolution was adopted that Eva be dismissed, and never again employed by the Union.[6]

Ray had something like a glow of triumph when she related how the committee had been forced to apologise to her. If she did not actually say she had them eating out of the palm of her hand, this is what she conveyed. She was, like Oscar, someone with a considerable ego. Even though I felt I was following in her footsteps, I could see it was unlikely Ray would interpret it that way. I was not, and could never be, her protégé. I also did not owe my position to her, as previous general secretaries of FCWU did.

Oscar was also not a protégé of Ray. On Oscar's telling, he had organised the workers at Laaiplek and St Helena Bay on his bicycle, and the Union was established there on his invitation.[7] One of the issues Ray had with Oscar came up apropos of my question about a married couple both working in the Union. Ray did not have a problem with that. She had a problem with leaders (men) having affairs with members (women). This was a gender issue, as she saw it, and there had been a debate in the Union at the time, in which Oscar had taken an opposing view to hers.[8]

I also knew from reading old minutes that Oscar had opposed Ray on another issue, with an eloquence no other official could have commanded: this was when FCWU had registered as a separate union for coloureds.[9] It would have been interesting to know why she had considered it necessary to register a separate union even before the Nats came to power. But the only issue of registration we discussed concerned the necessity at all costs to prevent the Vakbond from being registered.

It was a month or two before Minister Fanie Botha was to announce that government was not going to enforce the ban on trade unions recruiting contract workers. The question whether the emergent trade unions should register in terms of the amended legislation had not yet arisen, although it was bound to do so. SACTU's position on registration would have far-reaching repercussions for the emergent unions.

The question relating to SACTU that I raised was about forming a united front against the legislation among unions. Not that I was seeking her permission to do so: we had already called a meeting of trade unions in the Western Cape that were not affiliated to TUCSA, with this objective in mind.[10] Ray would have realised that forming a common front with other trade unions might lead to the establishment of a new federation or our joining an existing federation.

Probably it would have been of concern to SACTU if the Union affiliated to another federation. FCWU and AFCWU had been its most important affiliates, and this would have raised the existential question: did a federation in exile have reason any longer to exist? What would have sharpened their concern was that the only existing federation we could conceivably have joined was FOSATU, since FOSATU had by now acquired the reputation of being hostile towards the Congress movement.[11]

That would not, however, have been a valid reason for us not to affiliate to FOSATU. Political differences between organisations that did in fact represent workers should not stand in the way of unity, and there was no doubt FOSATU did have a significant membership. The question I had posed to the conference was whether it was 'in our interests to move closer and perhaps join this federation or not'.[12] The answer was that the management committee would decide, and there was a real possibility it would choose to affiliate. I did not tell her that. She could read my annual report for herself.

So what had my visit achieved? This was the question I mulled over on the road from Namaacha and afterwards. Ray had of course been a key figure in the history of the Union, and I was now part of this history, which was still unfolding. I had been curious, and wanted to fill some gaps in this history for myself. Meeting Ray had seemed something like a historical necessity.

I had also been seeking affirmation from the founder of the Union, and felt I had got it on two issues. The first was that there could be no compromise with corruption, and that this had always been the Union's position. The second issue was why it was imperative for us to form a common front with other trade unions. The implication was that it was up to those of us inside the country to determine how we did so. But these were all matters set out in various reports and minutes.

So I also had to ask myself the question: did the risk I had taken warrant the gains? Ray had apologised about meeting in Mozambique. The risk of her being kidnapped in Swaziland was too great. Presumably it was ANC security that had decided this, and it was also ANC security that was responsible for my return to Swaziland. Ray had discussed the arrangements in my presence with someone who I think was Jacob Zuma (we were not of course introduced). I had not been able to rely on these arrangements. Was it that I was seen as expendable, or the trade unions as unimportant?

My interchange with Indres gave me a glimpse of how much consternation the revival of the trade union movement, as well as the Wiehahn reforms, had stirred up outside the country: the reforms, because they could not be reconciled with the notion of South Africa as a fascist state; the revival of the union movement, because those in exile were bound to be suspicious of something they could not claim as their own, or that might represent an alternative locus of power.

If someone who was not privy to my visit had heard about it within 24 hours of my arrival in Maputo, what were the chances the Special Branch would not have heard about it from one of the many well-placed informers they doubtless had in Maputo? The Union was on a knife edge on three fronts. If something had happened to me, crossing over or on the road from Namaacha, the effect would have been incalculable.

So I started to have a kind of internal backlash, as the adrenalin gave way to anger: at myself, first and foremost, for seeking affirmation from Ray when I got more than enough from the workers, and for my egotistical belief in the 'historical necessity' of such a meeting. Angry also with the ANC, for its chaotic organisation, and with Ray for needlessly putting the Union at risk. Then the anger would give way to the realisation that in a polarised political situation you had to choose sides, and that there might be political dividends from my visit, even if these were not immediately apparent.

Years later, after I had left the Union, I had a chance meeting with the woman who had fetched me from the border with her partner. She told me how strong the suspicions of trade unions had been, suspicions she had shared, and that my visit had changed her own perceptions. Maybe it had also shifted perceptions towards emergent unions within SACTU, and although in many respects it was a different organisation, SACTU would on occasions claim the Union as its own.[13]

For my part, I continued to acknowledge that the Union had been an affiliate of SACTU and part of the same tradition. But this was increasingly looking like a sentimental evocation of tradition, which failed to confront the existential question. In a trade union federation, as much as a trade union, control should be exercised from below. The only role a federation in exile could play was thus a supportive one. It was 'we', the persons inside the country, inside the trade unions, who would have to reinterpret the tradition that SACTU had established, taking into account current conditions.

Paradoxically, the end result of my visit was that I felt more confident to pursue an independent line inside the country. I remembered a story Oscar Mpetha had told me, when I asked him about the relationship between the movement inside and outside the country, and who should determine the line to be taken. The issue had been the referendum on Transkei independence before it was granted. The object of the referendum was to show the world that the people of the Transkei wanted independence, so as to establish the legitimacy of a new 'state'. So the South African government took the unprecedented step of allowing people in the Transkei to form political parties and campaign.

Kaiser Matanzima's party was for independence and the party of Victor Poto was campaigning against it. Oscar argued that the movement should support Poto's campaign, because a 'no' vote in the referendum would be a major setback for the government, and could even have derailed the granting of independence. It was a similar argument Ferrus advanced for working inside the coloured Labour Party, but more compelling. 'Independence' was a one-off event. But the line from outside was that the movement should not participate in any form of Bantustan politics.

The person who held this line in the Cape was Elijah Lozah, but Oscar decided to defy it. The movement outside the country could not determine the agenda for those inside the country. He campaigned for Poto, who narrowly lost the referendum, and was regarded as a political outcast for doing so. Yet when Lozah was detained and died in detention, it was Oscar who turned his funeral into a political event and a demonstration of defiance. I remember reading about the procession through the streets of the township in the papers. According to Oscar, it was this that re-established his own credentials as a leader.[14]

Mediation

I took up the initiative of a united front on my return from 'leave', and was reminded how few options the Union had apart from affiliating to FOSATU. The two unions of which Bill* was the secretary illustrated our dilemma. Bill was another of those who regarded the secretaryship of a trade union as a profession, the objective of which was to serve as many unions as possible, to boost your income, rather than organising as many workers as possible. They occupied offices directly above us in Corporation Street.

The members of the one union could not have numbered more than a hundred skilled workers, in shops around the city; the executive of the other held its meetings on a Friday afternoon. You could tell there was a 'meeting' from the laughter and singing, or the pile of empty beer bottles at the bottom of the stairs on a Saturday morning.

The CTMWA had a significant coloured membership, but this was thanks to a closed shop agreement with the City of Cape Town. It had not organised workers of any other municipality, let alone African workers, and despite its general secretary having opened our previous conference, it had done what I could only interpret as a political about-face, and was no longer prepared to meet with us.[15]

The weakness of the emergent trade unions was also evident from the increasing arrogance of TUCSA. One of the TUCSA affiliates put forward a resolution at its September conference supporting the struggle of the Fatti's & Moni's workers, as well as the workers of Eveready in Port Elizabeth. The motion was defeated, albeit narrowly: there could hardly have been a more calculated slap in the face of the emergent unions.[16]

The Eveready strike was over the refusal of the bosses to recognise NUMAROSA, a FOSATU-affiliated union. Taking a leaf from Fatti's & Moni's, a boycott of Eveready products was called. It was not a success, but the fact that NUMAROSA was prepared to fight on behalf of its members made a favourable impression on the management committee. Several AFCWU delegates again suggested that the Union affiliate to FOSATU, but no decision was taken.

In the meantime, the war of attrition with Fatti's & Moni's continued, following our rejection of the offer of employment at Good Hope Bakery. However, our meeting with the Desmond Tutu had borne fruit after all. The SACC decided to take upon itself the role of honest broker, which we welcomed. Having tried without success to persuade the company to meet with us, it proposed facilitating a process of mediation.[17]

Mediation as a technique for resolving labour disputes was unknown in South Africa at that point. The profession of dispute resolution practitioner also did not exist, and there was no supposedly neutral list of practitioners from which the parties could choose. It would not have made a difference if there had been. If the company and the Union were agreed on anything, it was that it would be a fiction to suppose that anyone could purport to be neutral in a dispute such as this one. Inevitably he or she would have to take sides.

The proposal that was mooted, therefore, and eventually accepted, was that each party appoint a mediator who was acceptable to the other party.

The two mediators would then attempt jointly to facilitate a resolution. The Union nominated Mr Mandla, of the traders. We liked the idea of an African mediator, and we also wanted to keep the traders on board. However, he was not acceptable to Fatti's & Moni's, because of his involvement in the boycott, they said, so we settled on Alan Potash, who was part of the firm of attorneys that acted for the Union. The company then accepted as the second mediator a theologian the SACC had suggested, Dr James Leatt. The fact that the company had not suggested a mediator itself was not a good sign, and we did not have high hopes of the mediation process. It was nevertheless important for the workers' sake to keep alive the idea that there might still be a resolution.

Even with the prospect of mediation in the offing, there was no let-up in the war of attrition. After having held off doing so for so long, the company finally unleashed the inspectors of the Bantu Affairs Administration Board (BAAB). They raided the Nyanga hostels twice, bearing lists of names that management had provided. Several workers were arrested for being in Cape Town without permission. Wives of workers were also arrested. In each case, the Union instructed its attorney to defend them, and there was a further flurry of press statements, provoking further public outrage.

Stanford Booi's case, by way of comparison, was still wending its tortuous way to the Supreme Court. It would be nearly two years after the strike had ended before it was decided he had the right to remain legally in the urban area, and this was shortly after a worker called Rikhoto won a more famous case on the Witwatersrand. Rikhoto's case, in particular, has been held up as a shining example of how law could win rights for workers without their being organised.[18] Yet in fact none of the workers arrested went back to where they had come from. Long before the Supreme Court had recognised this right, workers were winning it in practice, through organisation.

There was still no sign of public interest in the strike flagging, and we continued to receive donations. One of the most generous, and surely the most unexpected, was from the international humanitarian agency Médecins Sans Frontières. The funds were to be disbursed for humanitarian aid, the accompanying letter said. We discussed what this meant. One interpretation might be that this was simply coded language for strike pay. Why else, after all, would anyone be sending money to a trade union for their members on strike? On the other hand, did strike pay qualify as humanitarian aid?

The outcome of this somewhat pedantic debate was that while the major portion of the money did in fact go towards strike pay, a portion was spent on something that would indubitably qualify as humanitarian aid and would also boost the morale of the embattled contract workers, in particular.[19] The Union would take a delegation of strikers to the Ciskei, with parcels of food and blankets, which they would deliver to the families of all the striking workers. Among the circle of friends and supporters that the strike had generated, a medical doctor volunteered to accompany the delegation. Medicines would also be provided to treat those who were sick. Another volunteer agreed to record the trip on videotape.

After much deliberation the general meeting elected five of their number: not the sort of persons who would be asked to represent workers at negotiations, but rather those best able to locate the workers' families, which were scattered over a wide area of the rural Ciskei. It was also proposed that Athalie accompany them. Amid much excitement, the delegation set off in a heavily laden van belonging to a striker called Council, who was one of the few with permanent rights, and with the doctor and others in a private vehicle.

Unlike almost everything else that happened during the strike, this trip was not publicised at all. I was worried that there might be some interference on the way. I also had another agenda, to which I did not want to draw attention. While in East London, I asked the delegation to find out what they could about the pineapple canning factories that were there. Publicity seemed inappropriate for another reason: this was something the Union was doing for the workers, and not just the workers with families in the Ciskei. It was also something the Union was doing for itself. We were overcoming a divide, between the rural and the urban, between those at work and those left behind, and ensuring that, whatever the outcome, there was understanding on both sides of the divide.

This was perhaps the most inspired decision of the strike: after more than five months on strike, and still reeling from the blow that the defections to Good Hope Bakery represented, something was needed to revive flagging morale. One could not fail to be moved watching the videotape of mothers and brothers and sisters addressing by name the workers who had sent them, and thanking them for remembering them.

Re-employment versus reinstatement

Fatti's & Moni's was on their third public relations consultant by the time the mediation process commenced. The object of the exercise, from our point of view, was not so much to persuade the mediators of the justice of our cause, which I think they both accepted, but that the stage had now been reached where the boycott had acquired its own momentum. Even if the Union were to accept re-employment at some other plant than the Bellville mill, it would not bring the boycott to an end.

The idea that the boycott had acquired its own momentum was consistent with our strategy of preserving an arm's-length relationship with the boycott, and there was some basis for saying so, in so far as the reputation of company had suffered. But I did not really believe it was true.

In the absence of organisation, consumers are fickle, and it was the traders' organisation that had given the boycott teeth. If pressed, I would have had to concede that I did not believe the traders or anyone else would think of carrying on the boycott without the Union's say-so.

For the company, the issue remained what it had always been: fear of the consequences of accepting a trade union in the workplace. This fear was less ambiguously expressed in the mediation than ever before, so we went to some lengths to try to reassure them. The workers would not get back pay. The Union would take steps to ensure the boycott was called off. At the same time there could also be no settlement without safeguards for the workers. Above all, there had to be safeguards to ensure that reinstated strikers were not again dismissed on their return to work.

Fatti's & Moni's was due to communicate its reply to our proposals on the Friday, but failed to do so. On the Saturday it issued a statement to the Sunday press, drafted by its newest public relations consultant. It was breaking off negotiations with the Union, as they put it.[20] The Union was trying to stretch out negotiations, the company said, in order to prolong the boycott.

This was a stupid thing to say. There could be no conceivable benefit for the Union in prolonging a boycott. It was also stupid to announce that it was going to settle with the workers without the Union, which in its own inept way it then tried to do. Foremen like Mr Christian and private detectives went to the homes of strikers, to urge them to return to work without the Union. None did. The Union now had more than enough ammunition to fire a devastating salvo in reply. The company was acting in bad faith. The workers would never return to work without the Union.[21]

The upshot was that Fatti's & Moni's succeeded in uniting the mediators against it. They went together to see Moni and Terblanche at the factory, and threatened to go public over what had in fact taken place in the mediation process. This is something that a mediator should never do, according to the disciples of mediation: this is what comes from elevating an intervention that should always be informed by the nature of the dispute to a discipline, with its own inviolable rules. The threat of the mediators to go public helped to break the deadlock.

The other thing that helped break the deadlock was that the company's latest public relations consultant had come to the conclusion it was fighting a losing battle. He told me so himself.[22] The mediation process resumed a mere five days after Fatti's & Moni's had announced it was breaking off negotiations. At about the same time Fanie Botha announced that even though he was not prepared to amend the legislation preventing migrants from belonging to a registered trade union, he would grant them a blanket exemption to do so – except those from 'foreign states'.[23]

This 'concession' was surely calculated to break the opposition of the emergent unions to the legislation, as it in fact did. It must also have made it politically more possible for Fatti's & Moni's to contemplate an agreement with a trade union representing migrant workers. No sooner had Botha made his announcement than Lucy Mvubelo of the National Union of Clothing Workers announced it would register. It was always regarded as the most pliant of the 'parallel' unions. Then FOSATU called a meeting of trade unions in Johannesburg to discuss what to do.

We had no hesitation in agreeing to attend this meeting, but had still to take a formal decision regarding Botha's 'concession', when Alec Erwin, the general secretary of FOSATU, came to see me. He seemed genuinely pleased when I told him that it looked as if there was going to be a settlement of the Fatti's & Moni's strike. What I might have told him was that Botha's 'concession' did not change the fact that the law provided for the racially separate trade unions and all that this implied, and there was no way we could register AFCWU as a separate trade union for Africans. It would have cemented the very division we had fought so hard to overcome.[24] There was freedom of association only in so far as the Minister allowed it.

The reason I did not put it as bluntly as that, I think, was that I did not want to anticipate what the Union would decide. Also, we were two whites with no worker representatives present, speaking the language of the well-to-do middle classes. This was also our first meeting and, as befitted our background, we maintained a degree of reserve towards each other. I doubt whether it would have made any difference if I had been more forthright, because I suspect the leadership of FOSATU had already opted to register. Alec's main concern was about the fallout that would ensue. Alec and I continued to preserve a degree of reserve towards each other over the ensuing years. I was surprised when, after the transition to democracy, Alec declared for the ANC and even more so for the Party.

Shortly after mediation resumed, on the evening of 6 November, the mediators phoned me from the factory, where they were in a meeting with Moni and Terblanche. The company wanted to meet with the Union and the mediators, they said. So Oscar and I, together with Friday and Spasie, went to find out what they had to say. Not knowing what to expect, Oscar and I went in first, to test the water, while the others waited outside.

I want to meet the general secretary of FCWU and AFCWU on his own, Moni said. This was the first time we had met in the flesh. I objected to meeting him without Oscar, but Moni was adamant. As managing director, he wanted a one-on-one meeting with his counterpart in the Union, with the mediators present. It was a status thing. Terblanche would also not be present. I hesitated, while the mediators, sitting around the boardroom table, watched to see what I would do.

It was not so much that it was against the policy of the Union for the general secretary to meet a boss on his own, because the mediators were there, and one of them was the Union's attorney. The problem was that they were all white men. The symbolism of going into a room of white men, and leaving my African colleague outside, was deeply objectionable. Yet that is what I did, and that is also how Oscar perceived what I had done. He told me so, as soon as I emerged. I felt deeply ashamed.

The meeting had been brief. The company was prepared to negotiate a settlement, Moni said. I am only prepared to negotiate with the others present, I replied, and that was what happened: Friday, Spasie, Oscar and I met Moni, Terblanche and the mediators around the boardroom table, late into the night, to hammer out the terms of a settlement. It would be a settlement covering all the workers who had rejected the offer of alternative employment in Good Hope Bakery and were still on strike. They were 56 in all.

It was not reinstatement, in the sense that the workers would be returning to precisely the same jobs. Reinstatement after this length of time was not practicable, the company argued. The Union also needed to make some concessions, in order to secure safeguards against the future dismissal of returning workers, and to secure the renewal of the contracts of the contract workers. In any event, what was more important than getting the same job back was getting back to the Bellville mill.

Yet even though it was not complete reinstatement, it was close enough: the workers would retain their seniority, and the period of the strike was to be treated as suspended service. The company agreed to try to place them back in their original jobs as soon as possible.[25] In order to renew the contracts of the contract workers, after reporting for duty the contract workers would have to go back to the homelands, where the local magistrate had issued the contract. Most had not seen their families in the homelands for over a year. The company also agreed to provide buses for them to go home at Christmas, and to grant one week's leave in advance, so they had some money.

The safeguard against dismissals proved the more difficult issue to resolve. Dismissal procedures or codes were at this point largely unknown in the South African workplace, and what the Union was seeking was to curtail management's power to summarily dismiss any of the returning workers. Terblanche fought against this with all the authority at his command. It was cutting into his managerial prerogative, he said. But it was his exercise of managerial prerogative that had caused the trouble in the first place, and Moni was in no mood to molly-coddle him. For our part there could be no settlement without these safeguards, otherwise the bosses would dismiss our members as soon as the opportunity presented, and we would be back to square one.

So it was agreed that none of the returning workers could be summarily dismissed unless for reasons of 'intoxication on the job or theft, for a period of one year'. The only other circumstance in which the company could dismiss was if it had issued a written warning (a copy of which was forwarded to the Union) and if the worker concerned failed to heed it. Even then, the company was obliged to notify the Union before dismissing anyone, and allow the Union the opportunity to make representations in this regard.[26] It did indeed cut into the managerial prerogative and was perhaps the final indignity for Terblanche.

The company had expected us to ask for formal recognition of the Union, which was something other trade unions regarded as significant at the time. But we did not particularly want a piece of paper stating the company 'recognised' the trade union. Recognition would be implicit in the agreement we arrived at, and we were happy to 'drop' a demand that we had actually never entertained. In any event, the factory had to be reorganised. There were the workers who never came out on strike in the first place, or who had turned their backs on the strike. There were also the scabs that had been hired at the start of the strike. It would be the task of the returning workers to organise all workers currently inside the factory, and they had undertaken to do so.

Certain provisions in the agreement were calculated to show workers currently inside the factory that those returning had not achieved much. During the first month after their return, for example, the workers were to earn the same wage as previously, even though wages had increased in the interim. But we were convinced that the one thing that would impress the workers currently inside the factory was the safeguard against dismissal. A year, we reckoned, should be enough time to get a majority again.

A founding strike

The next morning we reported back on what had transpired the night before. The workers were jubilant, but also understood the devil would be in the detail. The terms of the agreement were exhaustively discussed, and after a further round of negotiations, the agreement was signed on 8 November 1979. That evening we held a press conference at the Belmont Hotel. 'Strike at Fatti's is over' was the headline in the *Cape Times* the next day.[27] It hardly seemed conceivable that the following week the workers would be back at the factory they had left seven months before.

The strike was also the subject of an editorial in the *Cape Times*: 'the strike, and the subsequent consumer boycott ... was graphic evidence of new-found self-confidence among blacks. In different times in South Africa, people could be summarily sacked with little backlash from fellow workers or consumers. Now there is a new spirit of toughness in the labour movement. Workers are quick to pounce to the defence of their fellows. Country-wide consumer boycott can be organized in days, it appears ... South Africa is entering a phase in which blacks, newly conscious of their inherent worth and economic strength, are going to use their clout- whether they have political rights or not.'[28]

This was almost the polar opposite of the conclusion I had come to. It was because workers so seldom stood together that the strike had attracted so much attention. If there was any one lesson to be drawn from the strike, it would be that solidarity did not happen spontaneously, but was the product of painstaking organisation. Far from there being a 'new spirit of toughness' in the labour movement, I felt the Union's isolation keenly.

I was also anxious at the response of 'the community' and 'community activists' – concepts that had taken root during the strike. For the community activists, the strike had presented an ideal opportunity to agitate against the government and the bosses, and to promote whatever alternative they stood for or simply to promote themselves. The strike had opened up a political space outside the sterile bounds of parliamentary politics which many had not believed could exist.

For me, this was a space to organise from below. In order to build organisation from below, the difference between winning and losing was critical, but you could never expect to win a total victory. That only happened in games or in the mind. In order to make propaganda, however, winning or losing did not really matter. The agreement talked of re-employment and not reinstatement, and some would say this represented losing or that the capitulation of the bosses was not total enough.

There is a photograph commemorating the press conference at the Belmont Hotel. Against the backdrop of a banner stating 'Long Live FCWU and AFCWU' there are the clenched fists of the workers. Pendlani is standing in front, leading the salute: '*Amandla ngawethu*'. He is flanked by the newly elected president of FCWU, Van Graan, and an ecstatic Annie Mentoor. Everyone is in a celebratory mood, except me. The struggle was not really over, I thought at the time. Vormat would be telling people the Union had sold the workers out. Even though the traders fully supported the settlement, Terblanche would be looking for some pretext to scupper the agreement.

It was Hassan Howa, the self-important president of the South African Council on Sport (SACOS), who presented him with one.[29] The Union was misleading the workers, Howa announced, and threatened to continue the boycott without us. We invited SACOS to a meeting with the workers, so that they could hear from them in person whether they were being misled. Howa still maintained that the settlement was mistaken after the meeting, but grudgingly agreed to abide by the workers' decision. I was furious at the kind of middle-class arrogance this typified, and began drafting an open letter to him, which I never finished.

Our task now was to publicise the victory to the workers of Cape Town who did not read the *Cape Times*. We organised a meeting in the hall of St George's Cathedral in Cape Town, but it was naïve to suppose the unorganised would attend. We were speaking to the converted: the circle of supporters and friends the strike had generated, most of whom were activists drawn from the middle classes, together with members and officials of the Union, and a few other unions, notably the trade union formed by the Advice Bureau, the WPGWU.[30]

'The agreement reached is a victory of major importance for the workers and our Union. It is also a victory for workers everywhere, and for the organizations throughout the country and outside the country, who were prepared to support the workers cause,' I said in a statement distributed not only to the media but to the organisations that had supported us.[31] Indeed, it was the founding strike of the Union that emerged in the 1980s: this was in many respects a different trade union from that established following the Jones strike, in Ray's time. However, all this only became apparent in retrospect.

One of the first groups to realise the significance of this victory was the scabs. They were impressed no end when the strikers returned to work, and joined the Union almost immediately. Doubtless they were hoping to safeguard their jobs by doing so, but we had no problem with that. The Union had also not done anything to antagonise them, and this became for me an object lesson as to why violence against scabs was a misguided strategy.

On the other hand, the workers who had turned their back on the strike to take jobs at Good Hope Bakery kept their distance. Most of them had by this time been transferred from the bakery back to the Bellville factory, and must have feared the reaction of the other workers. Skilpad, who had led the split to Good Hope Bakery, had since been promoted, and was to become the implacable opponent of the Union.

The victory in the strike also went to show that trade unions were politically stronger than some supposed. Even an unregistered union representing contract workers could win recognition, regardless of government policy. To emphasise the point, we included Friday in the delegation that went by kombi to Johannesburg to attend the meeting called by FOSATU.

It was a sadly truncated gathering, considering how critical a juncture it was for the trade unions. WPGWU was there, and a delegation from our Johannesburg branch joined us. There was also a delegation from FOSATU. But trade unions organised in the Black Consciousness tradition, which then went under the title CCOBTU, were not prepared to associate with unions that had white officials. So they decided to have their own meeting. On the same day, in the same building, separate meetings were held.

The meeting started in the afternoon and went on into the evening, yet almost from the outset it was clear that there was little prospect of arriving at a common stand. The FOSATU trade unions had already decided to register. WPGWU and AFCWU had decided they would not. After a long discussion I suggested we try to draw up a statement of the principles that the organisations present seemed to share. This was a ploy to show FOSATU its position was untenable, and I think the joint statement on which we agreed did that. The unions would not accept registration which was not granted on the basis that unions were 'completely non-racial in their membership', it said.[32]

No further statements were supposed to be issued following the meeting, but FOSATU did in fact do so. While they agreed with the principles set out in the joint statement, FOSATU said, they were nevertheless going to register. I was annoyed, and issued a sharply worded response on behalf of the Union.[34] I don't think the media carried it, and it was probably as well they did not. Subsequent to the meeting Oscar proposed we join FOSATU, because they needed our leadership. However, I did not see how this would be possible, and the matter was left in abeyance.

14

The road to the crèche

THE FALLOUT FROM THE JOHANNESBURG MEETING of trade unions precipitated what became known as the registration debate. It generated much heat in political and academic circles, and raged on in journals and periodicals for some time, even after the circumstances prevailing at the time had changed. For ordinary workers, however, the issue could not have been simpler. If you believed that workers had the right to belong to one trade union, regardless of race, you could not go along with legislation that did not afford you that right. It was about having the courage of your convictions. Oscar had put it eloquently and tactfully at the meeting: there was nothing to fear from taking a stand.

This was not to say there could not be another clampdown on the trade unions, but that we should have confidence in the justice of our position. At a juncture at which the 'buy-in' of trade unions was essential for the reforms to be seen as legitimate, as I have said, the ability of trade unions to change the system depended on their staying outside it. Nothing happened after the Johannesburg meeting to suggest this assessment was incorrect. The Union had now entered a period of unprecedented growth. There was an upsurge in militancy in the industries we were organising except, curiously, in the aftermath of the strike, the industry of which Fatti's & Moni's was part. Workers did not care about the registered status of the Union, and one by one their employers were prepared to follow suit.

Does it matter that FOSATU and its supporters portrayed its position to register as tactical, and that of its opponents as a rejection 'in principle' of registration?[1] I do not want to fan any embers that might still remain from the registration debate. It remains important, however, to know the role trade unions in fact played if we are to understand how social change happens and what role a new generation of organisations might still play. There were trade unions that rejected registration in principle at a later juncture, when circumstances had changed, but at the time of the Johannesburg meeting, so far as we were concerned, non-racialism was the only issue of principle (although not the only issue).

Non-racialism was not about the race of trade union officials (which is what preoccupied the Black Consciousness group) but about overcoming divisions between workers. There was nothing abstract about our stance, because we actually had a non-racial membership and were trying to operate as a non-racial union. I do not think this was true of NUMAROSA, FOSATU's largest affiliate by some way, or any of its other unions. NUMAROSA was significant because there were large numbers of coloured workers as well as Africans in the motor vehicle assembly plants of Port Elizabeth. Yet there did not seem to be a semblance of solidarity between them, despite the 'parallel' union NUMAROSA had established, or because of it.[2] This came to light after the other high-profile strike of 1979, at Ford's Struandale plant, shortly after the Johannesburg meeting.

The president of the 'parallel' union was a worker at Struandale and also president of FOSATU. The plant might have been the flagship of the union, and the issue was victimisation.[3] But the organisation whose member was being victimised was not the union but the recently formed Port Elizabeth Black Civic Organisation (PEBCO). The workers believed that NUMAROSA or its parallel, or both, were on the bosses' side and the strike was settled without their involvement.[4] For us, the strike killed any prospect of taking further the idea of affiliating to FOSATU, at least for the time being.

PEBCO, as everyone could see, was a proxy for political organisation. It had utilised the concept of 'the community' to mobilise people, and activists in Cape Town were also intent on doing so. One such activist, Leila Patel, came to see us about establishing a newspaper to nurture community organisation in the Western Cape. The community had supported the Fatti's & Moni's strike, she said. Now the Union should support the community in their struggles.[5] We agreed. Soon afterwards, under her editorship, *Grassroots* newspaper was launched.

It was now accepted in struggle circles in the Western Cape that it was important for 'the community' to have the backing of 'the unions', but I don't think there was any consensus as to why it was important. The unions that 'the community' had in mind were only the WPGWU and ourselves: not FOSATU, and also not the behemoth of the established union movement, the Garment Workers' Union. Activists also referred to 'the community' in a very particular sense. The white Left were not really part of 'the community', and community activists were by definition black. Those we had to do with all had a better-than-average education and a command of English: among coloureds, this signified someone from the middle classes; among Africans, it was the urban insider. It was not long before questions of class began to sour this alliance.

My idea of community organisation was that it should focus on issues such as housing and the delivery of services, and develop the capacity to pursue these demands, which the Union did not have. The objective ought to be to draw in and unite ordinary working-class people, in much the same way as we were doing in the Union.[6] However, it did not ever seem likely this would happen if workers were not already united in the workplace. Two disputes at about this time illustrated the tangible impact the Union would have on two impoverished Western Cape communities: Saldanha Bay and Ceres.

Sea Harvest in Saldanha and Growers in Ceres both employed masses of ordinary workers, mostly women, and in the closing months of 1979 the temperature at Sea Harvest was close to boiling. In December, at the start of what used to be the builders' holidays, the workers went on strike. Less than two months later the workers at Growers did likewise. Both strikes were in factories and industries that had never been properly organised before.

In both towns, the factory concerned was far and away the biggest employer, so the whole of the community was affected, down to the Portuguese café-owner selling shrink-wrapped polony sandwiches at lunch breaks. Communities with no history of organisation were now gazing at workers streaming to the hall where the Union leadership was to report back on developments in the strike, rather than in their overalls to work. In Saldanha it was the community hall in Diazville, and in Ceres it was at a hall next to the crèche. Nowadays, militancy during strikes has become synonymous with violence directed at scabs. Then, the only violence was from the police, who were intent on breaking the resolve of the strikers. In both instances this was remarkable, but more so perhaps in the case of Ceres because it was so isolated.

Ceres is the name of a valley as well as a town, and this valley extends all the way to the mountains that border the Great Karoo, to the north-east, and the Tanqua Karoo and Cederberg, to the north-west. Another mountain range separates it from the valley of the Breede River, where Tulbagh, Wolseley and Worcester are located. These mountains are also the source of the river that runs through the town. At the bridge over the river is a road running to the right, parallel with it. Along this road are what look like municipal houses built for the working classes, although at that time the houses belonged to Growers. The bosses of Growers were farmers, and the management of Growers operated the factory in much the same way as a farm: the workers earned farm wages, and if they were dismissed, they were evicted from the houses summarily.[7] Growers also took the position that they were a farming operation, and therefore exempt from labour legislation.[8] The descent into the valley was thus also a descent into a different social universe. The strike helped transform social relations there, as well as labour relations in the industry of which it was part.

Direct negotiations

'Was it necessary for the workers to go on strike?' is a perennial question, whose answer is determined as much by class outlook as by circumstances. 'The workers are always right,' is what Bill Andrews had taught Oscar Mpetha.[9] The interests of the bosses and workers were in irreconcilable conflict and, whatever the circumstances, a trade union always had to take the workers' side. There was very little middle ground between this position and that of the government and the bosses. While the government might say the law allowed the workers to strike, there were virtually no circumstances in which lawful strikes could take place. All strikes that did take place were unlawful, and the outcome of any strike that continued beyond a couple of days was usually mass dismissal.

A negotiated settlement as the outcome of a strike, before Fatti's & Moni's, was practically unheard of. This negotiated settlement had also achieved significant gains, exemplified by the limitation of the employer's right to dismiss. Gains in the workplace were not going to resolve the bigger political question. Yet a negotiated settlement after such a protracted and publicised dispute was a potent symbol; even more so given that, in struggle circles at least, the bosses and government were seen as being in cahoots. On 'the Island', I was told by someone who was in a position to have known, people were impressed, and specifically Mandela.[10]

The limitation of the right to dismiss was also not a cosmetic reform. Without job security trade unions could not do what trade unions are supposed to do, namely organise workers and represent them. At a time when the United States subscribed to the doctrine of 'dismissal at will' (and still does), it shifted the balance of power in the workplace.[11] It also went to show there was a growing acceptance among the bosses of the need for reform. Yet it remained to be seen how far what they regarded as reform would go. What we wanted them to realise was that unless the genuine and deep-seated grievances of workers were addressed, strikes would become inevitable. That was also what we were trying to get Sea Harvest to understand.

'Conditions at [Sea Harvest] are worse than at any other big factory we have dealt with,' I reported to the 1979 conference.[12] A woman might be called on to work a 12-hour day, seven days a week, for weeks on end, and still risk being dismissed for being absent from work or for any other reason which her foreman felt warranted dismissal. There was practically no job security for ordinary workers as a consequence, and it was beginning to look as if top management, with a nod or a wink to lower management, had declared open season on the Union, simply because it was doing what a trade union has to do.

Relations had started to sour when the Hopefield workers joined the Union. This was not only because the Union now had something like a majority, but because management could no longer resort to a divide-and-rule strategy in terms of which the Hopefield workers were pitted against the local workers. Out of the blue, one of Mr Kramer's sidekicks announced that 'it was not fair to non-union members' for the Union to continue to have access to workers in the cloakrooms. This was a facility the Union had enjoyed since Johnny M's time, and it was arbitrary and unjust to withdraw it. But it was not set down in a formal agreement, and while litigating to secure access might have seemed attractive to the legally adventurous, it was not something that the Union would have spent its money on.[13] What was needed to consolidate organisation, we decided, was a paid official in Saldanha.

There were now enough workers paying subs to warrant employing one, and they elected Magrieta Wynand, an older worker from the inshore industry, as branch secretary. She proved to be an effective organiser as well as outspoken in the workers' defence. Before long, the management refused to deal with her, and established a liaison committee. As had been the case at Moberg's, the liaison committee was calculated to consolidate the authority of the foreman and supervisors over the ordinary workers.

The workers were outraged that someone who spoke their language, and in whom they had confidence, could be snubbed in this manner. There was a succession of walkouts over this and other instances of management high-handedness.[14] But the underlying issue was that the workers were fed up about their wages, and the large slice the company deducted from them for busing them to work. We did not apply for a Conciliation Board, as we might have done before Fatti's & Moni's. Other considerations aside, it took too long. Months had gone by since the Union had applied for a Conciliation Board at Growers, and still it had not been appointed. The Sea Harvest workers were not as patient.

Mr Kramer knew very well the workers were fed up with their wages, and there was nothing to prevent the company from negotiating about wages directly with the Union, although this was not something we had done before or that employers generally did. However, it seemed Mr Kramer was not prepared to negotiate wages in principle. Thinking he was probably afraid that one thing might lead to another, we tried to persuade him that one thing would indeed lead to another if the company refused to negotiate. All he was prepared to give us was an undertaking to review the wages, and to inform us what the outcome of this review was in three weeks' time.

Mr Kramer had a penchant for the colour blue, and I never saw him attired in any other. In his office, overlooking the turquoise bay, the curtains and carpets and upholstery were also all in shades of blue. It was there that we met, in early December, after the three weeks had expired, to hear what

Sea Harvest's proposal would be. But Mr Kramer did not want to tell us in Magrieta's presence. The solution to this particular impasse was to have Magrieta and her entire committee present, as well as the liaison committee.

The company wanted the liaison committee present to dispel any impression that it was indeed negotiating with the Union. For our part, we did not object: the Union might even be able to win them over, I thought, because we had a killer point. We had long been saying that Sea Harvest's wages were less than the canneries and the inshore industry, and its stock response had always been that they were not part of the same industry. I now knew its major competitor, I&J, was paying substantially more.

In fact, this was not quite a killer point, because the Union had not succeeded in making any headway in organising I&J's flagship factory in Woodstock, although this was not for lack of trying. But Mr Kramer was not to know this, and it was noticeable how incoherent he became as he tried to justify why Sea Harvest's wages were so much lower. My analysis of the situation, after the event, was that Sea Harvest had borrowed heavily to establish a plant of this size and a deep-sea fleet to boot. As a relatively new entrant in the industry, it was playing catch-up to I&J, and the wage level was a critical factor in its ability to do so. The upshot of the meeting was that Mr Kramer said he needed more time to respond to the workers' demands: he would give a definite answer the following Saturday.[15]

Next Saturday, reluctantly, we again travelled to Saldanha, to take our places across the familiar oak table in Mr Kramer's office. It was the end of a deeply draining year, and Athalie and I were desperate for a real holiday. This contributed to my sense of disbelief and despair at what Mr Kramer had to propose. This was, for many, no increase at all. For the remainder he was proposing individual increases on the basis of merit.[16] His justification went something like this: when the workers were employed, they entered into an individual contract of employment with Sea Harvest. No one was forced to work for a wage they did not want. So any question of an increase was strictly a matter between management and the individual worker. In practice, given the number of workers involved, this meant an increase determined by the supervisors and foreman.

It was such an extreme position that you might think Mr Kramer was naïve or badly advised. There would be no minimum wage at all. Instead of negotiations there would be communication. In fact, among manufacturing bosses, Mr Kramer was a heavy-weight. He was also forever on other continents, 'selling fish', as he would say. The board of his company were also not naïve people, as we would later find out.[17] The ideological position he was taking, which was implacably opposed to all that the Union stood for, was entirely in tune with the thinking that was beginning to gain currency in the UK, where Mrs Thatcher had recently been elected.

Evidently Mr Kramer thought he could ride out the strike that now seemed inevitable, and without exactly storming out, I announced our withdrawal from what were not really negotiations. I was fed up and exhausted, and resolved to let go of the situation. Athalie had arranged for us to go to a smallholding that belonged to a friend of a friend, near Stilbaai, where we could stay at no charge. It was agreed that Oscar would deal with the fallout at Sea Harvest, and that Oscar's grandchildren, Oscar and Prince, would come with us on holiday. My involvement in the strike that ensued was transacted from a public telephone. I was not able to let go altogether, I discovered.

At the start of what used to be called the builders' holidays, which I now think of as the December shutdown, Oscar and others held a meeting in Hopefield. The workers resolved they would simply stay away from work until Sea Harvest agreed to their demands.[18] A few days before Christmas, Kramer at last offered an across-the-board increase for all workers that could be described as realistic: but the workers would not accept anything that was not part of a signed agreement with the Union. On 7 January 1980, after a process that could at last be described as negotiations, agreement was reached. It was the first time the Union had negotiated wages directly with an employer outside the official system.[19]

The Union had taught Mr Kramer a lesson, the management committee meeting concluded. He had not expected workers to stay out on strike and forfeit their wages over Christmas. Yet the lesson would have been clearer if the middle layers in the factory and the African contract workers had come out as well.[20] The Africans lived in an isolated cove away from everyone else, in a hostel of surpassing squalor. They would not even be seen talking to the other workers for fear of their foreman, a Skilpad figure, who had recruited them in the homelands and, although ostensibly employed by Sea Harvest, spent more time hawking fish from the back of his bakkie. While the women had been at home, holding out for a negotiated agreement, the contract workers had carried on working.

'Die werkers wil weet'

The bosses of Growers, in contrast with Sea Harvest, were prepared to negotiate, even though they were farmers and had never done so before. They even roped in a Nat member of Parliament with links to the established union movement to advise how to do so, although he was probably more hindrance than help.[21] Predictably, the line they took was to offer a package that would have benefited the more skilled workers and those employed throughout the year. It included numerous wage differentials, an attendance bonus and a housing allowance.[22] This was precisely the kind of agreement the established unions were known for.

However, the Union's bargaining position was strong, because, as luck would have it, negotiations were taking place at the start of the season, something the canners were always careful to avoid. Our bottom line was to increase the minimum wages to a level comparable to that of the canneries, even though the process in which Growers was involved was different, as I have already mentioned, and so too was the market. Eventually Growers acceded to this. For some workers this represented an increase of over 100 per cent.[23] Once we had broken into a new sector, our next step was always to identify and organise the competitors. It was best to do so one at a time, to prevent firms from making common cause against the Union.[24] So while negotiations were in progress with Growers, we had already commenced organising its most immediate competitors, Elgin Fruit Packers (Elfco) and Krom River Apple Co-operative (Kromco), in the Grabouw valley. Like Growers, they were owned by local farmers and paid their workers farm wages.

One of the benefits of paying farm wages from the bosses' point of view is that there is not much need to monitor management's utilisation of labour. Because it is so cheap, it is possible to employ labour that is surplus to management's requirements, for sentimental or philanthropic reasons. Oom Klaas Marcus was one of several such workers, if the management of Growers was to be believed: he was 53 years old and approaching retirement age. Faced with the prospect of having to increase Oom Klaas's wages, management realised it could no longer afford him. Or so it said. From the workers' point of view, there was only one possible reason why Oom Klaas should be singled out, and that was because he had been a delegate at the negotiations. It was not so much what Oom Klaas had said, which in the broader scheme of things was of no consequence, but how he said it: he was a salt-of-the-earth kind of figure and had certainly not been deferential. I noticed the exchange of glances between the bosses at the time and the half-smiles they exchanged when he was speaking.

Oom Klaas had worked at Growers for 15 years. A week after the agreement with Growers was concluded, he was told by his foreman that his service was to end the following Tuesday. This was the very day the increases came into effect. As the workers saw it, Oom Klaas had placed himself on the line for his fellow workers, and would now not see any benefit from what he had achieved for them. So the Friday before, after tea break, the workers sent a deputation to management, while the rest waited for an answer.[25] The management refused to discuss anything with the deputation until the workers had returned to work, and the workers refused to return to work until they had received the assurance that Oom Klaas would not be dismissed. In effect they were placing themselves on the line for Oom Klaas. They would also not see any benefit from the increase unless he did so as well. So there was an impasse.

Annie Mentoor was at head office when I heard what was happening. The two of us immediately set off for Ceres. No one had been dismissed, Growers told us on our arrival. They had recently employed a particularly unctuous individual as personnel manager, and he did most of the talking for them. It was of course true no one had yet been dismissed, in that Oom Klaas was still employed. But management also did not deny what the foreman was alleged to have said to Oom Klaas. Instead we were asked to accept an assurance that no one would be dismissed without prior written notice, in which event the Union would be afforded the opportunity to make representations.

I had little doubt the plan had indeed been to get rid of Oom Klaas. Nevertheless, I was inclined to accept this position, as a way to allow management to back down while still saving face. But we could not agree to anything without speaking to the workers, and we were prevented from meeting them by the arrival of the Department of Labour, or the Department of Manpower Utilisation as they now styled themselves. The Department spent about three hours with the Union's committee drawing up a statement to the effect that the workers had gone on strike over the dismissal of Oom Klaas, which they wanted each worker to sign. The workers were not told why they needed to sign the statement, but almost certainly the Department was intending to use it as a basis for bringing criminal charges against them for striking illegally. However, the workers sensibly refused to sign and at the end of the day went home, without performing any further work.

We met with the workers over the weekend, and on Monday morning they gathered outside the plant in their overalls, ready to start working. They would do so, it had been decided, once they had the assurance Oom Klaas could start work with them. One of the managers appeared. You have three minutes to get to work, he announced abruptly. '*Die werkers wil weet ...*' one of the workers responded. The workers want to know whether Oom Klaas can start with them. It was a simple enough question to have answered, had management intended to honour its assurance, but he did not reply. The workers stayed put, and soon afterwards were told they were dismissed and should disperse. As they did so, the police, who must have been on standby, made their appearance, and laid into one of the workers for no apparent reason. Fighting broke out between workers and police.[26]

The decision to dismiss had clearly been a calculated one. The Union now had a major fight on its hands: over 700 jobs were at stake, with whole families dependent on wages earned at the plant and standing to lose their homes. The attitude of the bosses was also uncompromising. The workers can reapply for jobs, Growers said, in our first bid to negotiate a settlement, but there was no guarantee they would take everyone back. We have already employed new workers, they said.

We were pretty sure they were bluffing. It was now high season, when Growers usually took on additional labour. They needed the continuity an experienced workforce provided. What they were banking on was that workers would reapply for jobs, despite the Union. Then management would have a free hand to weed out not just Oom Klaas, but other 'troublemakers' as well. It was a reasonable assumption for Growers to have made. There was no memory of militancy in the valley. The nearest strike had been decades ago in Wolseley, when Annie Mentoor was a child. Many of the workers involved in that strike would have drifted towards the city, as Annie Mentoor's family had done.

Hardly any workers did reapply, and attendance at the meetings at the crèche was solid. As in the case of the Fatti's & Moni's strike, this was because of the many meetings held, and the slow process of organising the workers, and getting them to pay their subs. If there were workers in Ceres prepared to scab on the strikers, they were too few to maintain production. Growers went as far as the Karoo town of Touws River to recruit a workforce. But these were workers who knew about potatoes, and not about apples and pears. We smelt victory when Growers abandoned its endeavours to bus scabs from Touws River. After ten days, all the dismissed workers went back, including Oom Klaas. He worked there until he retired.[27]

The organisation of East London
It had been the noblest of strikes, because for solidarity's sake the workers had risked all that they stood to gain from the agreement reached, which was considerable. The integrity of their organisation mattered more than material gain: this was the statement they had made, and in the same year, 1980, there was a series of strikes in which the demand was simply recognition of the organisation. The first of these was a spin-off from our victory at Growers, but the most explicit was in East London.

AFCWU would be at the forefront of trade union organisation in East London. The process began with our trip to the Ciskei during the Fatti's & Moni's strike, when we established the whereabouts of Langeberg's pineapple factory. There was a follow-up visit at the end of 1979 by Oscar and Lizzie Phike, who took my yellow car. Lizzie had in the meantime been retrenched, and the Union employed her as an organiser.[28] This visit was to set in motion a train of events.

There were three other pineapple canneries in and around East London. One went under the name Western Province Preserving (WPP), which was curious, since it was located in the Eastern Province.[29] The reason, it transpired, was that it was owned by the Australian company that had founded Jones, and it supposedly felt some kind of attachment to the part of the country where it had first started out. Another factory belonged to Deepfreezing and Preserving, the English company with a factory at Macassar. The fourth was Collondale, which was privately owned. Almost all the workers were African and isiXhosa speakers, and most lived in the sprawling township of Mdantsane, reputed at the time to be the largest township after Soweto in the country. Organising in this area would clearly raise its own challenges, but fortuitously Oscar had a daughter living there, from a relationship 'on the side'. She was able to renew her acquaintance with her father, and Oscar and Lizzie had a place to stay.

The workers of East London did not regard it as any recommendation that the Union once had had a branch there. The only ones with any recollection of this were a handful of coloured workers at WPP, who were among the last to join, and the first to take fright, when the Special Branch began to show a particular interest in the Union. What did immediately catch the workers' interest, however, was that this was the trade union of the Fatti's & Moni's strike. Oscar and Lizzie recruited some 100 members at Langeberg, and promised to return to hold a general meeting.

Langeberg was the biggest of the pineapple canneries, and one of the biggest employers on the West Bank. This was the industrial area where most of East London's key industries were located, including the Mercedes-Benz assembly plant. Langeberg was also the employer in East London with which we had the most purchase, because of the other Langeberg factories we had organised, and especially the Boksburg factory, which also had an African workforce. However, relations with the Langeberg group were already complicated, because of the autonomy which management at its branches seemed to enjoy. Management at Boksburg had taken an increasingly hard line towards the Union, and relations with Neil Aggett in particular were strained. Since AFCWU's decision not to register, they had been unwilling to deal with AFCWU.

We are merely acting on the instructions of our head office, local management said, but it was clearly a position they wholeheartedly endorsed.[30] At about the same time management at Moberg's simply stopped recruiting African workers at the start of the apricot season. This was an alarming portent. Moberg's was the longest-organised factory in the Langeberg group, and the bond between African and coloured workers had always seemed strongest there.[31] It was also a bulwark against the Vakbond.[32]

Sensing the ground was about to shift beneath us, I pushed for a meeting with the managing director of the group, Dr Mouton. He was as far removed from the stereotype of the *kragdadige* Afrikaner as you could imagine, and he must also have commanded considerable respect within the Afrikaner business establishment. This was well represented in the Sanlam Building, where Langeberg had its head office, and which for a time was

Cape Town's tallest. All the factory managers were present, including the branch manager at East London. If we had not already secured more than a foothold at the East London plant, I doubt whether Dr Mouton would have agreed to meet us.

Dealings with AFCWU would be very much easier if it were to register, Dr Mouton said.[33] Yet he did not take issue with our explanation as to why AFCWU would not register, and the fact that representatives of both registered and unregistered unions were present, including Oscar and Lizzie, was an acknowledgement that our refusal to register a separate trade union was justifiable. We left thinking that our relationship with the biggest employer group we dealt with was on a new footing. In fact, the struggle within Langeberg was only beginning, and it was to lead to an intensification of the struggle on the other two fronts on which the Union had been fighting for the duration of the Fatti's & Moni's strike. A portent of what was to come was Oom Joe's failure to pitch at the management committee meeting, after he had lost out in the elections to be president.

At first, I fondly hoped he had accepted the outcome of the elections graciously. Then Oom Joe showed up in the middle of another meeting, and seated himself conspicuously at the back, in the same space the Jones group had occupied before the breakaway. He listened to my report of how Baba had not paid a cent of what she owed and had ignored the lawyer's letter.[34] Whether because he had been ousted as president, and thought I was somehow involved, or because I had shopped Baba, as he saw it, Oom Joe obviously had it in for me. The question was how far he was prepared to go: was he planning to do a Samson and bring down the whole temple with him? The opportunity to do so presented itself early in 1980, when the Vakbond applied for registration.[35] FCWU had the right to object, and it seemed our objection could not fail so long as we retained our membership at Moberg's and RFF. There were nevertheless a number of reasons for us to be extremely worried.

At Jones itself, Mr Woods had resigned. He was the only manager whose protestations of impartiality we believed, and there was speculation that he had fallen out with the bosses of Jones over the line they were taking towarsd the Vakbond. Juffrou H was still being granted access, and even before the Vakbond's application for registration, Jones began deducting their subs. This was illegal. At the same time, Jones refused to deduct subs by stop order for AFCWU.[36] On top of this, our organisation inside the factory was greatly weakened by the retrenchment of Lizzie Phike.[37] Although Liz Abrahams was now the official responsible for Jones, she had come from Moberg's. It remained to be seen whether she would be able to make any inroads into the Vakbond's membership.[38]

Then, at the very point on the agenda at which the management committee was to discuss the Vakbond's application for registration, there was a commotion at the back of the room coming from the space that Oom Joe and his followers now occupied. When Van Graan called the meeting to order, Mrs R interrupted him rudely and stormed out of the meeting.[39] It was an unprecedented and calculated slight. When called upon to explain her conduct at the next management committee meeting, Mrs R simply refused to do so.[40]

'Where are the financial statements?' Oom Joe asked pointedly at the same meeting. I had no idea what he was talking about, since statements of the Union's finances were presented religiously at every meeting, and when I asked politely which statements he was referring to, he remained tight-lipped.[41] He had not been asking a question as much as putting out an innuendo, I later realised, and financial statements featured in whatever plot he was hatching together with Mrs R. But who else was involved, and to what end?

The statements Oom Joe had been going on about, it later emerged, were in respect of donations received for the Fatti's & Moni's strike. These were handled by our attorneys – an arrangement we had made to insulate the Union against allegations that supporting workers financially was illegal – and the attorneys were responsible for preparing the statements. I myself was irritated at how long they had taken to do so, and as a result, in the closing weeks of the strike, we had opened a special savings account into which donations were deposited.

'The lawyers had better submit their statement this month,' Oom Joe said after yet another meeting had gone by without the lawyers having produced any statements, 'and they had better make sure that all the donations received are included,' he added darkly. Any possibility that he was merely being vigilant in matters financial was dispelled by Oom Joe's next line of attack. It was significant that it related to the organisation of East London.

Oscar and Lizzie had gone back to attend a general meeting, as they had promised. Almost the entire workforce of Langeberg were now members, as well as most of WPP, and the church hall in Duncan Village was packed to overflowing. There had not been a trade union meeting like this in living memory, anywhere in the country. The *Daily Dispatch* carried a story about it the next day. So did the *Cape Times* and the *Rand Daily Mail*. This was the earliest indication of the impact East London would have on the Union, and on the emergent trade union movement as a whole.

Buoyed by the extraordinary enthusiasm displayed by the workers, Oscar had decided to stay on and open a branch office. The problem was that he had again taken my car, and on the way to East London the oil-pump seized. By the time Oscar realised something was wrong, the engine had blown. We managed to scrape together the R853.92 it would cost to put in a new engine, and telegraph it to the Eastern Cape. This seemed like a substantial sum of money at the time.

This was the same yellow Cortina I had bought second-hand after the accident, instead of the bakkie out of the box which Oom Joe had wanted the Union to buy for me. But now Oom Joe seemed to be suggesting that I had enriched myself at the Union's expense by repairing it.[42] He had a no less absurd accusation to make about the organisation of East London. Athalie should have gone with Oscar, instead of Lizzie, he said, and should have used public transport to get there.

What was truly worrying was that among Oom Joe's followers was the charismatic Aunt Violet, and she seemed to find these accusations credible. Moberg's was being mobilised against me, at the very point at which unity in Paarl was imperative.[43] If worker leaders in the longest-organised factory in the Union could be turned so easily, what did this say about worker leadership? It did not seem like a coincidence that Vormat also chose this meeting to go on the offensive. Everyone pricked up their ears when she stood up to speak, since she hardly ever did so.

'There is something I want to bring to the attention of the meeting,' Vormat announced. 'I just want to tell the meeting that the workers of I&J Woodstock do not want Jan and Athalie.' It was a strange claim to make, since everyone knew we had no success in trying to meet the workers. No one at I&J can have known any of us. What remained to be explained was what Vormat had been doing by meeting workers from I&J.[44]

'When the vehicle is standing still, the dogs will piss on its tyres,' Alfred Noko never tired of saying whenever there was a report about an attack on the Union. 'When the vehicle is going forward, the dogs bark.' However, it was far easier to wax indignant when the attacks were external and took a form everyone recognised. It was more difficult when the attacks were from within. Some of the attacks from within, I have suggested, must have been masterminded by the Special Branch, and if Lennie's case was anything to go by, there were probably quite a number of members and officials on their payroll. But in Oom Joe's case, Langeberg management surely had a hand.

What would have exercised both the Special Branch and Langeberg in particular was that by organising East London, the Union had crossed the Eiselen line. This was of course the line that divided east from west, which constituted the Preference Area. In effect, we were encouraging workers on both sides of the line to fraternise. And while it is difficult to know in the case of Langeberg at what level management was involved, the complicity of the factory manager at Moberg's was clear. Straight after this management committee meeting, Oom Joe met with the workers in the cloakrooms, and the same week a letter arrived at head office. 'We as Langeberg branch,' it began. The upshot was that the branch would no longer pay affiliation fees to head office.

The letter was typed, but there was no typewriter in the branch office. I recognised the typeset and checked. It had been typed in the office of Moberg's manager.[45] At about the same time another letter arrived at head office. It seemed to be on a genuine Union letterhead from an unnamed member. There was mention of misappropriated money and, once again, of Athalie. But no member would have written such a letter, let alone have had the resources to type it on a Union letterhead. I did not know by what technology it had been produced, but obviously such technology existed and the Special Branch had access to it.

The decision no longer to pay affiliation fees was no small matter. In constitutional terms, it was one step away from a breakaway. In financial terms, it placed huge pressure on the head office, at a point at which its funds were already depleted: first, by the strikes at Sea Harvest and Growers, and now by the organisation of East London, including the R853.92 for the car. On top of this we needed to replace the kombi. The only way to do so was to get branches to contribute. These contributions were deposited in the same special savings account.[46]

Saldanha, of its own initiative, organised a dance to raise funds for head office, and Magrieta delivered the proceeds in the same week as the two letters arrived. All monies received had to be banked within three days of receipt, and in the interim were kept in the green safe in my office. It was Vormat's responsibility to do the banking each Friday, and that particular Friday she had a 'blackout'. That, at least, was her explanation about why she was not at work. So the money was not banked, and by Monday it was gone. Also gone was a file that I kept in the safe.

The Union was now in a full-blown organisational crisis. The disappearance of the file showed the motive was not simply theft. The file was my own record of the transactions relating to the special savings account. It could only be of benefit to someone wishing to prove there were donations that had not been banked or monies misappropriated. Unless, of course, it was I who was intent on destroying the evidence of my wrongdoing. Maybe the missing file would provide Oom Joe with the justification he was seeking, to break away altogether. That in turn would open the way for the Vakbond's application for registration to succeed, and defeat on the first front. What had previously seemed unthinkable, now seemed possible: that the enemy on the first front was making common cause with the enemy within.

Although this was the most serious organisational crisis the Union had faced in my time, I don't think anyone outside the Union had an inkling of how bad things were. Even inside the Union, it was only the branches attending the management committee meeting that knew. But the national executive committee was to meet in a few weeks' time. East London had already declared its intention to send a delegation. This would be my first encounter with the branch. Johannesburg and other far branches would be there. How could any of them have confidence in a head office where this kind of thing was happening? So we called a special meeting of the management committee, the second in a month.[47]

The last time I had the file in my hands, I remembered, was to show the treasurer there was a record of all the transactions on the special savings account. Betty was her name, and in all the years in which she had been treasurer, I had never heard her express an opinion, in public or in private. I had assumed it was because she was from Moberg's that she continued to be re-elected, but now I could see she was in cahoots with Oom Joe. The statements the branch wants, Oom Joe announced at this special meeting, were not those the lawyers were supposed to prepare after all, but those relating to the special savings account. But when it came to a discussion of how the theft had occurred, Oom Joe and his delegation were not interested in sticking around. These were not people who ever left meetings early, especially a juicy one like this. They must have already known, I thought at the time. Maybe they even had the file in their possession.

The details of the meeting, which started early on a Saturday afternoon and ended in the early hours of the next morning, are no longer important. Suffice it to say that I promised to prepare a statement in respect of the special savings account despite the loss of the file, and as it was getting dark, before the Dal delegation left, decided to go on the offensive. The working-class way of doing so is to pose a question, or series of questions, that exposes what is really going on. Why had their letter been typed on the same typewriter Moberg's manager used? Why did the letter refer to 'Langeberg' instead of Dal Josaphat branch? Why was there talk of a dance being organised by the branch where there would be a beauty competition to crown a Miss Langeberg?

I needed to know whether the worker leaders of Moberg were as gullible as they seemed, and whether anything I or anyone else could say would influence what individuals like Aunt Violet would do, once the collective had embarked on a predetermined course of action. They left without giving anything away. Afterwards, there was a long and painful discussion to try to get to the bottom of the theft. It ranged from questions as to who knew where the key to the safe was kept, to questions of motive. As the night drew on, the lines of division were clearly drawn, between those who believed that Vormat knew more than she was letting on, and the voices of caution, who were reluctant to draw any such inference and wanted proof of such an intention.

The voices of caution were those of coloured delegates, and it sounded to me as if racial solidarity was at play. It was already late in the evening when Moffat Manyosi, the African from Ashton who had been at my first management committee meeting, made an impassioned plea directed at the coloured delegates. There was a plot, was the tenor of what he said. It was not just a plot against the general secretary, but a plot against AFCWU, and the reason for it was that the Union was now organising branches where Africans were in the majority. It was East London he had in mind.

All the while Polly, the Jones worker, and a few others, together with Vormat herself, kept their thoughts to themselves. Polly was not a delegate, and was not in any way obliged to be there. She can only have been there out of concern for the Union and, it transpired, because she had something to say, although she alone knew how important it was. In the early hours of Sunday morning Polly got up to speak. Her concern for the Union must have outweighed any thought of adverse consequences, whether for Vormat or herself.

'Vormat came to me more than once', Polly said, 'to ask my help in getting rid of Jan. I asked her what Jan had done. She could not tell me. I then told Vormat I would help her get rid of Jan when he has done something which justified getting rid of him.' Here at last was the proof the meeting needed. Vormat did not even try to defend herself, and a resolution to dismiss her was adopted.

The next week Vormat came to collect her monies in lieu of notice. She was accompanied by a young man who I imagine was part of her cell. He looked at me as someone with a dark complexion might look at a white boss who, despite his protestations, was actually part of the problem rather than the solution.

The Grabouw strikes

I managed to prepare the financial statements Oom Joe had wanted despite the loss of the file, but there was no delegation from Moberg's at the next management committee meeting to hear the report.[48] In the interim the Union was able to warmly welcome, for the first time, a delegation from East London to the national executive committee.[49]

The management committee was captivated by the East London delegation because of their confidence and because they seemed to have no airs and graces. They, for their part, were enthused by meeting workers from places they had heard of but never seen, as well as places they had never heard of. This was also my first meeting with Bonisile Norushe, who had resigned his position as personnel officer of Langeberg to become the first elected branch secretary.

In the meantime, the organisation of two big pack-stores in the

Grabouw valley, Kromco and Elfco, was taking off. Grabouw is not close to Ceres, even as the crow flies, yet there were workers there who knew of the increases that Growers workers were getting, as did their bosses. They were, after all, part of the same industry. All three pack-stores marketed their produce through the same control board, which handled the export of their fruit. This control board was one of several which regulated the prices farmers were paid for their produce, and was in turn part of a broader system through which government controlled the prices of staples like bread and mealie meal.

Those who were not aware of the increases Growers workers were getting would soon be made aware. A situation in which three large pack-stores sold their fruit on overseas markets for the same price, but two of them paid lower wages, amounted to under-cutting. Our argument was always that any under-cutting by one employer of another was objectionable. In the case of Grabouw, this was the basis on which the workers were organised. But the corollary of this argument was that the Grabouw workers could not expect to get more than the Ceres workers.

Grabouw valley was then, and still is, the stamping ground of the urbane, well-to-do English-speaking farmer, and was not typical of farming communities elsewhere. But whereas the management at Elfco were quick to see which way the wind was blowing, and increased their minimum wages to what Growers were now paying, in order to forestall the Union, the manager at Kromco came from the mines and subscribed to a policy of law and order. When a group of workers stopped work on a Friday to demand a wage increase, he called the police.[50] We had not even had a chance to have a meeting to introduce ourselves.

The police arrived, and spotted one of the workers holding a placard demanding R40 a week – the same amount the Union had demanded at Fatti's & Moni's, although by now somewhat eroded by inflation. They tried to arrest him, and a tussle broke out, in which rocks were thrown. The police then summoned a specialised unit, the riot police, who arrived armed with guns, plastic shields and tear gas. When the riot police advanced, some workers stood their ground. Others fled, and tried to hide in the bushes and forest around the pack-store. Those that fled were pursued and arrested. The police let those that stood their ground alone.[51]

'We are ragged, we are hungry, we are dirty. We do not even get enough money to buy soap,' one of the arrested workers told me.[52] They were all young men, and most had come from towns in the Overberg, like Genadendal, Caledon and Riviersonderend. They had stopped work that Friday because the season was drawing to a close and, with it, their last opportunity to earn some money to take home with them. They were refused bail and held in police cells over the weekend. Some were beaten and tortured with a contraption the police kept at the station to administer electric shocks. A medical doctor whom we called on to examine them found burn marks on the workers' fingers consistent with their account. A 16-year-old had strangulation marks on his neck.[53]

So what had started as a comparatively small stoppage developed into a full-blown strike the Monday afterwards. Various attempts were made to force workers back to work, including threats to evict workers staying in factory houses or on farms belonging to the Kromco bosses.[54] The most vulnerable were those staying in factory hostels: workers from the Overberg towns, mainly women, and a contingent of 150 African contract workers, mainly from the Mount Fletcher district, who stayed in a different hostel, stuck away on the mountainside.

When the workers failed to return to work by the Tuesday, management issued an ultimatum to those in the hostels. That evening the police descended on the hostels and herded the workers onto buses to take them home. The same happened to the Africans the next day. So once again we found ourselves travelling to a community hall on a daily basis to meet with striking workers, only this was a grander facility than any we had encountered in townships elsewhere, the Gerald Wright Memorial Hall.

There was a prearranged meeting to introduce the Union at Elfco during that same week, so one afternoon we bid the striking Kromco workers goodbye and went to premises of their bosses' closest competitor. The meeting with Elfco was a genteel affair. Tea was served in fine china cups – I had by this point decided I did not agree with Liz's policy of not accepting anything from the bosses, when taken to the lengths of refusing a cup of tea – and one of the managers made a remark which was intended to convey, in a conversational way, how well he grasped that the cost of living was rising.

You can hardly afford to take your family away on holiday with a thousand rands, he said, looking at me, as though I would understand better than the officials accompanying me. I was amazed that anyone could contemplate spending as much on a holiday. It went to show how out of touch I was with people who perceived me as their social equal. That he perceived me as a social equal, even if I was somewhat shabbily dressed, went to show how the trade union endeavour was gaining acceptance among this class. No one mentioned what was happening to their competitor down the road, still less what was going on at the Memorial Hall. Yet it was certainly being closely monitored, not just by Elfco, but all the bosses in the valley.[55]

The settlement of the strike and a wage agreement were negotiated simultaneously with Kromco, over three days. The decision to negotiate represented a turnabout of note, and was clearly taken over the head of the factory manager. The chairperson of the board was present throughout the negotiations, which took place in its boardroom, as well as other board members.[56] They must have been worried at the negative publicity the strike had already generated and its impact on their markets overseas. They readily agreed to bring the wages in line with Growers. The problem was to get them to take all the workers back, and especially the African contract workers.

It is impossible to bring the workers back from Mount Fletcher, the bosses claimed, and listed all the reasons why this was so, including the difficulty of contacting workers who had already dispersed over a vast and remote area. In a final bid to secure agreement, they offered to pay each contract worker a lump sum, equal to half their wages for the remainder of the contract, at the increased rate we had negotiated. They would also guarantee the workers their jobs next season, they said.

Half the weekly wages at the increased rate represented almost as much as what the workers had previously been earning for a full week's work, and they were not required to do anything to earn it.[57] This seemed enough of a benefit to outweigh the risk of not resolving the strike, so ten days after the first stoppage we signed a settlement agreement. But management evidently could not bear the thought of paying so many to do no work. The sequel to the settlement was that Kromco discovered it was not 'impossible' to contact the workers after all, and brought back about half of them. When it came to rehiring the other half next season, they attempted to renege on their undertaking, and were only prevailed upon to do so after a further meeting with the board of directors.[58]

On the same day the settlement agreement with Kromco was being signed, and just before we were to depart from Kromco's premises, its management relayed an urgent message for me to contact Elfco. The workers were threatening to come out on strike, and the same afternoon wage negotiations at Elfco commenced. The problem was that because Elfco had already brought their wages in line with Growers, they had left themselves no room to manoeuvre. So we had nothing to report when we left Elfco premises later that night. The next day the workers came out on strike.

It was the most total strike at a large plant I had ever observed. The reason was that all the workers in the middle layers – the supervisors, foremen and skilled workers – supported the strike. There had also been considerable support from the middle layers during the Kromco and Growers strike, and it seemed the social divide in the coloured communities of Grabouw and Ceres was not as great as in Saldanha, and certainly not as great as in Cape Town. It was a divide that organisation could help overcome or exacerbate.

The strike placed the Union in a difficult position. We had organised the workers on the basis that there was no justification for one employer to pay less than its competitors, but in fact the only section of Elfco's workforce still earning less than Growers and Kromco was the African contract workers (you could always rely on the mean streak of the bosses to surface when contract worker were concerned).[59] It was dark, and we had to take back something to the workers who were now packed into the Gerald Wright Memorial Hall, still in their overalls, waiting to hear what we had achieved. We had not really achieved anything, except for the contract workers.

'What you are currently earning was thanks to the Union, and what it had won at Growers,' I told the workers. 'What we now have to bring you was not much more than what you are now getting,' and proceeded to read out the rates, including the piece rates for the packers, which enabled most to earn well above the hourly rate. This was a repackaged version of what they were already getting, and I was exhausted and filled with apprehension. We had never been in the position of having so little to bring to so many.

When I stopped speaking there was an uneasy silence. I would not have been surprised if someone had stood up from the floor and virulently denounced the offer. Instead, one of the committee members, a worker from the middle layers, came up to me where I was standing next to a microphone, and warmly shook my hand. The hall burst into enthusiastic applause. The strike had not been so much about wages as about teaching the bosses a lesson. The lesson can only have been about respect for the organisation to which the workers belonged. There were clearly deeper issues between workers and their bosses than could be resolved over china teacups.

15

The Graaff-Reinet road

A SUCCESSION OF VICTORIES in a succession of strikes went to show, we thought, that we did not need to rely on a strategy of negotiating at Conciliation Boards. Instead we would engage in what we now termed 'direct negotiations'. Where it made sense to get different employers around the same table, we could simply establish our own forum to do so, as in fact we later did for the three pack-stores. There was also no cause for us to consider the possibility of forming an Industrial Council. So from 1980, the year in which FOSATU and its affiliates began trying to enter the system, FCWU began withdrawing from it.[1]

This new strategy did not mean we could afford to lower our standards of organisation. New factories were organised in the same way as before, by getting a majority to sign applications forms and pay their subs. Our best prospects were in sectors in which the Union already had a commanding presence. On the other hand, the fiercest resistance could be expected in unorganised sectors, as our experience at Fatti's & Moni's and Growers had taught us. 'The Union will take up whatever you demand', we used to say whenever workers made demands. 'But do not blame us if we do not succeed, when there are other workers doing the same jobs as you, but who are unorganised, and are willing to put up with much less than you are demanding.'

Where we could we would name the unorganised workplaces belonging to the same company, or producing the same product, that weakened the Union's bargaining position. Workers everywhere understood this logic. Often there would be someone present in a meeting who had worked for the company we had named, who would remind workers what it was like to be without an organisation. In this way workers came to accept responsibility for organising the unorganised in the industry of which they were part. We had not emphasised this enough when organising Fatti's & Moni's at Bellville, and it proved too late to do so at the end of the strike. Although the workers nodded in agreement when implored to organise the unorganised mills, in fact they were happy to leave this to the organisers. It was different matter when we organised the Isando plant of Fatti's & Moni's outside Johannesburg. This, once again, went to show how much it mattered that workers were organised at the outset.

The Union had already established contact with workers at the Isando mill by the time the boycott committee was established in Johannesburg, and had been meeting a group of them regularly; but the workers were understandably cautious about the Union showing its hand at this stage. Soon after the settlement was reached, a meeting with the workers was arranged. By February 1980 the Union had a majority and had introduced itself to the management.[2] This was the company's flagship factory, and it was located in a new industrial area, along with a number of other food factories. Almost all the workers stayed in Tembisa township, and most were on contract from the northern reaches of the country and spoke Sepedi and Shangaan. The prospects for growth were promising, but the branch was struggling. It seemed to be the fault of the paid officials.

Beatrice*, whom the Jones workers had elected from among their number, spent an inordinate amount of time doing her hair in the office. A second organiser had now been elected, from among the members at Langeberg Boksburg. She was bright and keen, yet I feared it would be only a matter of time before Beatrice's indolence rubbed off on her or else she would be defeated by the magnitude of the task. Already the Union was covering an area that stretched from west of Johannesburg to the East Rand, and from Isando to the north. It was too large an area to operate without a vehicle, and the branch did not have the money to buy one. It also did not have money to pay affiliation fees to head office or another salary.

So Neil Aggett continued to work without a salary, and subsidise what he did for the Union through working sessions at Baragwanath Hospital and later in Tembisa. The branch could not have done without someone like him if it was to reinvent whatever tradition of worker organisation there once had been on the Reef and navigate all the shoals on which such a venture could easily be wrecked: an intellectual with understanding and vision. It was not realistic to expect someone whose only experience was working in a factory to play this role. So Neil in effect became a regional secretary, although there was not yet a region in the Transvaal. He was also the driver who would take the officials to the factories in his car.

I was not comfortable about Neil working full-time on a voluntary basis, and did not believe he could continue to do so much longer. Apart from anything else, Beatrice used the fact that he was neither elected nor remunerated by the Union to avoid being accountable for how she spent her days. Neil, on the other hand, was worried that being on the Union's payroll would make it easier for the Military Police to track him down and arrest him for avoiding conscription into the army. He also keenly felt his isolation, and the isolation of the branch, from the rest of the Union. So in April of that year a delegation from the Western Cape, including workers from Langeberg and Fatti's & Moni's, accompanied me on a week-long visit to the Reef, during which we held lunch-hour meetings at as many factories as we could. However, the right to hold lunch-hour meetings, which by this point was taken as a given in the Western Cape, was more controversial on the Reef.[3] Permission for the meeting at Fatti's & Moni's was only granted after intense pressure from the workers. It was the first the Union held in Isando.

Petrol restrictions were still in force, and I needed a permit to drive back to Cape Town on the weekend. 'My father has recently had heart surgery,' I told some bureaucrat in whatever office it was that granted permits in downtown Johannesburg. It was true that I was concerned about his condition, which was serious enough for him to have taken early retirement. But there was no suggestion it was worsening at the time, and I doubt whether he would have been prepared to say it was, if the police had asked. Obtaining a permit under false pretences was a criminal offence.

My father was still my reference point for what the white ruling class were thinking. Although to others he would advance a perfectly rational defence of his son's involvement in the trade unions, he always felt obliged to uphold the status quo during my visits, which would invariably degenerate into a political argument. Bishop Muzorewa had been trounced by the Patriotic Front formed by ZANU and ZAPU in Zimbabwe's first democratic elections, yet he kept on being taken in by a succession of Muzorewa clones: stooges that the PW Botha government sponsored as 'responsible' leaders of African people.

Soon after our visit Neil phoned to say they were being followed, and that the Special Branch had questioned Beatrice.[4] Later he wrote to provide more details: the Special Branch were 'following me and even people who visited me ... on two occasions when we went to meetings in the township, a fleet of five cars followed us. There is also evidence they have been into the office (e.g. urine in the water jug).'[5]

Another ominous development was the detention of most of the officials of WPGWU during the Red Meat strike. This was a strike over recognition of the WPGWU, which had succeeded in organising just about all the plants processing red meat in greater Cape Town. It represented a significant shift from organising generally to the development of an industrial strategy, but rather than taking on the meat bosses one by one, as we would have done, all its members came out in a one-day strike. The response of the bosses was savage: 800 workers lost their jobs.[6]

If there was another clampdown, it was obvious that the trade unions that refused to register would be targeted. Our newly elected branch secretary in East London, Norushe, was detained on a Friday, but we only got to hear of it on the Tuesday after it had happened, because Monday was

16 June, the anniversary of the start of the Soweto uprising, and no one in the Union was at work. There was also no one we could get hold of in East London to tell us more, so Oscar and I left by road the same evening to find out.[7] To get to the Eastern Cape from Cape Town you can travel along the coast, through Port Elizabeth, or the Karoo way, through Beaufort West. The contract workers usually took the Karoo way in their buses and taxis, and that was also the route Oscar preferred. It was not the shortest route, and I had yet to be persuaded it was the safest.

The detention of Norushe seemed to be in response to events at Western Province Preserving (WPP) that week, we later established, but it was the situation at Langeberg which occupied most of our visit. After a series of meetings we set out on the return trip, taking the way we had come. But to travel at night I again needed a permit. My younger brother Martin's wife had died tragically while I was in East London. I had hardly known her, but I certainly had to get back for the funeral. That was the reason I gave the East London bureaucrats, but if I had not had this reason, I would have thought of something else. There was a management committee meeting on Sunday and I had to prepare for it.

From East London you take the King William's Town Road, and shortly after crossing the Buffalo River, as you leave the town behind you, there is a turn-off from the N2 on your right to Alice, and, by way of a gradual ascent to the central plateau, you pass through a series of frontier towns with 1820 Settler names: Fort Beaufort, Adelaide, Bedford and Somerset East. Beyond Somerset East you are in the Karoo proper and on the way to Graaff-Reinet, the site of an old Boer republic and the birthplace of Robert Sobukwe.

'We are on our way back to Cape Town, from a business trip in East London,' I told the police at the first roadblock, shortly after the turn-off to Alice. They shone their torches intently through the passenger-side windows of the kombi before they came up to the driver-side window. An African man in a suit next to a young white must have aroused suspicion. The next roadblock we encountered was somewhere before Somerset East. Political tensions were obviously running high so soon after 16 June, but again we were allowed to proceed. The roadblock after that, on the Graaff-Reinet road, was unquestionably set up specially for us.

A phalanx of police surrounded the kombi, some with rifles. Oscar and I were made to get out and were body-searched, while the Special Branch officer in charge searched the interior. There he found the permit for us to be travelling at night and the permit for the Johannesburg trip as well, which I had carelessly left in the back seat and forgotten about. He gave something like a cry of glee as he looked at the permit by the light of his torch: so, he said, you have been misusing permits to travel around the country. A more junior officer was in the meantime going through the contents of Oscar's bag in the boot, where he found another titbit: a June 16 pamphlet, calling for a stayaway.

I was standing next to Oscar at the boot while the junior officer completed his search, and after finding nothing else of interest, he took a closer look at the pamphlet, which was written in struggle-speak. I then observed an extraordinary exchange between them. Look what I have found in the old man's bag, the junior officer cried out to his superior in Afrikaans. 'No, you did not!' said Oscar, with such emphasis and conviction that the junior officer became perceptibly flustered and was no longer sure where he had found the pamphlet. All my middle-class conditioning militated against the self-preserving lie because, I think, the middle classes generally have no need to tell such lies. Honesty is the best policy, but not for Oscar in these circumstances or for the working class in many other circumstances.

We were taken to what must have been the offices of the Special Branch in Graaff-Reinet, where there were more questions, about the pamphlet and the permits, and what we were doing in East London. But there was not much they could ask about the pamphlet, because they now did not know who had it, and anyway it was only a single pamphlet. The officer in charge made several phone calls in an adjacent office, one of which must have been to a superior, who told him to release us. He was clearly not pleased to do so. Since Steve Biko's death in detention, the Security Police of the Eastern Cape were generally regarded as the most vicious in the country. It was already dawn when we set off on the last leg of our journey.

The East London strikes

If the strike at WPP was not the first, it was one of the first of a wave of strikes in East London that started in 1980. It was also one of less than a handful that achieved its goal. However, it was not reported in the press, probably because it was resolved the next workday.[8] The strike at Langeberg, which commenced on the day of our arrival, was also not reported, although it was one of the others that also achieved its goal.[9] This was the disbandment of the liaison committee and the recognition of the Union. In the interim Norushe was detained, along with several other workers at WPP, including the chairperson of the committee, James Mpemvushe. The workers were released after two days, except two shop stewards, who were charged with participating in an illegal strike.[10] Although these were ordinary workers, doing the same jobs as their counterparts in the Western Cape, none of them seemed to have been intimidated by the experience. It was also no shock to them that their branch secretary was now in detention. This was the frontier of the old Cape Colony, where this sort of thing, and worse, could be expected.

Mpemvushe spoke a beautifully articulated, deliberate isiXhosa, but no

English. Since he wore the khaki jacket of someone in the middle layers of the factory, isiXhosa must have been used by lower and middle management as well. The middle layers at Langeberg, with the exception of the members of the liaison committee, had also all joined the Union, but Welile Mzozayana, who was the chairperson of the Union at the factory, as well as the East London branch as a whole, represented a kind of worker intelligentsia that I had not previously encountered. He used the word 'orthography' in a meeting with Langeberg management, in its correct context and with something approximating a BBC accent.[11] Wages would not have been his primary reason for joining the Union.

The strike at Langeberg began after the morning tea break, with the workers demanding the resignation of the liaison committee. They also wanted the committee to resign in front of them, as had happened at WPP.[12] Since the liaison committee was for them the workplace equivalent of a Bantustan government, it had to be a public resignation. The workers suspended the strike that afternoon, when they saw that management were prepared to meet with us, but set a condition: they should be informed of the outcome of the meeting by 9.30 the next morning. This was the time of the morning tea break. So negotiations went on into the night, while Langeberg rushed its group personnel manager to East London. They resumed early the next morning.

As the 9.30 deadline approached, management were still clinging doggedly to their liaison committee, whose existence, they assured us, in no way affected their willingness to deal with the Union. So a statement was drawn up, which committed Langeberg to negotiating with the Union, but fudged the issue of the liaison committee.[13] Oscar and I stood to one side, while the branch manager read it out to the assembled workers. It was in English. Many would not have understood a word. 'What does the statement say about the liaison committee?' someone asked. The workers were clearly not satisfied. 'We want to see the liaison committee resign,' someone else shouted. There and then, they meant. The situation was now on a knife edge, and management was not able to defuse it. This much was clear to Brand, the group personnel manager. He came over to where Oscar and I were standing, imploring us to do something. After a period of judicious hesitation, and with a shrug of his shoulders, all of which the workers were able to observe, Oscar relented.

'Let me see if they will listen to me,' he said to Brand, and stepped forward. There must have been more than a thousand, completely quiet. 'The liaison committee is finished,' he told the workers, in isiXhosa, and so it was. It was more satisfying theatre by far to be addressed by a genuine African leader, with the management looking on, than a public humiliation of the liaison committee would have been. Long after the workers had streamed back to work, management was eating out of the Union's hand, but it remained to be seen for how long.

The East London strikes were to become the most significant wave of strikes since Durban 1973. Unlike in Durban, however, the strikes did not happen spontaneously. They also did not happen simultaneously. They were triggered by organisation, and apart from the strikes at Langeberg and WPP and two other factories that AFCWU organised, the union involved was SAAWU, and the demand was recognition of the trade union. When Oscar and Lizzie had first visited East London, SAAWU was still trying to establish itself. Oscar had an important influence on its charismatic and charming young president, Thozamile Gqweta.[14] So, from the outset, relations between our branch and SAAWU were close, and the two unions had adjacent offices in an old building near the town centre.[15] SAAWU offered to help keep our office going while Norushe was detained, and we gladly accepted. When SAAWU officials were detained at a later juncture, I and others negotiated on their behalf.[16] We saw each other as allies, and were identified with one another by the East London public.

However, the two unions represented very different constituencies, and operated in very different ways. The first indication of this was that although Gqweta's position was president, he worked full-time for SAAWU and seemed to be a paid official. He had also not been a worker before becoming a paid official, if 'worker' is understood to mean a worker in a factory such as those SAAWU sought to organise. In fact, it was difficult to tell from the outside who was a worker in SAAWU and who was a paid official or simply a hanger-on: their office had become a magnet for charming young men, some of whom were workers and some of whom were not. The distinction between a paid official and a worker was inviolate in the Union's tradition. A president could not be accountable to workers if he or she was a paid official or worked full-time for the trade union. Workers were the decision-makers, and a president also symbolised that.

Resolution on the third front
If proof were needed as to why the distinction between paid officials and workers was so important, it was provided on our return from East London. The financial books of Dal branch had been submitted to head office, as each branch with its own bank account was required to do at the end of every financial year, and they were in a mess. This, I now knew, was the way a corrupt official tries to camouflage dishonesty.

Upon further investigation, it emerged that Mrs R had been paying herself more salary cheques than she was entitled to, and that there were other payments made out to Mrs R for which there was no apparent explanation. The signatures on certain cheques looked as if they had been forged. Mrs R had been the paid official whom I had trusted more than any other when I started out in the Union, and from whom I had also learnt more than from any other. Lilian had been right: no one could be trusted. The excess salary and unexplained payments must surely be the explanation for her recent outrageous behaviour.

I reported to the management committee meeting what I had found, but since the FCWU delegates from Moberg's were now boycotting the meetings, the workers were none the wiser. The signatures that appeared forged were those of Oom Joe and the branch treasurer, and the Union needed to know where they stood in this matter. So a meeting was convened at the auditors' office with Oom Joe, Mrs R and the branch secretary. Oom Joe blandly denied his signature was forged, and claimed all the payments to Mrs R were authorised. This was not much of a shock. I was pretty sure the workers would not have authorised additional payments to Mrs R, and it was already clear that she and Oom Joe were in this together. It was different matter, however, when the treasurer was confronted with what had to be a forgery of her signature.

'That's my signature,' Gladys* told Mr B, the auditor, softly and without conviction. Gladys was one of the kind faces I knew well from management committee meetings. Maybe I should not have been as astounded as I was. She had to go back by train to Paarl, together with Oom Joe and Mrs R, to a factory where Oom Joe was in a position to make her working life a misery. It was just another self-preserving lie, and no less astounding than Mr B shrugging his shoulders and saying there was nothing to be done in the face of such a denial.

I then convened another meeting with the branch to ask what it was going to do about the financial irregularities.[17] It did not seem it was going to do anything, and Oom Joe and one of his sidekicks took the opportunity to launch a series of barbs at me, with the evident approval of Aunt Violet. There was no indication that any of the members would not follow wherever Oom Joe was leading them. While all this was going on, the three-year agreement for the canning industry expired, and the news came through that the Vakbond had been provisionally registered.[18]

The implications for the Union were momentous. Legally, the Department could not have granted full registration to the Vakbond over the objections of the Union: the Union still represented a majority of workers in the three factories that fell within the magisterial district of Paarl. Provisional registration was nevertheless a huge boost for the Vakbond. Already the Vakbond was telling workers at Jones that their organisation was registered. It also meant the Vakbond could apply for full registration again in a year's time.[19] If Oom Joe and Mrs R were envisaging a breakaway, they would have time enough time to make common cause with the Vakbond. If that happened, the Union would be sunk in its heartland.

Our plan had been to negotiate directly with the canners for a new agreement, for the first time.[20] It would now be too risky to pursue. Instead, for the last time, negotiations took place at a Conciliation Board, and for the first time there was a delegation from AFCWU as observers. You might describe this as a shameful climbdown for the Union, and humiliating for the African members. But the fact that the chairperson, the Department's divisional inspector, prevented the AFCWU from speaking made a lasting impression on the FCWU delegates present.[21]

Annie Adams, for one, was convinced Oom Joe and Mrs R were planning a breakaway, and discussed the matter with Liz. I met with them at their request one weekday evening, not long after my apparently unsuccessful meeting with Dal branch. What they had in mind was a 'compromise'. In order to avoid another breakaway, and keep Oom Joe and Mrs R in the Union, we should overlook the financial irregularities. My surprise at this proposal gave way to consternation when Liz went on to mention that she had been approached the same day by her next-door neighbour, an ordinary worker at Moberg's, who had also asked to meet me that very evening. Liz had thought a meeting with Annie was more important, and put her neighbour off.

I was furious, although I tried not to show it. All the while we had been trying to get through to the members at Moberg's, so that they could understand what their leaders were up to. Even if it had been a meeting with a single worker, it was an opportunity to grab with both hands. The worker might since have developed cold feet, and the opportunity lost. In fact, the opportunity had not been lost, and the meeting took place in Liz's *voorkamer* the following evening. It was not just Liz's neighbour who was there, but a group of Moberg's workers, including several who had sat in the management committee meeting, listening and saying nothing. Also present was Aunt Violet and Gladys, the treasurer who had so recently denied her signature was forged.

'*Ek het altyd 'n sambreel oor [Oom Joe] se kop gedra*,' said Aunt Violet. I always held an umbrella over Oom Joe's head. She was not prepared to do so any longer. Other members of the committee had been speaking to her, because of the respect she commanded among the workers, and the penny had eventually dropped. Gladys explained how Oom Joe and Mrs R told her to acknowledge the signatures on the cheques as her own. There was a keen air of excitement among them. All were women, and they were challenging an overbearing man, Oom Joe, who also personified male authority in their workplace. The plan agreed upon was to hold a lunch-hour meeting at the factory with a view to asking the workers to attend a general meeting after work. This was the only factory where the factory manager had always agreed to meetings as a matter of course, but that was about to change.

Since Oom Joe was still chairperson of the branch, I had to tell him of our intention to hold a meeting at the factory. He was vehemently opposed to it. I nevertheless arranged a date with the factory manager, and when we arrived at the factory we found Oom Joe waiting for us. The management wanted to see us, he said, and at the manager's office we were kept waiting until the start of lunch break, at which point the manager and Oom Joe began to question why we wanted to hold the meeting. The manager then asked to meet with me alone. I refused, point-blank, and we proceeded to the cloakroom, defying him, as it were, to stop us. He did not do so.[22]

The cloakroom was packed. Even before a word was spoken, it was clear from the reception the workers gave us that the hold Oom Joe and Mrs R had once had over them had been broken. We told them we planned to hold a general meeting at the office to conduct elections in three days' time. The workers were not expecting Oom Joe to attend that meeting or, for that matter, Mrs R, and decided in the interim whom they wanted to assume the chair: a good-looking young man called Jeff, who wore a white coat.

A resolution of no confidence in Oom Joe and Mrs R was adopted. There was another resolution adopted to reunite with Paarl branch, and bring a formal close to a wound that had been festering for more than 15 years.[23] But the hopes the Moberg's workers had placed in Jeff proved short-lived. Like other coloured men with some education, he opted to leave the factory for an administrative job. Oom Joe, on the other hand, worked for Langeberg until he unexpectedly died.

In the meantime, the funeral of my brother Martin's wife had taken place, and sometime afterwards a Special Branch officer rang the bell of his Sea Point flat. Martin looked down from the balcony and asked what he wanted. The officer introduced himself and made some snide reference to Martin's wife supposedly having died. Martin was outraged. Realising he was on a wrong tack, the officer left, and I heard nothing further of the matter of the permits. Fortuitously, they had chosen the wrong one to investigate.

Someone whom the Special Branch had in their sights for far longer than me was of course Oscar. I had by now realised that Oscar's habit of dropping asleep at the steering wheel and other unlikely places was not just because he was old. Almost every time I took him home to Nyanga there would be people waiting for him: whether from the women's group that had recently been formed, or the residents' association, or the youth. So his return from a Union meeting would be the start of another meeting, which went on into the night. No doubt there would be more such meetings on the nights he was not accompanying me to branches.

In the winter of 1980, the issue around which the community in Cape Town was being mobilised was the steep increase in fares that Golden Arrow bus company was seeking.[24] The buses were subsidised by government, which also determined the fares. When the company was granted something like the fares it was seeking, there was a call for a boycott. For the boycott to succeed there had to be an alternative for workers to get to work. Oscar was involved in organising an unofficial taxi service, and was now busy in the early mornings as well, monitoring the cars packed with workers traveling to Claremont and Mowbray stations. It was then that the police started to intervene, and the boycott started to spiral into violence, culminating in the ambush of a motorist on Lansdowne Road near Nyanga. Oscar was detained shortly afterwards. After two eventful years, his second spell in the Union had ended. He would not return.

PART 4
Recognition

*'One knows the good people by the fact
That they get better
When one knows them. The good people
Invite one to improve them, for
How does anyone get wiser? By listening
And by being told something.'*
– BERTOLT BRECHT, 'SONG ABOUT THE GOOD PEOPLE'

16

The road to Kidd's Beach

THE FIRST DEMOCRATIC ELECTIONS in Zimbabwe in 1980 were a watershed.¹ The people of South Africa would surely not accept anything less than Zimbabweans had, and sooner or later there would have to be elections in which everyone could vote. But no one in the circles I moved in thought that this would come about in the same way as it had in Zimbabwe, through a protracted guerrilla war. South Africa was, after all, an industrialised society with a predominantly urban population.

That, as I have said, was why people like me had opted to organise workers. I had no qualms about that decision at the start of 1981. The experience of the previous five years had also shown, I thought, that a trade union was the best way to go about organising workers. This was above all because it was an autonomous form of organisation, which could sustain itself, because it was financed by its members.² Workers and the class they came from needed organisations for the long haul.

I doubt whether anyone in the trade unions supposed that they would be capable of bringing about social change on their own, but over the previous five years we had shown that trade unions had a role in bringing about change that no one else could fulfil. The Union had become a non-racial organisation at a time when few others were. It was in the process of winning recognition for its right to be a non-racial organisation even though the law had yet to recognise this right. Recognition of non-racial democracy in the engine rooms of the economy would represent the faltering first steps toward the institution of democratic rule in the country as a whole.

The Union was itself a model of the kind of democracy it would like there to be in the country, founded on respect for local autonomy. Although it had been touch and go, it had thus far managed to see off serious threats from within without compromising this model. It had also managed to stop the rot that had set in because of a leadership that had been corrupted. Another year would pass before the last of the 'leaders' from Johnny M's time was found to be dishonest despite her honest face.³ By this time it was already well understood within the organisation how corruption manifested, and why you could not have corrupt leaders in any kind of democracy.

The debilitating power struggles that were the manifestation of a corrupt leadership clinging to power were also at an end, and the sense of relief members felt was palpable. The mood among the delegates who packed into the Paarl office from near and far for the management committee meeting was celebratory. Despite the shadow cast by the detention of Oscar and Norushe, our most articulate African leaders, things were looking up. The task at hand was now to consolidate our organisation nationally, and join hands with other trade unions committed to organising all workers, regardless of race.

Naturally, we hoped that other trade unions would be persuaded to adopt the same model of organisation as we did, if they had not already done so, but we also had to acknowledge a contradiction in our claim to be one trade union. The fact of the matter was that we could not truly be one union so long as we were constituted as a registered and an unregistered union. Since it was not possible for us to be one registered trade union, in terms of the legislation, the only alternative was for FCWU to forfeit its registered status, as Oscar had argued it should do in 1947.

So long as the Vakbond was provisionally registered, however, this was not a possibility we could even consider. The Department of Labour could be expected to try to help the Vakbond establish itself in whatever way it could, and when the Vakbond's provisional registration expired later that year, it was bound to renew it.[4] If the Union forfeited its registered status, it would not be able to object to the Vakbond becoming a fully registered trade union. There was an additional complication, which would profoundly affect the balance of power in Paarl. Dr Mouton summoned us to a meeting to inform us that Langeberg was taking over Jones.[5]

This situation in Paarl was one of the reasons why I could not consider making way for someone else as the fifth anniversary of my becoming general secretary approached. However, the strike at Langeberg had brought home to me the contradiction of being a general secretary who could not address the workers of East London in their own language. Now I was in East London again, at the start of 1981, because of the fall-out from a strike at Collondale Cannery and because our local official was in jail. All eyes were on what was going on in East London at the time.

You would not have expected a small cannery like Collondale to respond as ruthlessly as it did when the two largest pineapple canneries in East London were negotiating with AFCWU.[6] However, the majority of bosses in East London had resolved to adopt a common front, in terms of which they would quash any attempts by unregistered trade unions to organise their workplaces. The proof of this, it seems to me, was a well-publicised meeting that Minister Fanie Botha himself held with the captains of industry in East London, where he urged them to do precisely that.[7] Botha's new job title was Minister of Manpower Utilisation.

I would be sitting in the office in East London as workers began trickling in next door, where SAAWU had its office, to report (yet again) that an entire workforce, or an entire shift, had been dismissed. I do not remember any getting their jobs back. There were some bosses who broke rank and recognised SAAWU, such as Chloride, but for the most part their common front held. It was not clear whether SAAWU would survive this onslaught.

Collondale, so far as we could establish, was a private company, owned by one of the local captains of industry, Corder Tilney. If it had any links with a larger group with which we might have had purchase, they were not visible. I phoned Tilney to try to set up a meeting. He was not prepared to meet with the Union, he said, but eventually offered to meet with me at his farm: man to man, as it were, or white to white, or factory boss to union boss. This was not our policy, I told him, but would he at least meet with us so that we could explain this policy to him? I flew to East London without getting a clear answer from him, to meet about having a meeting.

Tilney also owned a pineapple farm, or rather a suite of farms, on the road to Kidd's Beach, a small seaside resort where my mother used to go on holiday as a child. From the Union's office in the old part of town you would take the bridge over the Buffalo River, now named after Steve Biko, and pass through the factories of the West Bank in the direction of the airport. Collondale Cannery was just on the other side of the airport. Beyond that the commercial farms start, and the turn-off to Kidd's Beach. A little further on was the barren, over-grazed territory of the Ciskei, dotted with huts and pot-bellied toddlers.

Someone from the white Left gave Sisa Njikelana and me a lift. Sisa was an official of SAAWU, and had a hand in organising the plant on our behalf. We were dropped off next to the neatly ordered rows of pineapple plants at the turn-off to the farm. We walked the last stretch ourselves and knocked on a door marked 'office'. Tilney looked up when we entered but did not offer us a chair, after we had introduced ourselves, and remained seated while I tried to find some hook with which to draw him into a process of negotiations.

'I had said I would meet with you,' he said, addressing me. 'I did not say I was going to meet with a delegation', referring to Sisa. I tried to engage him on the humanitarian consequences of 400 workers and their dependants losing their livelihoods, but he was not biting. He was simply not prepared to discuss this. It was at this point that his secretary interjected. She was sitting at the far end of the office and had evidently been listening to what was said.

'Mr. Tilney is a good man,' she said passionately. 'Ask him to show you what he keeps beside him at all times', she challenged us, 'it is there in the right-hand drawer of his desk.' We did not respond to the challenge, but

Tilney obliged her by reaching down to open the draw in question. I was half expecting him to produce a gun. Instead, he took out a black leather-bound Bible, which he clutched meaningfully. We left without anything we could report back to the workers: not even Tilney's justification for dismissing them.

Njikelana had trained as a teacher, he told me while we waited by the pineapple fields for our lift back. However, the only posts that were available were in the Ciskei, and you did not get on the payroll of the Ciskei government unless you were acceptable to the Sebes: Lennox, his brother Charles, and their cronies.[8] Now he was working as an organiser for SAAWU. Like Gqweta and Welile, our branch chairperson, he belonged to an intelligentsia. The calibre of the East London intelligentsia made organising there different from any other urban centre. It was far closer to the working class than its Soweto-based equivalent on the Reef, and far more open to collaborating with the white Left than the intelligentsia rooted in the teaching profession in Cape Town.

I stayed with Sisa that night in Mdantsane. He did not stay in one place, out of fear of harassment from the Ciskei special forces, and took me to a stand-alone structure adjoining someone's house. It was made of the cement blocks that township houses are usually made of, and had nothing inside it but the two beds on which we slept. The person whose house it was locked us in, from the outside, in case the special forces came in the night and tried to abduct us. The door was metal, and I might have preferred to take my chances with the special forces, but was not asked. As a white in the townships, you had, as always, to trust the judgement of the people you were with.

Independence

There was a vast gap between what workers in the canneries and in other 'low-wage' industries earned and the higher-tech plants of East London, like Mercedes-Benz: greater, I suspect, than in any other industrial centre in the country.[9] Competition for places in the higher-tech plants must have been intense. The only other prospect of material advancement for the relatively well-educated African was to get onto the payroll of the Ciskei government.

Mdantsane graphically illustrated the absurdity of having a separate government for what was to all intents and purposes the township of East London, and a resolutely self-seeking type of person was required to partner government in a fraud like this. The Sebe brothers fitted the bill nicely. It was not difficult to see why an intelligentsia would become radicalised in this context. Perhaps it was inevitable in the circumstances that SAAWU would increasingly assume the role of the official opposition.

As far as the Sebes were concerned, AFCWU was simply an extension of SAAWU, but to me the differences were becoming clearer. The SAAWU members were mostly from the higher-tech plants: men in pressed jeans and fashionable tops, who used English and isiXhosa interchangeably. The Collondale women wore the long dresses and traditional headdresses of rural Xhosa, and some had their faces painted white. There was not even someone among them who could translate what I wanted to say. It would have been rural people like them who had been employed as scabs, to replace the striking workers at Collondale. It would also be people like them who would soon be herded into some stadium to applaud Ciskei independence.

That made it all the more important to ensure they were properly organised, but it seemed that some in SAAWU regarded them as country bumpkins, and that the leadership had not absorbed any lessons from AFCWU about proper organisation. Collecting subs was obviously not something they considered important, because they relied on donor funding.[10] Also, although it had the structure of a trade union federation it did not have any affiliates. Instead it organised all kinds of factories generally into one big union, without regard to the industries in which they were engaged.

The head office of SAAWU was supposed to be in Durban, where it seemed SAAWU did operate as a federation of industrial unions. Gqweta told me that its food manufacturing affiliate, which was called the Food and Allied Workers' Union, wanted to transfer its members to AFCWU in accordance with the agreement to collaborate entered into between SAAWU and AFCWU in East London. I had no idea what to expect when Gqweta and I flew to Durban to discuss the transfer.[11]

We were fetched from the airport by Sam Kikine, the general secretary of SAAWU, who was accompanied by Griffiths Mxenge, an attorney from King William's Town, now practising in Durban. Kikine was a short man with platform heels and the gift of the gab. Stopping at the barrier on our way out of the parking area, he rolled down the driver-side window and made a voluble attempt to recruit the parking attendant into the Air, Sea and Road Workers' Union. This, I later found out, was one of the 'affiliates' of SAAWU, and his behaviour showed how he operated. Griffiths did not talk much but listened intently to what I had to say on our way to the house of an Indian comrade where they had arranged for me to be put up. Evidently he had come to size me up. About a year later he was brutally murdered.[12]

If we were to be a truly national union, we needed to establish a branch in Durban.[13] It would be difficult to do so without local political support. Kikine seemed to have that, so I reserved judgement, and had a close look the following day at the records of the members he proposed transferring. They were located in a string of plants in Mobeni, the industrial area south of

205

the city. Apparently workers from Mobeni had been prominent in the 1973 Durban strikes, but now it seemed the factories were mostly unorganised. There were not more than five members at any one plant. Kikine had not thought to consult with them about transferring them to another union. We agreed to leave the matter in abeyance until he had done so.[14]

The Collondale workers had been out of work for some three months when I told them that the head office could not keep on paying them strike pay any longer.[15] The workers understood this, but when I suggested they accept the strike was over, because there was nothing more the Union could do for them, they would not hear of it. They resolved not even to collect their arrears wages at the cannery.[16] It would be more than a year later that I accompanied about 60 workers to the factory to collect their wages. The other workers, it turned out, must have gone to the factory individually to collect their wages long before then.[17]

The same kind of thing happened at the SAAWU plants. The spirit of no surrender was contagious, and none of the SAAWU members who were dismissed were ever prepared to admit defeat in meetings, even if it was clear there was nothing their union could do (or was doing) to get them back into the factory. Once the meetings were over, the dismissed workers would vote with their feet, as workers have to do, and accept whatever jobs they could find elsewhere. Consequently there was no opportunity for the union to regroup or rebuild. Within six months of his speech to the captains of industry in East London, Fanie Botha was to change his line on the unregistered trade unions.[18] If the strikes were a factor, it was not because they had been successful. Every workforce that was dismissed with no apparent repercussions for the bosses represented a setback for organisation and strengthened the hand of the bosses.

Gqweta did not seem to see this, because the wave SAAWU was riding was too alluring. If you had chanced into one of SAAWU's general meetings, it would have been obvious why. Meetings were not possible in the Ciskei, so SAAWU took the bold step of hiring the City Hall, an old building from the era of the Cape Colony. From the stage Gqweta, Sisa and the other leaders looked down onto a wood-panelled interior packed with men, singing in unison of the imminence of change. The trumpets outside the walls of Jericho could not have matched an assembly of Xhosas singing.

Recognition
I was hoping that in time the differences of approach that had emerged with SAAWU would be overcome, and that SAAWU would come round to adopting our approach. However, the leadership of the branch, I noticed, did not share my optimism. They realised long before I did how deep-seated our differences were. Although these differences were not aired at that time,

it was clear they related to the political role of a trade union, as well as the fact that its leadership was not accountable to workers.

For SAAWU, as I saw it, the political role of the trade union was to mobilise workers on what I call the political front, in support of the Congress alliance and its proxies. Although this could not be openly acknowledged, the basis of our informal alliance with SAAWU was that we were on the same side. For the Union, however, the proper organisation of workers itself represented an important and necessary political task.

Proper organisation had to start in the workplace, and a critical component of this task was to prevent a situation in which Africans in 'higher-level skilled jobs', once organised into trade unions, would lord it over the ordinary workers in the same way as whites and coloureds in the established union movement had done and were doing still. This is the situation which I believed the government's labour relations reforms were calculated to bring about.[19] The issue that brought the difference between the two unions to the fore was recognition. Recognition, as we understood it, entailed a commitment to negotiate with a trade union. The agreement that was the outcome of these negotiations was confirmation enough that it was recognised. We could not see the point of a separate recognition agreement, which is why we had not sought one when we settled the Fatti's & Moni's strike.

A matter of months after the strike was settled, Fatti's & Moni's had tested the Union's resolve by proposing to retrench workers, but we told them that this would be a breach of the agreement, and they backed down.[20] For our part, it was critical that the Union be seen to abide by the agreement, so we honoured the undertaking not to pursue any wage demands for a year. And the returning workers honoured their undertaking to the Union to organise the unorganised workers in the Bellville factory. When the settlement agreement expired, only the members of the liaison committee and a few of their hangers-on were outside the Union.

The liaison committee included people like Christian, the foreman who had eaten the workers' money, and Skilpad, who had led the breakaway from the strike. Keeping the liaison committee going was a way of telling these workers and others who inclined to follow their example, 'So what if there is a union in this factory: we the management, will stand by you.' Now, at both Isando and Bellville, workers were demanding the liaison committees be disbanded, and for a time it seemed that Peter Moni had been deliberately avoiding meeting us to discuss the issue.[21]

Not so, Peter Moni assured us, when he eventually met with a delegation from the Union. Fatti's & Moni's was 'working towards recognition of the Union'.[22] All the company had to do, we replied, was to respect the workers' wishes as to who to represent them. But Moni had something more elaborate in mind, it emerged a few weeks later, when we received his letter enclosing the draft of a 'formal recognition agreement'. This, he explained, would 'clearly be to the benefit of both parties as our relationship will then be firmly established and defined'.[23]

'It is agreed and accepted that both the management and the union have a common objective in ensuring the efficient running and continued prosperity of the company ...' began a page and a half of 'general principles'. Certainly it was in the interests of the Union's members that the company kept on running. But you could only view the 'continued prosperity of the company' as a shared objective if you thought there was no conflict of interests in the labour relationship between workers and their bosses.[24] How could a trade union be expected to endorse such a position, you had to wonder.

The only 'principle' to which workers could relate was the so-called freedom of association. This, as I have said, was what the Fatti's & Moni's strike had been about: the right of all workers to associate, regardless of race. It was a right they had won. But I very much doubted the bosses had the same understanding of this principle, and the way they framed the right was calculated to limit its application: to protect the right of the individual *not* to associate, and to prevent the Union from imposing membership as a condition of employment. As it happened, the Union did not support the closed shop. But if we were not asking for a closed shop, why should we have to enter this debate at all?

The Union was not yet ready to concede the need for a separate recognition agreement, and we countered with a draft of our own, focusing on wages and conditions of work. The first issue to be overcome at the negotiations that followed was whose draft would take precedence.[25] Moni's justification for preferring his draft was that before negotiating wages 'we need to establish the "rules of the game"'. It was clearly intended to be a long game. Foisting the company's own ideological premises on us was only the start of what was bound to be a process of trying to incorporate the Union into their chain of command in the workplace. Their draft pretty much affirmed management's right to carry on as it had done before, including a statement that 'the parties to this agreement accept that the mobility and flexibility of labour are desirable for the achievement of the company's objectives'.[26]

The compromise eventually struck was that we would negotiate 'their' draft, provided that at the end of the three days we had also arrived at a wage agreement. So negotiations proceeded by a protracted process of stripping away from the company's draft, taking us through the night into the early hours of the morning of the fourth day. What was left was nothing we particularly wanted, but also nothing we particularly minded.

At about the same time as these negotiations were taking place, SAAWU met with Chloride and concluded an agreement within a matter of hours.[27]

SAAWU had accepted with minor variations the draft Chloride had presented them, and saw nothing wrong with provisions to which the Union had objected strenuously. Some of these provisions were identical or very similar to the Fatti's & Moni's draft. Perhaps this was because Chloride relied on the same consultant or the same precedent. I saw it as evidence of a concerted strategy on the part of the bosses.

Establishing labour rights

Fatti's & Moni's accused the Union of wanting to make up the rules as we went along. This was not an unreasonable thing to want to do, from the workers' point of view. The 'game', to adopt the Fatti's & Moni's metaphor, had only just started. The only rules the workers can be expected to respect, we argued, are rules that the workers understand and see the need for, based on their own experience.

But the labour relationship was anything but a game, of course, and what a trade union agreed to in these negotiations would have longer-term implications, or the bosses would not be putting so much time and energy into them. What does the phrase 'the mobility and flexibility of labour' actually mean, we asked, and why should the Union accept it as a desirable objective? Fatti's & Moni's ummed and ahed, and could not give us a coherent answer. It was obviously a phrase that emanated from some employer's think tank, and represented their aspiration about how the labour relationship should be structured.[28] Eventually they agreed to drop it.

Over the next few years I was to hear the same tired metaphor about the 'rules of the game' from a succession of personnel managers, some of whom were in the process of redefining themselves as human resources (HR) practitioners. All of them would argue from the standpoint that the rights workers have are conferred by the text of the agreement. That was also why they did not want to negotiate wages without a formal recognition agreement. It was an ideological reason. They did not want to accept that workers are capable of establishing rights in the workplace through their own organisation.

Our standpoint was that labour rights are won primarily through organisation, and as a general proposition, the less said in an agreement the better. The 'strikes and lockout' clause we eventually agreed on with Fatti's & Moni's still had a few redundant words in it, but had been whittled down to a sentence: 'Both parties reaffirm their fundamental belief in dialogue, discussion and negotiation as being the method of conducting industrial relationships in all circumstances.' It exemplified what I would call the minimalist approach to a formal agreement.

A clear example of how rights are won through organisation concerned the level at which we were bargaining. Undoubtedly Fatti's & Moni's would rather have been negotiating at plant level. Were there plant-level agreements, it would be easier to shift production between plants and realise the 'mobility and flexibility' that was evidently becoming so important for the bosses. The only reason we were negotiating at a company level (with more than one plant of the same company) was that the union was organised nationally.

Interspersed with the wrangling about recognition, an agreement on wages had been comparatively straightforward. General workers were to get R40 a week in the case of women and R45 for men. Although the gender differential was problematic, we had accepted that it was justified on occupational grounds: the work the male general workers did was, by any definition, extremely taxing. There was also no system in terms of which the different jobs were graded, and we had not had time to devise our own proposal for grades of work.[29]

The symbolism of a R40 minimum was not lost on either the workers or the company. The workers had achieved their demand some 18 months after their employers had refused even to entertain it. Even though the cost of living had risen in the interim, I myself had not expected us to get there so soon, and thought it politic to be generous in our press statement. The agreement reached 'put Fatti's & Moni's in the vanguard of progressive employers,' I said, although I doubt this was where it wanted to be. I also doubt whether the media were interested any longer with what was happening there.

This was not our first company-level negotiation outside the official system – we had already negotiated at company level with SAD. But in dried fruit, SAD had no competitors. There were serious competitors in the milling industry, as Fatti's & Moni's was at pains to stress, and the danger of negotiating at company level was that the Union inadvertently advantaged unorganised competitors or competitors that refused to recognise trade unions, like Collondale. The only solution seemed to be industry-level negotiations, and this is what we now proposed in the case of the two pineapple canning factories in East London. It would have been short-sighted to attempt to play the one off against the other.

We met with the bosses of WPP and Langeberg in a beach-front hotel.[30] Some consultant must have warned Langeberg days before of the perils of negotiating wages without a recognition agreement, and Langeberg management got the wrong end of the stick. Their opening gambit was that they wanted a 'negotiation agreement' rather than a wage agreement, but soon realised it was far too late to raise this kind of objection. So it was that AFCWU concluded its first industry-level wage agreement without having negotiated a recognition agreement.[31] So far as I know, it was also the first such agreement negotiated with an unregistered trade union.

However, the date on which Ciskei was to become independent was approaching, and the situation in Mdantsane was deteriorating. The day before the negotiations were due to begin, Welile, the chairperson at Langeberg, was detained by the Ciskei police, together with a number of people from SAAWU. A week or so later Van Graan, the president of FCWU, and Noko, the AFCWU vice-president, together with Kallie, a member of the management committee from Tulbagh, were travelling between Mdantsane and East London when they were detained, along with a cohort of young men, including members of AFCWU and SAAWU.[32] For the young men from East London, this kind of harassment was nothing out of the ordinary. When they were all released after a few days, they were buoyant and joking. However, I suspect the two coloured men (more than Noko) were deeply shaken. Van Graan was then 60-something. Not long afterwards Kallie was promoted in the factory and lost interest in the Union.

17

Malta Road

'ARE YOU HOLDING DISCUSSIONS with other milling companies?' Mr Moni asked us with great seriousness during our negotiations.[1] We always paid close attention to these kinds of remarks. Although we did not accept everything the bosses said at face value, it was in their interests to get us to understand enough about their business to buy their arguments. Sometimes the only way to check out a particular proposition was with another lot of bosses, in another set of negotiations. In this instance we knew what he was getting at.

Nowadays, the way the bosses like to explain their businesses is in terms of a 'value chain', which begins on a farm (in the case of foodstuffs) and ends with the consumer buying the product in a retail store. This is to focus on a vertical relationship between entirely different kinds of business, rather than the relationship between businesses of the same kind. It seems to me it is also a consequence of the way in which the bosses have restructured their different operations. This began at about the time I am now talking about.

At that time, the horizontal relationship between the different manufacturers was far more important, at least in food manufacturing factories. In fact, in the milling industry, as we had learned during the Fatti's & Moni's strike, the vertical relationship was of secondary importance. The price of wheat was controlled, and so was the mark-up the millers charged on their flour price, as well as the price of bread.[2] The 'control board' for this industry was part of a network of control boards the government had appointed to control the marketing of most agricultural products.

As a consequence of price controls, we did not need to concern ourselves with what the farmers were paid or the price at which bread or flour was sold. On the other hand, the bosses could not compensate for an increase in wages simply by increasing the price of their product. If we pushed Fatti's & Moni's further than 'other milling companies', who were not organised, we might place our members' jobs at risk. Once we began to organise these other companies we realised how vulnerable the position of Fatti's & Moni's in the industry had all along been.

This was because a significant amount of capital was required to establish a mill, and milling was dominated by large conglomerates for this reason. Foremost among them were Premier Food and Tiger Oats, which in turn was part of the largest manufacturing conglomerate in the country, the Barlow Rand group. These conglomerates had extensive holdings in other sectors of food manufacturing, notably the fishing industry.[3] Indeed, ownership in food manufacturing was probably more concentrated than in any other sector of manufacturing.[4]

Premier, as it was known, has a mill in Salt River, halfway along a band of industry located between the railway lines from the city centre to the northern and southern suburbs: it starts at Woodstock, where I&J, another conglomerate, had its flagship factory, and runs through to Maitland. To get to the mill from the Union's head office you would take the lower of the two 'main' roads, running parallel to the Southern Suburbs line. Past the traffic circle you get Malta Road, running through the heart of Salt River, and a stone church. Then you come to Attwell's Bakery, on your left, and the bridge over the railway line.

The mill is besides the bridge, on what must have been the banks of the river that once gave Salt River its name, but that now finds its way via stormwater drains and covered canals into the Liesbeek. The South African Milling Company, it was called. I would imagine there was no flour mill in the country to match it when it was built. The interior of the mill was more like an ocean liner than a factory, with wooden stairs and gleaming brass fittings, and chutes of canvas framed in teak.

The entrance to the mill is under the bridge, and the organisers would wait for the workers by the pillars, at lunchtime or after work, to invite them to meetings at the church hall. It was the same strategy we adopted at I&J at about the same time, except that the meeting place for I&J was the Anglican church hall in Woodstock, and the entrance to the factory was much more exposed: it was safer to wait at the subway leading to Woodstock station, rather than in the view of supervisors and foremen.

The two workforces reflected the difference between a labour-intensive operation and one that was not. Bar a handful of women cleaning offices and working in the canteen, the workers at SA Milling were African men, and there were fewer than 200 of them. Almost all of them were contract workers from the Eastern Cape. Our selling point was not so much the wages we had negotiated with Fatti's & Moni's as the fact that the Union had been able to get contract workers reinstated in their jobs. A majority joined and were paying subs when we wrote to the company early in 1981.[5]

There were more than 1,000 workers at I&J Woodstock, coloured women from Bonteheuwel and Retreat and other gang-ridden areas on the Flats, as well as closer at hand, from places like Kensington and Woodstock. The task of getting a majority to fill in application forms and paying subs was, by any standard, daunting.

Verification

We received a polite and timeous reply to our letter from SA Milling, and before long were in a meeting with Mr Wolly Wolthers, the joint managing director of Premier, who came from Johannesburg to discuss how they would go about verifying our membership. It was unprecedented for us to be having a conversation about something so mundane with this level of management, and an indication of the sensitivity with which it regarded relations with the emergent trade unions, which was in no small part a consequence of the Fatti's & Moni's strike.[6] It was also because, it would later transpire, Premier saw itself as being in the vanguard of progressive employers.

This was doubtless because it had a wheat or maize mill in every major centre and bakeries in every town of the country, as well as plants producing other staple foods. It understood very well that not only its workforce was overwhelmingly African, but the consumers of its products as well. It is also worth noting that it had never been approached by one of the emergent trade unions before.[7] The same was true of its main competitor, Tiger, and of I&J.[8]

A firm of auditors was appointed by Premier to conduct what it termed a 'verification exercise', to confirm that more than half the 'weekly paid' workers were paid-up members, according to the exacting standard we had set.[9] By the time the verification was completed, Premier had appointed its first group human resources manager, Dr Human*. Before he was able to assert his authority in the relationship with the Union, Wolthers had agreed to negotiate wages without having negotiated a recognition agreement.[10]

To make matters worse from his viewpoint, the Union's wage demands were linked to proposed grades of work. At the time, the workers were all paid the same wages, regardless of the work they did. They wanted some acknowledgment of the role they actually played in the production process, which is why grades of work, as we saw it, had to be part and parcel of any wage negotiation. What Dr Human held out for, however, was what he described as an 'objective' job-grading system.[11] Essentially, this meant grading jobs by way of a bureaucratic process, controlled by management.

Other branches in the Union, in the meantime, were taking their own initiatives to organise milling plants. The next breakthrough was in East London, where the branch organised two plants producing animal feeds.[12] One was Epol, which was part of a separate division of Premier. The other was Meadow Feeds, which was part of the Tiger group.[13] Milling, we had in the interim learned, included the milling of maize, not only for human consumption, but as a key ingredient of animal feeds.

It was a barely literate contract worker on the committee at Fatti's & Moni's, Neil told me, who had done more than anyone else to recruit unorganised workers from other plants in the Isando area. Epic Oil was the first to be organised. It belonged to another subsidiary of Premier, producing cooking oils from sunflower and other seeds. By the time the negotiation of the recognition agreement at Salt River began, the Union was organised at SA Milling in Isando, the flagship of its milling division.

The negotiation of the recognition agreement at the Salt River mill took place over more than four months, during which time the Union had verified membership at the other plants, and it was agreed that the agreement would cover the Isando mill as well. At Isando, however, the company proposed a speedier method of verification. Instead of having auditors check on the payment of subs, they would conduct a ballot: for or against the Union.

Speed was not a good idea when it came to proper organisation. We also did not want there to be such a low threshold as to let in trade unions that did not properly organise workers. In the case of the Isando mill, however, workers were already paying their subs by hand when management proposed a ballot, and they went ahead with it, whether in good faith or not, without the Union's say-so. The outcome was a resounding endorsement of the Union. True to the tradition of boycotting any vote about which the electorate had not been consulted, only four out of 400 workers in the factory voted.

Realising it had blundered, management asked the Union if it would agree to a fresh ballot. We agreed. There was something compelling about a ballot in the workplace to determine who was to represent the workers in a country in which, as citizens, they were disenfranchised. Some 85 per cent of the workers voted to be represented by the Union, and from then on the ballot was the method of verification in Premier plants.[14]

Premier had been making up rules as it went along. Dr Human was the first of a succession of managers that it was to hire, both at its group offices and the different plants, to help it do so, but whereas Mr Wolthers was a genial man, who more than any other manager we had encountered was receptive to the Union's narrative, Dr Human had quite fixed views, acquired, I would guess, at some university in the American Midwest, where he had gone to refine his understanding of 'human resources'.

Job grading, it transpired, was one of the central planks of the 'human resources' school of management, because it concerned upward mobility in the workplace, and Dr Human communicated his views in a pedantic, if not patronising, manner. However, he was at least sincere and not devious. The same could not be said of some of his counterparts at the other conglomerates. While the organisation of the milling industry was gathering steam, there had also been progress in organising I&J.

I&J produced frozen vegetables as well as processed fish, and we knew the vegetables were grown under irrigation in Marble Hall, in what was then the Eastern Transvaal, because Table Top claimed this gave them a competitive advantage. What Table Top did not tell us was where their processing plant was located. I asked Neil to find out. What he discovered instead was a plant of I&J in Benrose producing preserves and pickles.[15] The local management was happy to deal with the Union until their head office in Cape Town got wind of this.[16]

I&J, it seems, was not an adherent of the 'human resources' school. It had a group personnel manager, and below him was Mr J, the group industrial relations manager. Mr J wrote us a letter posing a string of questions about the Union. It was obvious from his belligerent tone that they were looking for a reason to delay recognising the Union. In the meantime I&J had invited another union into the factory producing frozen vegetables that Neil had been looking for all the while. It turned out to be in Springs.[17] I&J had been playing for time. If it had to recognise trade unions, it clearly preferred to deal with a multiplicity of them. It was a divide-and-rule strategy.[18]

It was no secret, at the time, that the Union was recruiting at I&J's fish processing plants at the time, and had already secured a majority in the Cape Town docks. This was where the trawlers that supplied the processing plants berthed, but the workers were African and on contract. When their contracts expired, the company simply did not renew them.[19] This was a 'business decision', Mr J assured us, which had absolutely nothing to do with their joining the Union. Business decision in management-speak meant 'none of your business'.

The turnaround at its flagship fish-processing plant in Woodstock happened quite suddenly. After two years of trying, a majority joined, and would stream to the church hall after work each week, before catching buses and trains to their various destinations on the Cape Flats.[20] We met with the group personnel manager and Mr J at our Corporation Street offices, and agreed to verify our paid-up membership the way we were accustomed.

The same management team came to our offices a week later, armed with sheaves of computer printouts, which were checked against the factory number of each worker, as recorded by the shop stewards in the little blue book. It was the first and only occasion we let management itself verify our membership in this way. To our alarm, it soon became apparent that there were many more workers on I&J's printouts than we had dreamt possible. On page after page of their printouts there was not a single tick to indicate a factory number that matched the Union's records.

I looked at these pages more closely and noticed the code 'MBH'. This, I guessed, referred to Marble Hall. The bosses had included printouts from an unorganised factory in another part of the country to inflate the numbers. Mr J flushed red and admitted his 'mistake' when I pointed this out. The group personnel manager feigned ignorance. Also present was an IR manager they had recently appointed at the plant. He was palpably shocked. But I&J had agreed to abide by the outcome of the verification exercise. There was no going back on that.

A new breed of manager
Discretion was the watchword among the bosses when it came to disclosing who owned whom, or to whom managers were ultimately accountable. We had to work out for ourselves that Oceana, which owned the fishing factories on the West Coast, in turn belonged to Tiger, or that I&J was part of the Anglovaal group. At about the same time we attained a majority at I&J, with the help of East London branch, we organised Table Top's other competitor, Land Harvest.

The Land Harvest factory was in Port Elizabeth, and verification in this instance took place by show of strength.[21] Pendlani and I were in Port Elizabeth trying to set up a meeting with management, when we were hastily summoned to the factory. The workers refused to go back to work after their tea break until we had addressed them. I have a vivid memory of being escorted into the factory on a grey winter's morning by young Bernard Mahlakahlaka, whom they had elected as chairperson, and being welcomed by ululating women in gum-boots and a multitude of coloured overalls.[22]

When we found out that Land Harvest was owned by another conglomerate, Imperial, we were able to put two and two together.[23] Imperial also owned a half-share of Sea Harvest. The clue had been on its letterhead all along, on which a company was obliged by law to list its directors. The chairman of the board was also the chairman of Old Mutual.[24] Old Mutual, the insurance giant, provided financial backing to Barlow Rand, and both Tiger and Imperial were part of the Barlow Rand group.[25]

Ostensibly Premier was a South African company. It traced its origins to a family-owned business, running trading stores that supplied the mines, as did Tiger. In the case of Premier, it had been the Bloom family business, and like a farm handed down to the eldest son, the family connection had been preserved in the person of its chairperson, Tony Bloom, a suave man with liberal views. We had to find out for ourselves that a UK-based food conglomerate, Associated British Foods, owned half the shares of the group.[26]

Trade union organisation was to expose differences between the conglomerates over the extent to which plant managers could be regarded as being their own bosses, which none of them of course were. A related issue over which they also differed was the Union's drive to centralise negotiations and standardise agreements. Compelling the bosses to behave more rationally, and regulating the terms on which they competed with one another, was part of a trade union's social mission, as we saw it.[27]

These differences corresponded to the different political stances of the conglomerates. Wolthers would adopt a somewhat conspiratorial tone whenever relations with government came up in Premier's dealings with the Union, as though we were both on the same side of the political fence. Its stated aim was to keep the government at arm's length. Tiger's position was to deny it had any position at all. This boiled down to an apolitical stance, coupled with a dogged insistence that our relationship with management was at a plant level, and that the head office could not tell plant managers what to do. Other companies in the Barlow Rand stable adopted the same approach.

The different political positions of the bosses no doubt influenced whom they appointed to assist them in their dealings with the trade unions. Wolthers was a practical man, with a background in accounting. The nearest thing to an intellectual in the ranks of management would be someone like Dr Human or Mr J. These were part of a new breed of managers, who were young and university educated. They were not union-bashers as much as people who liked to think they 'understood' the unions, and why people like me had been drawn to the unions. Their job was to anticipate what the unions would come up with. It was also to talk down the more reactionary factory managers, and encourage a move away from apartheid-style controls. This new breed of manager saw the need for management to become racially inclusive, and believed government was not moving quickly enough. It was also obvious that they believed that negotiating with an organisation whose members were Africans was symbolically important. For the Union, however, the question was whether they had the power to impose changes in the workplace.

On this point, there was unanimity among the bosses. Whatever their title, this new breed of managers were expected to behave as though they were consultants, whispering in the ears of the 'line manager', as they now preferred to call the plant or factory management, or passing them a stream of notes. 'We are only here to advise,' they would say, when their role was queried. 'We have no power to make decisions on our own.'[28]

The draft recognition agreement Premier sent us for SA Milling covered much the same ground as the one Fatti's & Moni's had proposed, and by the time negotiations were concluded, some four months later, negotiations were under way with plants belonging to the other conglomerates: with Tiger in East London, with Imperial in Port Elizabeth, and with I&J in Cape Town.[29] In each case, the content of the employers' drafts, and even the style, were similar. Identically worded phrases would crop up time and again. The bosses were clearly attending the same seminars and drawing on the same pool of expertise.

The trade unions, however, were not. In order to give effect to its minimalist approach to recognition, the Union now had its own standard draft. It strongly affirmed the autonomy of the Union, because it would only be possible to maintain democracy in the workplace with the backing of an organisation that could not be manipulated by the bosses or, for that matter, the state. We could see from the bemused reaction of the bosses, however, that other trade unions did not share these concerns.

In East London we deadlocked over the election of trade union representatives, which for us concerned how we conducted our own affairs and which was an issue of autonomy. However, SAAWU had already conceded to elections being held in terms of the agreement and being subject to management scrutiny.[30] It was an early example of how the lack of a common approach among unions undermined our bargaining position.

The language question
The outcome of the different agreements that were being negotiated by ourselves and other trade unions was the evolution of a labour relations system for the workplace parallel to the official system. But if the art of negotiation was to seem more reasonable than the other side, in wage negotiations the terrain was rands and cents, and you knew how many workers stood to benefit from the trade-offs. In the case of recognition the terrain was rules and principles and the written word. The trade-offs, if there were any, were far from obvious.

'The company recognizes the union as the representative of the company employees who are members of the union,' the bosses would routinely state in their drafts. It was this kind of play on words that exemplified the problem. 'We know we are representative of our members,' we would routinely reply. 'We do not need an agreement to tell us what is self-evident.' The issue was that if a trade union was required to prove it represented more than 50 per cent of the workers in order to be recognised, as was invariably the case at that time, then it should be recognised as representing all the workers in that workplace.[31]

'That is not at all what was intended' is another phrase we would routinely hear from mouths of management, referring to this or that phrase in their draft, to discount what were often realistic concerns that workers had as to what it actually meant. Usually the workers present did not have a command of English to argue the point. This highlighted another problematic aspect of the negotiation process. It was partly a problem of translation.

Premier was the only company that laid on translation as a matter of course, and their translator of choice was Mr Mlokoti, whom Wolthers referred to as Clarence. Clarence's job, someone told us, was to 'put out fires' wherever in the group they broke out, and he bore himself with an air of chiefly authority. As a translator, however, he was apt to soften the stark terms in which disagreement was expressed to the point of misrepresenting what was being said. Athalie's command of isiXhosa was by now good enough to correct him, and he had the good grace to accept this.

The other companies did not provide translation, probably because there was no African in the management hierarchy that the bosses trusted to put their position. In East London, Tiger brought in a white worker, an artisan. It was not uncommon for whites in the rural Eastern Cape to speak isiXhosa, but it was a competency you did not trumpet in the apartheid era, lest it betray your lowly class origins. He soon proved to be at a loss for words, because he did not understand what we were getting at.

This was not always obvious. It took me a while to put my finger on the fact that whatever concessions the bosses made were more often than not framed in a way which asserted the authority of management. So instead of saying 'general meetings may be held with the permission of management', they would say 'no general meetings will be held without the permission of management'. This was even clearer with the procedures they proposed. These invariably had lots of stages, with each stage corresponding to a level of authority in the management hierarchy.

Language, of course, shapes how people perceive reality and, in this instance, how workers understood the labour relationship. At root and base the problem with the process was that the bosses, through their more sophisticated command of the English language, coupled with the fact that they were in a position of power over the workers, tried to impose their own version of the labour relationship on them. It seemed as though only someone with my kind of education would be able to expose this. Since every clause that compromised the Union's position would set a precedent for others, I felt obliged to be personally involved in each negotiation.

Since negotiations were now taking place all over the country, I was in an aeroplane every other week, if not every week, and under increasing pressure of time. This in turn compelled me to be more comprising, and doubly so in the case of Land Harvest in Port Elizabeth, where we had agreed to commit ourselves to reaching an agreement in one day, which was itself a compromise that would inevitably force us into compromises we would otherwise not have been willing to make. What made the occasion the more poignant was that Liz Abrahams, who had been a general secretary without the benefit of a university education, accompanied me.

We travelled by car and spent the night in Zwide township in the homes of workers. The negotiations began very early in the morning at the factory, which was located in a new industrial area close to Motherwell township, which was then in the process of being established. There was not a café or shop in sight. So we were wholly dependent on the bosses for food and drink, or we had to travel some distance to fend for ourselves.

I had no problem accepting the sandwiches and tea they provided, nor did the committee. But Liz refused to touch any of it. You had to respect her steadfastness, particularly since she did not seek to draw attention to it in any way. On the other hand, we needed sustenance to focus on the issues at stake, as the negotiations drew on through the afternoon and into the night. Liz had almost nothing to contribute on the issues themselves, and I wondered what she was thinking. For me, her silence signified how far the Union had moved from the time when she was general secretary.

A parallel dispute resolution system

If I ask myself now whether the inordinate amount of time and energy we put into these negotiations was worthwhile, defending a particular line, I would say the procedures we established made a difference, not only in the day-to-day in the workplace, but in establishing certain labour rights.

'What you need', we would tell the bosses (and this was our killer point), 'is procedures that actually work, and are seen to work.' We knew what worked, because we had been involved in more major disputes than many, and many of them had been resolved. The bosses knew this, and we had more success in persuading them to adopt our procedures than in respect of other provisions.

The procedure of most immediate relevance to workers was the disciplinary procedure, and our draft was an elaboration of our original settlement agreement with Fatti's & Moni's: a system of warnings, culminating in a final written warning, before dismissal could be contemplated, and with very few exceptions.[32] The bottom line was that no worker could be dismissed without the Union being consulted beforehand. This later evolved into 'no dismissal without a hearing'.

All the conglomerates we dealt with agreed to this. Employers could no longer dismiss at will, and this represented a major gain. As the same conglomerates entered into agreements with other trade unions, a right not to be unfairly dismissed became established in practice. This was, as a matter of fact, well before the Industrial Court recognised such a right.[33] So far as we were concerned, it was thanks to the workers of Rainbow, Fatti's & Moni's, Growers and elsewhere who had put their jobs on the line to defend their fellow workers.

This was why we could not accept that dismissals could not give rise to a dispute, another hotly contested issue in what we called the disputes procedure. Disputes, Premier argued, arose out of negotiations. Discipline was non-negotiable. It could only give rise to an appeal. All the other conglomerates said the same.[34] It was as if they were at pains to deny that the right they were in the process of conceding, the right not to be unfairly dismissed, was due to pressure from below.[35]

The solution for disputes that the bosses pushed very hard for us to accept was compulsory arbitration. This meant appointing a 'third party' to determine the dispute, usually after a first attempt to mediate a solution.[36] We did not have a problem with mediation, although we argued for joint mediation, where each party appoint its own mediator, and the two jointly attempt to resolve the dispute. This was of course based on our experience at Fatti's & Moni's. We could even accept arbitration where it was voluntary. However, we would not agree to compulsory arbitration under any circumstances.

This was because, linked to the dispute procedure, was an undertaking to which Premier and others gave the portentous title 'Labour Peace Obligation', but which we preferred to call 'strikes and lockouts'. In its bare outlines, it was that the employers would not lock out, and the trade union would not go on strike, until the dispute procedure had been exhausted.[37] If compulsory arbitration was the last step in that procedure, it would mean we could not strike at all, since the outcome of arbitration was supposed to be binding. We would in effect have renounced our weapon of last resort, although not in so many words.

The irony, in our case, was that what the bosses were proposing was what the official system required us in any event to do in the industries in which the Union was originally based, where strikes were illegal. It was in fact a replication, in a parallel system, of the most objectionable features of the official labour relations system. What we held out for, and eventually persuaded the conglomerates to agree to, was a parallel system which represented a real alternative to the official system, although time would tell whether this alternative was sustainable.

We did not presume to demand a right to strike in this parallel system.[38] The time was not ripe to do so. There did, however, need to be an acceptance that strikes happen. Where they did, as a matter of course, the Union would 'make every endeavour to ensure that normal working conditions are restored'.[39] But it was not in a position to command workers, even if the workers had not followed procedures before going on strike. Indeed, procedures could not always be followed, and the example we always gave was where workers were victimised. The outcome of the argument, in the case of Premier, was a clause that read as follows: 'Should a dispute have arisen from the dismissal of an employee(s), such employee(s) shall be reinstated until the dispute settlement procedure has been exhausted.' To the extent that rights can be won through negotiations, this seemed an outstanding example. 'Reinstatement', it was clear, meant reinstatement in the job at the workplace. It did not mean the kind of 'reinstatement' that would later be ordered by the Industrial Court, which amounted to ordering the employer to continue to remunerate the employee, and was to all intents and purposes indistinguishable from suspension on full pay.

Dr Human was deeply unhappy about this clause. When we later extended the agreement to other Premier factories, he always tried to change it. 'That stupid clause', I heard him mutter. It will mean, he argued at the time, that workers will go on strike each time a worker is dismissed, to get the dismissed worker back. The workers will only go on strike, we countered, when the dismissal is unfair. You are always asking the Union to accept assurances that managers will behave reasonably, but you are not willing to accept an equivalent assurance that workers will behave reasonably.

The proof of the pudding was that the workers did behave reasonably. Workers were dismissed, not only at SA Milling, but at other plants to which the agreement was extended, and there were no strikes over unfair dismissals until a few years later, when five workers were dismissed at an Epol plant in Pretoria West, as I will presently relate.

'The flexibility and mobility of labour'

Of all the identically worded phrases that cropped up time and again in the bosses' drafts, the one that has the most resonance today is the 'the mobility and flexibility of labour'. Fatti's & Moni's had not been able to explain why they wanted this phrase in an agreement, nor had the other bosses, but it was to some extent self-evident. They wanted carte blanche to utilise labour in whatever way suited them.

This did not just mean carrying on as the bosses had done before there was an effective trade union presence in the workplace, it would transpire. It had to do with the way employment was structured, and how workers were deployed. We had already encountered an early prototype of a 'flexible firm' at SAPCo. What was flexible about it was the way it utilised the seasonal workforce. The distinction between them and the relatively minuscule 'permanent' workforce was rigid, as I have already mentioned.

It was surely no coincidence that SAPCo was American-owned. Now, in the year in which Ronald Reagan was elected President of the United States, a UK-controlled company was trying to get us to accept a similar distinction between a 'permanent' workforce and the rest. As Premier would have it, only permanent workers would be eligible to be members of the Union, as well as the other benefits associated with this status.[40]

'What is a permanent employee?' we asked. It was a 'monthly paid' employee who also belonged to the company's pension fund, we were told. It had nothing to do with a worker's legal status, in other words. Rather, it was a status assigned to you by management, by way of reward. Alongside these 'permanent' workers, doing the same work month in, month out, were 'weekly paid' workers. The Union had already recruited them, and the fact that some had worked for years without being made permanent was a burning issue.

'I have given instructions that the practice of employing workers who are not made permanent must stop,' Wolthers told us to the factory manager's face. The factory manager muttered in his beard, and a week or so later some weekly paid employees were made permanent. Not long after that, more 'weekly paid' workers were employed in their place. As well as 'casuals' employed at the gate, local management continued to practise 'flexibility and mobility' in this way, despite what its head office said.[41] The only difference a clause in the recognition agreement might have made was to make it more difficult for the Union to oppose it.

Another category of worker Premier wanted out of the Union was security. Wolthers was adamant about this. It had obviously been discussed at the highest level. Premier was in the process of establishing its own security company, he told us, which would be a separate legal entity.[42] If the term 'outsourcing' had already been coined at this point, it had not yet filtered down to South Africa. There was also no mention of the rationale usually given for outsourcing, that security (in this instance) was not their 'core business'. In fact, it was precisely because security was 'core' to their business that Premier wanted a separate company. Wolthers was quite frank about this. If there was a strike, the bosses wanted to be sure their property would still be guarded.

So in 1981 Premier had already begun fragmenting its existing workforce. I don't think this was simply in response to unionisation, because during a break in the recognition negotiations, Wolthers passed on a snippet of information that I had cause to remember. My mother had been an enthusiastic patron of a supermarket in Belmont Road ever since it opened, little more than a decade before. Now Pick n Pay had supermarkets and hypermarkets all over the country, and was turning the retail sector upside down with its approach to retailing.

Manufacturers are in dread of Pick n Pay, Wolthers said. You do not tell them what price you want for a bag of flour, and or what it costs to produce it. They are not interested. They tell you what they will pay, and if you don't agree, they remove your product from their shelves.[43] I was later to hear other employers confirm this. The fear of being blacklisted by South Africa's dominant retailer meant they were not willing to risk passing on price increases to the consumer. The balance of power between retailers and manufacturers was shifting.

Pick n Pay's self-serving sales pitch was that it was driving down prices in the name of the consumer. It was also bound to drive down the profits of the manufacturers. The beauty of it was that 'the consumer' was a class-neutral concept. The worker was, after all, also a consumer, and in the UK and Europe it was the working class that bought the sugar-laden product the canners produced. But consumer tastes were changing. At the same time, the industry now had competition from Greece, which had preferential access to the European market.[44]

This was no doubt why Jones had sold out to Langeberg.[45] It was inevitable that this would lead to further consolidation of production and retrenchments. What we did not foresee was which plants would be most affected. I had silently hoped it would be Jones, because the Vakbond still had a majority of the coloured workforce there. Instead, it was where the Union was the strongest: in Montagu, and at Moberg's.[46] Both stopped canning fruit, and it was only a matter of time before they would close altogether.

The canning industry was now in crisis, and the next to close was RFF. Amongst those retrenched was the president of FCWU, Van Graan.[47] This closure, following the closure of Moberg's, might have spelled triumph for the Vakbond six months earlier. But after four years of trying, we had at last won back the majority of coloured workers at Jones, including, to my great surprise, my most formidable (and seemingly implacable) foe, Grootvoet. When the Vakbond did apply for full registration, their lawyer had to invoke free market principles rather than numbers. 'It is accepted in all spheres', it said, 'that competition is to the sole benefit of those who eventually stand to gain from it and that the same principle applies in respect of trade unions.'[48] What he was trying to say, I think, was that it would be beneficial for the workers to have a competitor trade union.

Six months after RFF, Oakglen Canning in Wellington closed.[49] At about the same time, Deepfreezing and Preserving in Firgrove closed its cannery, and retrenched the larger part of its workforce.[50] Although the vegetable-canning part of the industry was not affected by the difficulties on the export market, a year later Jones in Industria also closed.[51] Gants was next. In the space of a few years the industry in which the Union was founded, and a major employer, was reduced to a handful of factories.

It seemed as though there was little but solace the Union could offer the workers being retrenched. But you could also argue that it could have done more to support them, and did not do so because at the same time as it was haemorrhaging in its rural heartland, it was making significant gains in the urban centres. Something had shifted in the awareness of workers in general. With the backing of workers of SA Milling, I&J, Fatti's & Moni's, and others, it had been possible to hold a proper general meeting in Cape Town, the outcome of four years of persistent effort.[52] Cape Town was now set to become one of the most important branches in the Union. The re-establishment of the Port Elizabeth branch a few months later was another landmark.[53]

These urban branches were already far more extensive than anything that had existed in Ray's time, and were grappling with issues of which the Union had no experience. Because of the geographical spread of the organisation, I discussed with Neil and others dividing Johannesburg branch into two, in order to accommodate the growing membership in what was to become the Kempton Park branch. This necessitated the establishment of a Transvaal region, of which Neil was already de facto secretary. What we still had to work out was how to accommodate a regional structure in the Union's constitution. The inner circle in Paarl, of which Liz had been an integral part, was too far removed from these questions to offer advice, and was falling apart. Paarl was no longer the centre of the Union.

18
Off Bhunga Avenue

NEXT DOOR TO THE MILL in Salt River was a bakery that also belonged to Premier. Not long after the wage negotiations at the mill, a bright-eyed young man came to see me at the head office. He was neatly turned out, in a jacket and pressed pants, and clutched a pork-pie hat in his hand: a migrant worker, obviously. 'The workers of Attwell's Bakery have sent me,' he said, and introduced himself as John. I made sure to ascertain his surname, which was Pici, so as not to be 'just another white' for whom African surnames are too difficult to pronounce.

This was the first time a worker had come to the head office asking to be organised and I told him we could not help him. John, however, was a determined man, and this was the beginning of an unusual conversation, stretching over many visits to the head office at regular intervals for the best part of a year. At a much later juncture, John and I would start another conversation that would take place over an even longer period. My point of departure in 1981 was that the bakery was organised by another trade union, or rather the coloured workers were.

The Union had grown from its base in Paarl to become a national organisation without ever being accused of poaching members from another trade union. It could not afford to be accused of poaching now. We had been talking about the need for a united front among trade unions for some time and needed to do something about it urgently. Labour Minister Fanie Botha had announced the law was to be amended, and my mandate was to convene a conference of trade unions which would, we hoped, anticipate what the government was likely to come up with.[1]

The idea of a conference was, of course, a borrowing from our own tradition, in which the conference was the workers' parliament. The question was how to ensure that workers were actually represented there. The Bakery Employees' Industrial Union epitomised the problem. It was not affiliated to TUCSA and could therefore be regarded as 'independent'. I had, however, met its secretary on the occasion of a visit by the general secretary of the International Union of Foodworkers (IUF), to which his union was affiliated.[2] He had intimated that the reason why it did not organise African workers was that the law stopped it from doing so.

'It is a union for the bosses,' was John's view of the Bakery Employees, as he would refer to it.[3] 'The African workers want nothing to do with it.' That might be so, would be my reply, but the Union was still in the process of organising the milling industry. There were about half a dozen bakeries producing standard loaves in Cape Town alone, and we were not ready to take them on. We could also not organise the African workers without also organising the coloured workers. Come back when you have spoken to the coloured workers and heard what they have to say, I told him.

'The women are scared to be seen with me,' John told me a month or so later. It was a familiar situation. The coloured workforce was female, except for a few skilled workers, and the supervisors and foremen composed the leadership of the Bakery Employees. It was one thing for workers to be up against a white manager. It was quite another when it was 'your own' who were in authority over you: more especially when 'your own' claimed to be your leaders.

As was the case with the cones section at Fatti's & Moni's, the women were 'members' by virtue of a closed shop clause in an Industrial Council agreement. John and the African workers knew nothing about the Industrial Council, and I doubt whether many of the coloured women did either, even though all their wages were set by the agreement the Bakery Employees entered into there. This agreement had been extended to all the bakeries in the Western Cape; there were other councils in other regions. Employers had vested interests in this system and would do whatever they could to help the Bakery Employees retain its hold. For me, this was a significant complication. John did not know what he was letting himself in for.

Even if John had known, I doubt whether it would have made a difference. There were a number of factors that explain why John and others like him were so much more confident than the coloured women. The law's promise that African workers could join trade unions might have been one, but it was surely of secondary importance.[4] The factor that seems to me decisive was what he had observed at the mill next door: a model of organisation that worked for workers like him. It was not a model Fanie Botha would have approved of, not so much because it brought together workers of different races, but because it brought together ordinary workers. Government was undoubtedly still hoping only those in 'higher-level skilled jobs' would join unions.[5]

Race was also still the primary basis on which trade unions were categorised when Botha invited trade unions to a briefing about the proposed amendments. There were separate sessions: one for the 'whites only' unions, one for unions with mixed (coloured and white) or coloured membership, and yet another session for the 'black' unions that had applied for registration, such as the FOSATU unions.[6] FCWU was invited to the same session as the Bakery Employees at the Department's head office in Pretoria. Trade unions that had refused to apply for registration like AFCWU were of course not invited.[7] AFCWU nevertheless supported a decision that I attend.

The essence of what Botha had to say had already been leaked to the press, and having separate sessions obviously enabled him to tailor his message for each audience. The government was going to drop the requirement that trade unions had to register in respect of a specific racial group – although the government would also not stop them from doing so.[8] At the end of the briefing, each of the trade unionists present got to shake his hand. He was flanked by his adviser, who was Wiehahn himself. As each came forward, he whispered in Botha's ear. Whatever he whispered when my turn came did not prompt the same beaming smile that others were given. What kind of Theron speaks English, Botha said to me with a scowl. It might have been my Uncle Johan speaking.

Dropping the requirement that trade unions register in respect of a specific racial group meant the goalposts had shifted since the unsuccessful meeting between FOSATU, WPGWU and ourselves in Johannesburg.[9] We therefore could not allow the question of registration to be an obstacle to the formation of a front.[10] This was what I said to David Lewis, who was by now general secretary of WPGWU, our first port of call on questions to do with other unions. David was sceptical about a front with FOSATU, because of the fallout over an article on the registration issue published in WPGWU's name.[11] The union was nevertheless willing to go along with us. So was SAAWU in East London.

There was also some scepticism about FOSATU within the Union. Apart from our disagreement on the registration issue, it seemed to be controlled by a close-knit group of individuals, from the top down. According to Neil Aggett, who was close to members of MAWU, there was disaffection within its own ranks about this.[12] If the front we were proposing was to precipitate a realignment in the movement, leading to a new federation, it needed to be formed from below. Yet there was no bad blood between FOSATU and us at a national level, and it readily agreed to a meeting to discuss our proposal. These talks about talks would be critical if the conference was to avoid being seen as the initiative of any one organisation.

We especially wanted to avoid a repetition of what had happened the last time, when different union groupings held different meetings over the same issues. The lesson we had drawn from that meeting was that the unions in the Black Consciousness tradition would not come to a meeting called by FOSATU. These unions were now affiliated to CUSA, which had a reputation for being soft on the bosses. It was nevertheless important that it be included, not least to prevent one section of the emergent movement being played off against another, as I&J had contrived to do at its Springs plant. This was also the only time our organisational paths could be said to have crossed with CUSA, but we had never spoken to its leadership. So I asked Neil to enlist the help of Emma Mashinini, the feisty general secretary of CCAWUSA, to set up a meeting with CUSA's general secretary. CCAWUSA had until recently been affiliated to CUSA. A few months earlier we had helped it establish a branch in Cape Town.

It was difficult even then to have to wade through the flotsam of elaborate names and acronyms that each wave of trade union organisation would wash up. The CC in CCAWUSA stood for 'commercial and catering' and its members were in the retail sector. It would in time become SACCAWU. To simplify matters for our own members, we referred to the food affiliates of FOSATU and CUSA as Sweet Food and Food Bev respectively. WPGWU, which had been the Advice Bureau, was about to become the General Workers' Union (GWU). To simplify what each organisation stood for, you would have to understand its model of organisation. Although the name of the GWU indicated a general organisation, it was in the process of developing an industrial strategy based on its organisation of stevedores in the ports.

Neil wrote to me about his meeting Piroshaw Camay, the general secretary of CUSA. 'I suggested the meeting could not have the aim at this stage of dissolving all the differences between the various unions but on deciding on certain common positions ... so that the state would be faced with a united position from the unions before it made its decisions on legislation.'[13] This was the line we had discussed we would take, and Camay was present at a meeting in Johannesburg in February 1981 to discuss the issues on which a common position might be possible.[14] There were six in all. Registration, we suggested, should be dealt with under the heading 'trade union recognition and negotiation procedures'.[15] The idea was to de-emphasise it.

'Will the conference proposed be some kind of propaganda exercise, or will it be followed up by some kind of action?' Joe Foster from FOSATU had asked at the start of the meeting. He did not spell out what kind of action he had in mind, but we understood him to mean FOSATU was open to a realignment of the trade union movement and the formation of a new federation. 'We were certainly hoping that something more concrete would emerge,' was my response, 'although even if it were no more than a propaganda exercise it would still have some value.' These talks about talks were for me the start of a process known as the 'trade union unity talks' that would culminate in the establishment of a new federation of trade unions more than four years later.

Each of the organisations present agreed to submit a statement of its position on the six issues, but only FOSATU, WPGWU and FCWU/AFCWU in fact did so.[16] These three organisations would also be the most committed participants in the unity talks; however, if there had ever been a possibility of their going it alone, it had passed.[17] The venue for the follow-up meeting was Port Elizabeth.[18] As I recall, this was at FOSATU's suggestion, which struck me as strange at the time, because in the aftermath of the strike at Ford's Struandale plant a breakaway union had been established. This was the Motor Assembly and Component Workers' Union (MACWUSA), and FOSATU did not want it at the conference.[19] From the outside it looked as if there were now separate unions for separate races in the city's most important industry.[20]

A date for the conference was set, but days after the Port Elizabeth meeting Minister Botha published the amendments to the legislation he had promised. It was no longer a question of anticipating what government would come up with, but responding to a Bill that had been tabled in Parliament.[21] The Bill confirmed that the goalposts had indeed shifted, and trade unions were entitled to have a non-racial membership. This was, in my view, the first decisive step towards dismantling apartheid.[22] It also recognised the right of migrants to belong to a registered union, rather than by courtesy of the Minister. Government could now credibly claim for the first time to have recognised freedom of association.

The Bill also said unregistered trade unions would be required to submit the same documentation to the Department as registered trade unions did. No one expected that, and it was a complete about-face on a position Botha had taken only six months before.[23] There could now be no doubt that it was legitimate for employers to deal with them. This was also another reason why it made no sense for unregistered and registered trade unions to be at odds with one another. The fact that the most draconian controls that the Bill proposed applied to both registered and unregistered trade unions alike made the argument even more compelling.[24] There was one stick intended just for unregistered trade unions. Stop orders for unregistered trade unions would be illegal.[25]

At this point, much of SAAWU's East London leadership was detained. Our differences with SAAWU boiled down to differences over its model of organisation, but Gqweta and Njikelana represented an articulate African leadership that the emergent trade unions needed. SAAWU in East London had also inspired a new wave of unregistered trade unions. Of these, GAWU in Johannesburg was typical. G stood for general and A for allied, to emphasise its intention to organise generally. So the date and venue for the conference were settled as soon as the SAAWU leaders were released. Sydney Mufamadi, the general secretary of GAWU, must have been barely in his twenties when he pitched up at my Waterkant Street cottage the day before, together with Gatsby Mazwi from the BMWU and someone called Monde.[26] Neil had brought them in his VW Beetle and they all stayed the night. Sydney and Gatsby typified an urbane Soweto intelligentsia. I am not sure what they made of the shower in the kitchen.

The venue was in Langa, Cape Town's oldest township, and the nearest to the city, squeezed between the industrial band that ends at Ndabeni and the newer industrial area of Epping. Bhunga Avenue is the main access road from the N2, and leads to what used to be a stop street but is now a traffic circle. Across the stop street, set back from the sandy verges, you get a row of palm trees, unusual for a South African township, and behind that the familiar single-storey township houses. But to get to the meeting you had to turn left. In those days, except for the access roads, township streets had no names or, if they did, the people living there did not know them. Street signs of any kind were a rarity. Nowadays its name is inscribed on the kerbstone: Ndabeni Avenue.

St Francis Centre is part of a complex at the end of Ndabeni Avenue that is dominated by a Catholic Church and a large acacia. It must still have been a sapling when the meeting took place in some prefabricated buildings where people wanting to learn could attend night classes after work. Gqweta, who had a sense of the big occasion, proposed dropping the more mundane term 'conference' and calling it a summit.

'The toiling masses of Africa, SA section'
'Summit' was too self-important a title for the Langa conference for my taste. However, we had defused the issue of registration (we thought) despite a few barbs directed at FOSATU and CUSA.[27] We had also made it clear that we would resist what were seen as controls on trade unions proposed in the Bill.[28] Above all, we had managed to present a united front. Many had been looking for just such a sign from the emergent trade union movement, and the meeting attracted a great deal of attention. As a propaganda exercise, it had some value. Personally, I had hoped for more substance. The only position adopted that went further than the fairly superficial consensus already reached in the talks about talks was one that would soon be abandoned, on Industrial Councils.[29] While it was important to resist 'control' by government, it was no less important to be mindful of what I earlier referred to as 'incorporation', whereby trade unions that ostensibly represented workers became a means of exercising power over them.

The irony was that while rejecting 'controls' by government, the emergent unions were not interested in discussing how the recognition agreements they were entering into might be herding them into a similar kraal. Recognition agreements were instead regarded as badges of accomplishment, to be worn proudly, as though indicating that the union concerned had arrived, no matter how it had got there. Gqweta cut an impressive figure, but I suspect his mind was elsewhere. The attitude of the FOSATU delegation seemed defensive. It included a new recruit, Jay Naidoo, who had been appointed as an organiser of Sweet Food and would before long replace its ineffective general secretary. All in all, there was something staged about the occasion, like a photograph not everyone is happy to be part of.

The one resolution that might have brought a federation closer was that trade unions form 'regional solidarity committees'. It was implemented in Cape Town, but that was only because there were already strong ties of solidarity between WPGWU and the Union. It was not implemented anywhere else. We had presented a united front, but it did not really exist in practice. So while the Langa 'summit' is generally seen as the start of the 'unity talks' that led to the formation of a new federation, it really only raised the prospect of one. This had to be a more inclusive federation than any of the existing ones, representing workers of different races and trade unions from different political traditions.

To understand the halting process that would follow, I must emphasise that it was not only on account of the government and the law that a new federation was important for us. It was also because we were increasingly aware of the fragility of our organisation in areas far from our base. We had grown to a point at which co-operation with other trade unions had become imperative. A series of strikes during 1981 underscored this. Entire workforces had been dismissed before we had even taken stock of the fact that they were being organised.[30] At Table Top in George we were all of a sudden engaged in a desperate struggle to retrieve what had taken years to build, following a mass dismissal in bizarre circumstances.[31] It was a similar scenario later in the year with an ill-considered strike at WPP in East London.[32] The issue these strikes raised was one of capacity, the lack of which could only be overcome if unions pooled resources and co-operated in other ways. If we were not able to co-operate, the inevitable consequence of growth would be a decline in the standard of organisation.

The affiliates of FOSATU were able to co-operate with one another, it seemed, because they had more or less agreed on the same model of organisation. Our experience in East London had taught us that co-operation was not possible without this kind of agreement, but it had become a lot more difficult to reach because trade unions in South Africa were now fashionable.[33] A swathe of people who would have viewed trade unions dismissively in 1976 now saw them as 'relevant'. These included political romantics hoping to be at the scene of the action, careerists looking for opportunities for personal advancement, and colourful characters of all kinds, as well as those who were seriously committed. I would count politicos who believed in the need for a vanguard party as being in the last-mentioned category, including current and future members of the Party. Importantly, workers in 'higher-level skilled jobs' had also begun to see trade unions in a new light.

And all the while new trade unions were being established: some would be bogus, some would have no clear model of organisation, and some would still be developing one. In my book, a trade union with no clear model of organisation had no imperative to co-operate with others. This was one reason why a conversation about the different models trade unions were adopting was never on the cards. Another reason was that it would have forced into the open the extent to which the emergent union depended on foreign funding, which was almost total. Trade unions were competing with one another for dollars. Funders who had no interest in organisations based on solidarity were happy to make an exception in the case of South Africa, because of the Polish equivalent of an unregistered trade union called Solidarity. In Poland, organised workers had been the battering ram that forced a repressive regime to open its gates: a regime installed by the Soviet Union. The irony about this was that unions whose leadership would have been in sympathy with the Soviet Union were being funded by the ICFTU or the international trade secretariats affiliated to it, of which the IUF was one.[34]

So the trade unions that made up the emergent trade unions were in competition with one another at the same time as they were in competition with the established union movement in the workplaces and industries where it was present. Mostly these were in industries in which there were Industrial Councils. The established trade unions in cahoots with the bosses would utilise their position on these councils to try to keep these newcomers out. Up the road from the offices of the Bakery Employees was our prototype of a 'union for the bosses', controlled by a father-and-son dynasty, the Petersens. In a few years activists would establish a new trade union in a bid to wean workers away from the Garment Workers' Union.[35] They would be defeated, as successive attempts to oust the Petersens over decades had been defeated.

If government had been paying heed to anyone when it tabled its Bill, it would have been to the bosses and the established trade unions, especially TUCSA. TUCSA unions were handicapped by being prevented from forming non-racial unions, and so was the Bakery Employees. It showed its true colours when African workers at Attwell's were given forms to sign, authorising deductions for a union they had never joined. This must have been done with the complicity of local management, but probably not its head office, which had already voted with its feet by recognising AFCWU and also acknowledged the justice of our demand to become one non-racial trade union. Premier and others like them must surely have told government its position was untenable. The suddenness of government's about-face showed it felt it had to move quickly.

Government was responding to what was happening on the ground, in the workplace. The law had to change as a result of what had been won through organisation. That is what we believed at the time, and I still do. The demand by FOSATU trade unions for non-racial trade unions must certainly have contributed to the pressures on government from below; but I did not take seriously its threat to challenge the government legally if it did not register its affiliates on a non-racial basis. It was only much later that I became aware of what I regard as a myth: that it was in response to a legal challenge by FOSATU that government changed tack.[36] It is easy to see why those who wished to promote the regulation of labour relations by courts might give credence to such a myth.

None of this explains why, despite the Langa 'summit', divisions between registered and unregistered trade unions became sharper. This had to do with what was happening on the ground, in the community; more specifically the political contestation going on in the townships. The trial of Norushe, our branch secretary in East London, was indicative. After eight months in detention he was brought to court to testify as a state witness against someone accused of belonging to the ANC and was sentenced to one year in prison for refusing to do so.[37] The charges Oscar was now facing, together with a group of Nyanga youths, included murder. It was claimed he had masterminded the crime, but the real objective of the trial, we had no doubt, was to remove someone seen as key in the internal revival of the Congress movement.[38]

The point at which the internal revival of the Congress movement became public was a funeral in which the Union was intimately involved. That Easter, the Labour Party leader from Worcester was killed in a car crash.[39] The management committee met on the day of his funeral and decided to go en masse to Worcester, where we met up with Liz and Virginia, who had been part of the committee that organised the funeral.[40] So had Johnny Issel, who by this time had resigned his post in the medical fund. Johnny was the one who had reviled Hennie Ferrus as a stooge during the Rainbow strike. On arrival at the church I happened to be present when Oom At greeted Liz Abrahams.

Oom At was better known for the sonorous voice he assumed when he opened meetings with a prayer than for his leadership of Ashton branch. He was wearing his Sunday best. Liz was togged out in khaki with green, gold and black epaulettes: the uniform of an ANC volunteer in the golden fifties. So were all the marshals.[41] What, I wondered, did Oom At make of this? Was this the way to draw conservative elements of the coloured working class into the political struggle? The interior was also festooned in ANC colours, as was the coffin. We marched to the graveyard following an ANC flag. The *Sunday Times* later carried a claim from the family that the funeral had been hijacked.[42] I would say it was. On the other hand, someone who had said he would have worn skins for the struggle would surely not have minded.

The internal revival of the Congress movement pertinently raised the question of our relationship with SACTU. There had never been an open discussion within the Union about this, but clearly the federation in exile could bring its influence to bear to support or retard co-operation among unions and the formation of a new federation. I was hoping SACTU would be supportive, for our sake, and because what was happening inside the country was more important than what was happening outside. SACTU was also still my historical model for the federation that I hoped would be established.[43] Then a letter arrived, as if to remind us there were other claimants to the SACTU tradition. It was an invitation to AFCWU from Sam Kikine, the general secretary of SAAWU, to a 'special workers' congress' of unregistered unions. It was to take place in Port Elizabeth, and MACWUSA was to make the arrangements.[44]

'What has to be borne in mind', Kikine wrote in his inimitable style, 'is the fact that the present workers' struggle being waged by the toiling masses of Africa, SA section, is the struggle that prolongs the same workers' struggle waged by the oppressed people of South Africa. In 1919 and 1954 respectively when the first non-racial federation of toiling masses was formed, that is SOUTH AFRICAN CONGRESS OF TRADE UNIONS, sons and daughters of peasants, you are therefore being cordially invited.' Although the congress he was proposing never took place, the intention was clearly to sabotage the Langa 'summit'. This was the last straw for the Union, and soon afterwards we moved our Durban office.[45] It may also have been the last straw for SAAWU in East London, as its leaders ceased any further pretense at being part of the same organisation from about this point.

At the time, I did not think SACTU could possibly have backed an initiative aimed only at unregistered trade unions, which was calculated to be divisive.[46] But the only way I could have found out was by contacting Ray. The last time I had done so was through Henning, the German intermediary. I am not sure I would have risked using him to get a message to her, but the possibility of doing so was now closed to me because, at about this time, I received a visit at head office from a balding man from Johannesburg whose name I no longer recall. I often received unannounced visits from people like him, claiming to be on 'our side'. Generally, I would listen to what they had to say, and while I was doing so on this particular occasion he slipped in a casual and entirely unexpected reference to Henning. Inadvertently, my reaction betrayed that I knew whom he was talking about. I could tell from his response that this flicker of recognition was what he had come for. Afterwards, I heard my visitor was suspected of being an informer.

This was also the period when, every so often, our home phone would ring in the middle of the night. When Athalie or I answered, there would be the sound of protracted heavy breathing at the other end.[47] So when Henning contacted me some time afterwards, and we met in Sea Point, I was more than a little guarded, if not paranoiac. I am sorry, I cannot discuss anything with you, I told him. He could not understand why not. I liked Henning, and did not want to offend him, yet I felt unable to say anything further without exposing myself, and perhaps him, to further risk. I left and did not see him again.

The next time the flag was flown was at an 'Anti-SAIC' meeting, to mobilise a campaign against the elections for the South African Indian Council.[48] We had no inkling this would happen when we decided to send delegates to the meeting, and it was our offices and the offices of the WPGWU that were raided the following week. The Special Branch were armed with a search warrant, and removed documents, some of which related to the campaign.[49] Since no other organisations were raided, the message could not be clearer. We were not to cross the line between trade union organisation and involvement in political activities.[50]

The Labour Relations Act

Needless to say, we did not accept the validity of the line the Special Branch sought to enforce, and were going to push it. But the dilemma the Union faced in doing so was that its political influence, like the influence of any membership-based organisation, depended on maintaining its base. It could not afford to adopt positions that would create disunity in its own ranks. At the conference that year it became clear that this was also the dilemma confronting the Union, once the Bill became an Act, and we had to decide how to become the non-racial trade union we had always wanted to be. It was no longer a simply matter of becoming one registered trade union.

It was because FCWU was registered that it was able to prevent the Vakbond from getting registration, as Polly from Paarl pointed out. Through relying on their own organisation and going on strike, Monica from Johannesburg countered, workers at Langeberg Boksburg had secured a better agreement than the registered trade union had in the Western Cape. In the course of this debate it became clear that the African delegates were not prepared to countenance becoming members of one registered trade union.[51] Registration for them was now akin to collaboration. So the issue over which the Union had originally split into two was one over which workers were still divided.

The political contestation taking place in the townships was also taking place in sections of the coloured community but not, evidently, among working-class sections in the Western Cape. It seems to me that coloured members from about this time on began to display increased deference towards African members on questions of a political nature. However, the fact that the African workers were taking more radical political positions did not mean they were more advanced, and, in any event, we could not simply be guided by what was said in meetings. I could not for the life of me see coloured workers accepting a decision to deregister FCWU. Someone like Oom At would vote with his feet, and ribbons and flags would not stop him.

The 'solution' the conference arrived at was to defer a decision, to allow more time for deliberation at the branches.[52] It was the first of a succession of postponements of a decision that looked as if it could split the Union, over a period of more than three years. The Bill then became an Act, with a new name, emphasising that this was the most far-reaching overhaul of the law since the first Industrial Conciliation Act: it would now be the Labour Relations Act (LRA).[53] Government had dropped the grosser forms of control proposed in the Bill, but moved almost immediately to stop those few employers that were deducting the subs of AFCWU by stop order.[54] 'This is a small matter,' Pendlani had said, when this came to light. 'The Union was not built up on stop orders, but on collections by hand.' However, this was not entirely true. The revival of AFCWU had in large part been made possible by the resources of FCWU.

At the same time as the issue of registration was threatening to drive African and coloured workers apart, John was making headway at Attwell's Bakery. By the time the Bakery Employees' Union formally complained to us that we were poaching their members, we had also been approached by workers from other bakeries. These included Good Hope, the bakery belonging to Fatti's & Moni's whose bread the traders had boycotted, and at which those who had turned their back on the strike were initially placed before returning to the mill.[55]

'We have not poached any of your members,' we were able to tell the secretary of the Bakery Employees' Union truthfully. 'We cannot stop workers from speaking to other workers, or from asking to join our union.'[56] It was to take another six months or so before the coloured women at Attwell's were able to muster the courage to resign en masse from the Bakery Employees and send a delegation to the head office requesting to join the Union.[57] They were accompanied, of course, by John.

19

The road to the cemetery

EVEN IF THE LANGA SUMMIT had not done much more than raise the prospect of a new federation, it gave all those who had opted to organise workers a new sense of purpose. This was what we would tell our children about what we did during the struggle. However, a new federation would be a paper tiger unless the unions that formed it had properly organised workers. Neil Aggett understood that as well as anyone else, and had utilised the space there was to organise in the workplace to good effect, under the most difficult circumstances: nowhere more so than at Langeberg's Boksburg factory.

The Union had got its foot in the door at Boksburg by virtue of its agreement with the Canners' Association, but it was not realistic to expect workers to continue to accept what amounted to a rural wage in an area in which wages (and expenses) were much higher.[1] The workers went on strike to secure a higher wage than had been agreed at an industry level.[2] Langeberg was bitter about being compelled to pay a higher wage, and now accused the Union of 'political incitement'. This was because Neil would open lunch-hour meetings with an *Amandla!*, which was at this point practically routine at all our meetings and the meetings of other trade unions.[3] *Amandla* was the call, and *ngawethu* the response: 'Power is ours', a particularly apt statement for workers in production, as well as being a rallying call for people oppressed by apartheid rule. I also suspect management blamed Neil for foiling its attempts to establish a special relationship with the chairperson of the Union, which was very reminiscent of the relationship Langeberg once had with Oom Joe. This was the kind of 'big man' democracy the powers that be were comfortable with.

The kind of reaction the *Amandla* salute elicited was also because the East Rand was one of the preserves of the Transvaal Nats, who were always more hard-line than their counterparts in the Western Cape. Also, the factory was slap in the middle of the industrial heartland of the Reef, at a time when organisation on the Reef was also lagging behind the coastal centres. The Special Branch would surely have told the bosses that SAAWU's leadership were coming. This would have been alarming news so soon after the East London strikes. We agreed to let SAAWU use our office as its base, and Neil would be our pointsman in dealings with them.

Coupled with incidents like the Special Branch urinating in a water jug in the Wanderers Street office, Langeberg's blaming of the messenger suggested that Neil was being demonised, and that management was in cahoots with the Special branch. Neil himself believed the HR manager at Boksburg was their pointsman at the factory, and hated him. This hatred can only have been because of what Neil represented. The rational, quietly spoken manner in which Neil would argue the workers' position did nothing to lessen their animosity, or the fact that Neil had moderated his image: he no longer wore his hair as long, and wore a tweed jacket to meetings instead of just an open-necked shirt.

It was not just because of the HR manager's smarmy attitude that Neil believed he was in cahoots with the Special Branch. One day a member of the committee was called to the factory's reception area and offered money to spy on the Union by someone who said he was from the Department of Manpower Utilisation but who must have been Special Branch. This could not have happened without the HR manager's say-so. No less telling was that the worker concerned was in financial difficulties. Only someone in the HR manager's position could have known this.[4]

Neil welcomed SAAWU's move to Johannesburg, because he thought it would strengthen the position of the trade unions on the political front and counter the political timidity of FOSATU. But even though we still considered ourselves as being on the same side as SAAWU, I felt it necessary to warn him about the differences between SAAWU and ourselves that had started coming to the fore in East London.[5] We had a reasonable opportunity to talk in the closing months of 1981. Neil was in Cape Town for a week and stayed with us in our Waterkant Street cottage.

He was still the intensely committed person he had been when we first met, but a lot wiser than the person who had secreted a copy of *What Is to Be Done* under his couch. He had also moved in the same direction as I had, not because of me, but in response to similar experiences. I would say we had become close, and he was also close to Athalie. We did not just talk shop. One of the things I had in common with Neil was an overbearing father, but Neil's father problem was incomparably worse than mine. They were not speaking at all.

If Neil did not say so in so many words (and I am fairly sure he did), he regarded his father as a racist and a fervent supporter of white minority rule. However, I do not think the issue between them was just about politics. It was also about respect, and I could see that standing up to his father had taken its toll of Neil as a person. After his return to Johannesburg, towards the end of 1981, he phoned to say he had moved out of the house where he lived with his long-standing partner, Liz, into a flat with someone called Doug Hindson. I assumed that meant the break-up of his relationship, but did not ask.

This was one of the last conversations we had, because, on 27 November 1981, he was detained, along with a number of others, including Gqweta, Njikelana and Kikine of SAAWU, and Emma Mashinini and Alan Fine of CCAWUSA.[6] Detention now served the function that banning orders formerly had done, in maintaining a line between trade union organisation and politics.[7] So the detention would not have been unexpected.

'He knew that he might one day be jailed, as so many other trade unionists have been detained,' a pamphlet later put out by the Union stated. 'But he had nothing to fear from detention.' Neil might even have welcomed the respite detention offered initially. He had never had a paid holiday since he began working for the Union, and I imagine he was as near to the end of his tether as I was at the time. It was not altogether true, however, that Neil had nothing to fear from detention. He was afraid it would come out under interrogation that he was dodging military conscription.

'Obviously the easiest way in which they will get hold of you is through the SB,' Neil had written to me. 'They' were the Military Police. Although there seemed to be little coordination between the Military Police and the Special Branch, this might change. 'The main problem arises in interrogation [when] the SBs [go] into your history, education, money background etc, but this is unlikely to catch one if you have a good story (e.g. exemption).'[8] If it came out he was dodging conscription, Neil would either have to do his military service or go to jail. The alternative, if the opportunity presented, would be to skip the country.

I do not remember ever advising what he should do in such circumstances, and it would have been presumptuous for me to do so.[9] What we were both concerned about, rather, was the effect on the organisation of a sudden departure, and the kind of stories the Special Branch were likely to spread if that happened. Neil suggested that he inform the leadership of the branch that he was dodging conscription, even though there were obvious risks in doing so. I don't know if he ever got round to this.

At the time of his detention, Athalie and I were on the Transkei coast, where there were no newspapers and phones. This was meant to be our opportunity to reclaim our personal lives. Instead, I reverted to a childhood hobby and obsessively scoured the shoreline for shells. A party of holidaymakers arrived not long after Neil had been detained, including people I knew from circles he moved in. They said nothing to me about the detentions. Maybe they thought I knew about them, and there was nothing anyone could do anyway.

So determined had I been to take a full three weeks' holiday, my first since I had started working for the Union, that I hardly took note of another Bill. This was to amend the law government had just amended, and concerned the Industrial Court.[10] The court already had the power to determine something called an 'unfair labour practice', which had a definition so open-ended that the Industrial Court had carte blanche to do so.[11] But it could not do so without litigants, and up until this point there had been hardly any. The object of the amendments was to make litigating in the Industrial Court more attractive to trade unions in particular. Although the amendments would in time affect us all profoundly, the Industrial Court was to remain a topic which we did not discuss with other trade unions, or they with us.

There is always something you can do, was my reaction, when I first learned about Neil's detention on my return to work. What is happening outside must have an effect on what is happening inside, even in John Vorster Square.[12] So, apart from making representations to all and sundry, we resolved to take legal action to recover the bank deposit books, receipt books, minute books and files on our dealings with various employers.[13] The documents which the Special Branch seized went to show that it was on account of his Union activities he was being detained, we argued.

'Those that saw him in detention can testify he was in good spirits, and very confident that however long he might stay in jail, the authorities had no charges to bring against him,' said a leaflet the Union later brought out, referring to the 14-day period in terms of which he was initially detained.[14] When that period expired, he was detained under the notorious section 6 of the Terrorism Act. This provided for indefinite detention.[15]

In the early hours of Friday, 5 February 1982, Neil died a lonely death in his cell. A labour reporter I was friendly with phoned to tell me about it. According to the police, he had been found hanging from the bars, with a kikoi cloth around his neck. He had killed himself, they said. My immediate reaction was that this was a lie. Neil would never kill himself: he had been killed. The Security Police were responsible, but the government as a whole was implicated.[16]

'He took his own life, so the police say. We say: He was killed. Nothing the authorities say to clear themselves of complicity in his death can convince us otherwise,' read the leaflet the Union put out. 'The trade union struggle is an open struggle. Everything a trade union leader does or says is put to the workers ... That is the way our Union works and that is the way Neil conducted himself. That is why we can never be expected to believe Neil had any reason to kill himself.'[17]

We struggled to find any photographs of Neil for the leaflet. His parents did not have any photographs of Neil as we knew him. Quite deliberately, he had left no trace of his identity that the Military Police might pick up. However, acquaintances of mine had filmed part of the proceedings of our last conference.[18] The image by which he is most often remembered was taken from that. He is listening intently to the proceedings, with his left forearm across his chest, and a pen clutched in his right hand. The handle of the movie camera partially obscures his head, and he is wearing his tweed jacket.

The Union, and the emergent trade union movement, were now the focus of national attention. I was to address a series of memorial meetings, starting at St George's Cathedral in Cape Town on the Wednesday following his death.[19] Lizzie also spoke movingly, in isiXhosa, and Athalie translated.

We had already decided that the next day, 11 February, all our members would stop work for half an hour, between 11.30 and 12, as a mark of respect for Neil and an expression of our outrage. We also called on workers everywhere, and unions and employers, to abide by this call. There were undoubtedly factories that would have stopped for longer; half an hour was what we thought all could achieve, including those that were not so well organised or had recalcitrant bosses.[20]

The stoppage was intended not to coincide with lunch hours. The bosses should feel some of the pain we felt, by sacrificing a portion of their precious production. We also wanted to demonstrate the breadth of our organisation and that of the emergent unions. That would also send out a signal to government and the bosses. All the bosses to whom we spoke about the planned stoppage – the head offices of the conglomerates – indicated they would not stand in the way of the stoppage, including Langeberg.

On the day of the stoppage, as reports trickled in to head office from the different branches, I travelled to Durban. There was a memorial meeting in the cathedral that night. I travelled to Johannesburg the next day, a Friday. A press conference had been arranged for the afternoon. There had already been a story in the *Sunday Times* that Neil had been engaged in subversive activities, doubtless planted by the Special Branch, which was known to have close links with the paper. As though to confirm the Special Branch version, a pamphlet was then distributed in the city claiming Neil as an ANC member.[21] Among the people I met upon my arrival, including those who had been his friends, there was a mood of fatalism. I was dismayed.

'Trade unionist hangs himself,' was the caption of a pamphlet issued by FOSATU, which had a line drawing showing a limp body hanging. I was appalled. Not only was it in bad taste, it contradicted the very point that the Union was trying to make. Privately, I might have conceded the possibility that his suicide had not been staged. However, there was no reason whatever to make this concession publicly, and to do so could only lessen the pressure on government.

As someone who was part of a struggle, I believed the appropriate response was outrage. The press conference was packed with foreign and local media and other hangers-on. I gave short shrift to the suggestion from an American journalist (from the *Christian Science Monitor*, I think) that Neil must surely have done something to be detained and, by implication was the architect of his own demise.

The funeral was to be on Saturday morning, and the buses bringing delegates from the Western Cape branches of the Union were already on their way. That evening I met with the organising committee at Doug Hindson's flat. I had not met Doug before, but I knew that he was one of the people to whom Neil had become close in the period prior to his detention. There must have been some 20 people present. Virginia, who had been sent from head office a few days earlier, had already told me there was an issue over which the committee was divided. I did not yet know what it was. It turned out to be about the flag that would drape the coffin. The Union had already made up a red flag with the Union's logo of hands clasped in unity, in white. Some wanted an ANC flag on the coffin instead.

Absolutely not, I said. Neil was detained because of his union activities, and it was on account of his union activities he was now dead. To put an ANC flag on the coffin would have been tantamount to saying he was detained because he was involved in underground activities, as the government had all along maintained. It must have been clear that I was not going to brook any opposition on this point, and I was confident the leadership of the Union would back me. Clearly, however, some of those present were not happy.

One of those who had wanted the ANC flag on the coffin asked if they could see the speech I intended to deliver. It was typed out and I did not mind leaving it with them. Sipho, who had been an organiser for MAWU, brought back my speech to me the next day, half an hour or so before the service was to begin. I was in St Mary's Cathedral in the city centre, where people were already gathering for the funeral service. The buses from the Western Cape had arrived and I remember greeting Aunt Hester from Drakenstein in her red union T-shirt in the aisle.

The speech does not really say anything, said Sipho. I was quite put out. My intention was to say what Neil stood for, what he had achieved, and why he had died. I thought I had not pulled any punches. What I was trying to do in the Union was to extend a space we had demonstrated existed, to organise. This is what I would say Neil was trying to do. We had never disagreed about this. So it was because of his trade union work that he had been detained, and it was the government that had killed him.

All this, I afterwards realised, was not what Sipho wanted to hear, but he would not say what this was. I added a few sentences in a bid to satisfy Sipho and his circle: concerning Neil's stand on registration and what he saw as 'reformist' tendencies in the unions. His position in this regard was far more uncompromising than my own.[22] But I was not going to say that the work he was doing in the Union was part of some ANC master plan, when it was not. In retrospect, I think such a statement was what Sipho's circle were looking for.

The mourners that streamed into the cathedral were from all races and classes, and some were attired in church gear and some in union T-shirts. The group that made the biggest impression, perhaps because they arrived en masse, was the nurses from Baragwanath in their blue and white uniforms. After the service, as everyone streamed out of the cathedral after the coffin, there was a trumpet blast. I looked up and saw the green, gold and black flag.

We had already resolved to walk with the coffin to the cemetery, and we did so following that flag, borne by a series of youths I did not recognise, running ahead of the hearse. Periodically, one set of youths would dash to the front and swap with the bearers. Taking turns in this way must have been a strategy to avoid being identified on camera. Themba Mpetha, Oscar's youngest, told me sometime afterwards he had brought the flag from Cape Town on the bus. The boldness of the statement could well have been Oscar's conception.

West Park cemetery lay to the west, over the ridge that runs from west to east, separating the city centre from the Northern Suburbs. From St Mary's, the way out of the city centre was along Jeppe Street, past the office blocks and shops, which in those days drew customers of all races on a Saturday midday, towards the outskirts of the city centre, where the cityscape flattens out into garages and nondescript warehouses. Thirty years afterwards I am not able to reconcile a memory of the gasworks and walking up Jan Smuts Drive past the University. Was it because a group of marchers took a different route?

What I am sure about is the stately block of flats, painted white, as Empire Road becomes Kingsway, and the curious onlookers on the balconies. It must have been an extraordinary sight: mourners of all races, from all sections of society. Above all, I remember the gum trees towering over the road in the summer afternoon, as you descend the hill from Melville, along what has become Beyers Naudé Drive. It was from here that you could see both the front of the march and the tail-enders. The gum trees are still there. So are the flats with their balconies, but you now have to look through razor-wire to see them.

Outside the law
At first I was anxious the flag would provoke a reaction from the police, but as the march proceeded I became more relaxed about it. It did not compromise the autonomy of the Union to follow the flag: it was not, after all, something we had laid on. It also did not compromise the political position we had taken, that the detentions were an attack on the emergent trade unions. I also do not think there was any disrespect to Neil's memory in flying the flag, even if members of his family and others saw it that way.

Just as our trade union work was establishing and preserving a space to organise, flying the flag was a way of creating political space internally. It could even have been regarded as marking a shift away from a position of overt hostility towards the emergent trade unions on the part of the ANC, and an acknowledgement of the work they were doing inside the country. That would also have been the statement Oscar would have wanted to make, although it would be stretching it to suppose this expressed the view of the ANC as a whole.

It is therefore not correct to say the funeral was 'hijacked', although it might have been.[23] It had not in fact been an ANC funeral, as Hennie Ferrus's had, and the flag had been spirited away by the time the procession arrived at the cemetery. It was messages and speeches from unionists and friends that predominated at the graveside.[24] Joe Foster, the general secretary of FOSATU, was among the unionists present, and he told me how well its affiliates had supported the stoppage. There was a perceptible sense of solidarity among those present.

If the flag was an acknowledgement of the work Neil had been doing, he would have welcomed it. 'The people I am working with', he had written some two years before, 'feel it is reasonable to continue working in the open, and increasingly it will be necessary to combine legal and illegal work in this country. They feel that to confine one's activity to purely legal work is letting one's limits and direction be defined and decided by the state.'[25] I would not have asked who 'they' were, but 'they' almost certainly all supported the ANC. Put it this way: in those days we all supported the ANC, or nearly all of us, because the ANC seemed to be the only organisation capable of changing the existing order.[26]

Among the activists he had engaged with was Barbara Hogan, who had been chairperson of the Fatti's & Moni's support committee in the Transvaal. We now know that Hogan was an actual member of the ANC, and regarded herself as being 'under discipline'. When 'Lusaka' asked her to compile a list of people she worked with, she complied. We now know she was being set up by Special Branch agents to do so. Neil's name was on that list, as well as the names of other detainees, although she did not categorise him as a 'close comrade'. It was this list that had precipitated the detentions.[27]

We know from the inquest that Neil denied being what Lieutenant Steven Whitehead regarded as a card-carrying member of the ANC. Whitehead, the Special Branch officer in charge of his interrogation, seems to have accepted that. However, it was enough for Neil to be a supporter or sympathiser to warrant his detention.[28] Yet Neil's support would surely have been conditional, since it was not clear how far the ANC was prepared to go in advancing the interests of workers and the working class. He would also not have been naïve about the activists with whom he had contact, or the danger that trade unions would be hijacked. He had already had to do with just such an attempt, by members of the Fatti's & Moni's support committee.[29]

Whether you were a member or a supporter, the nub of the matter, as I saw it, was what this implied when you were working in a democratic, membership-based organisation like the Union. Was your primary loyalty to the ANC or the organisation you worked for? My own view was that what I was doing was consistent with being a member of the ANC. However, I did not consider myself to be 'under discipline' or accountable to the ANC in respect of what I was doing in the Union. I was accountable to the workers who employed me, through its structures.

I am quite sure that is also how Neil would have seen it, because that is how he acted in the Union, with a strong sense of accountability to both the workers and higher structures, including myself. It was also clear from what he had written and said that he saw the building of an autonomous workers' organisation as a necessary step for workers to be able to develop a consciousness of themselves as a class, and to be able to engage with other classes in political organisations. He would have been suspicious of a view that saw this work as being subordinate or secondary to the primary struggle, the 'political struggle', as though what he was doing was not political.

The fact that Neil did not believe the line between legal and illegal was sacrosanct did not mean he was guilty of a crime. That was Whitehead's problem. It might have been enough for Neil to be a supporter or sympathiser to warrant his detention, but it was not enough to charge him in the kind of show trial the Special Branch were evidently cooking up.[30] One line of endeavour seemed to be to establish a link between him and SACTU, in the person of Hogan. But Whitehead's particular mission seems to have been to get Neil to admit to being a communist.[31]

'I support Marxist ideology and am therefore a communist,' Neil wrote in a statement to the Security Police. Surely Neil's ideas about the organisation of workers and the working class drew inspiration from the ideas of Marx, as did many among the loose agglomeration that was the white Left. Perhaps, like me, he didn't like labels, since I never heard him describe himself as a Marxist. Even if he had, however, he could not conceivably have believed this was the same thing as being a communist, which, in the context of South Africa, meant being a member of the Party.

There would, in time, be those among the white Left who would declare themselves for the Party. But at that time the Party had not even begun to come to terms with its dogged support for Stalin and its wilful blindness to what the Soviet Union had become. Most in the white Left, in my estimation, saw this as a serious problem. Neil was particular about what words meant. Coming from this tradition, you could not think you could be a communist without being a member of the Party.[32]

I am on trial for my ideas, was what Neil was really saying. The people who are interrogating me are blinded by their own ideology. They are putting words in my mouth and trying to put their own nonsensical labels on me. To convey this message most clearly, he added the following sentence: 'I am also an idealist.' Very likely, at one time or another, Neil would have heard himself being as described as an idealist, as I have, perhaps by a benign relative trying to come to terms with the choices he had made. How naïve, he would perhaps have thought, without taking issue with this description. He was, after all, inspired by the idea that things could be different.

But Neil had read philosophy. He knew that in philosophical terms it was a contradiction to say one was a Marxist and also an idealist. Idealism was the converse of dialectical materialism. I also do not think he would want to be remembered as an idealist, as much as for what he actually did and the difference he actually made. We make history, as Neil was to confirm, but not in circumstances of our own choosing.

Whitehead must have hated Neil, in much the same way the HR manager of Langeberg did, unconsciously, because of what he thought he represented. I was hoping to get an idea of what that might be on the day I attended the inquest, but had to content myself with the images of Whitehead in the press: an earnest young man who would have looked at home on any Afrikaans university campus at the time. He would have found Neil's austere lifestyle, which he had been observing over a period of three years, particularly disturbing. Evidently Whitehead himself preferred the good life. He was described as a natty dresser, wearing 'fashionable teardrop glasses' and Gucci-like shoes.[33]

Interrogating someone for 62 hours on account of their ideas is the sort of thing you might do to a traitor to the class that you aspire to belong to.[34] The statement Whitehead eventually did extract sounded much like the 'confessions' that Stalin's police extracted from those that deviated from his line: 'confessions' so out of character and improbable that they had to be extracted by torture. The irony of the analogy would have been lost on Whitehead.

Sleep deprivation, relentless humiliations and pitiless interrogation would be enough to get anyone to sign a statement. But 'to live outside the law you must be honest': shame attaches to someone outside the law, as Neil was, who makes a statement that is to be used against your comrades. It is the kind of shame others in struggle circles have never been able to live down entirely. Sleep deprivation, relentless humiliations and pitiless interrogation were also calculated to destroy Neil's sense of integrity as a being.[35] It seems they also caused him to lose perspective on the significance of a 'confession' that in the broader scheme of things did not amount to much.

Auret van Heerden, who was in the cell opposite Neil's, was the detainee who saw him last, but was only released later that year, as the inquest was drawing to a close. Neil told him he had been broken. Auret's account was a strong indication that Neil had killed himself. What finally persuaded me this was what had actually happened was when Norushe told me about his experience of detention. However, this did not mean that any of the accusations we had levelled against the government were unwarranted or untrue.

Solitary confinement could drive anyone to kill himself or herself, was Norushe's point. But Neil was not anyone for those who knew him. For me, the clue to understanding why Neil would have killed himself was the very high value he attached to the respect he had earned in the organisation, and how very alone in the world he must have felt, at the point at which he believed he had betrayed it. Curiously, it did not seem the Special Branch were bothered about whether he had done his military service.

Aftermath

Workers in factories across the country stopped work, sometimes with the consent of their employers, and no loss of pay, and sometimes despite their employers.[36] Even newly organised plants stopped work. However, there were two branches where workers did not participate. In Durban, the Union could at least say a branch was still being established. There was no such excuse at Ashton. The workers at Langeberg had asked their bosses whether they could stop work, and waited for a reply that never came. At Ashton Canning the answer was no, and that was that.

We described the work stoppage following Neil's death as a political strike. If that is an accurate characterisation, it was the only one of its time which had a clear demand: to end the detentions and harassment of trade unionists.[37] It seems to me that from this point on elements in government did begin to adopt a more careful approach towards the emergent trade unions. While I do not want to exaggerate the impact of Neil's death and the stoppage that followed, it was a turning point in relations between the emergent unions and the state.

The fact that the police left the funeral procession entirely alone was an indication of this. There were some banners seized after the funeral, outside the cemetery, but that was the only time I was aware of the police.[38] Another indication was the coverage on the SABC news that night. It reported the funeral as its lead story, including my accusations against the government. Not surprisingly, it studiously avoided any footage of the flag being paraded through the commercial capital of the country.

On the other hand, if there was a change of attitude towards the emergent trade unions, it did not mean an end to 'dirty tricks' by the security forces. The phone calls to my home in the middle of the night stopped, but soon after our return from the funeral there was again a theft from the safe of monies to cover the costs of the funeral. The system had been tightened up since Vormat's departure. I had no doubt the Special Branch was responsible, and neither did the management committee meeting.[39] We suspected it was the Special Branch who, not long afterwards, tampered with a Union vehicle in George, but perhaps it was the doing of one of the shadowy security structures set up under Prime Minister PW Botha as part of his 'total strategy'. George was after all PW's constituency.

I was about to take the long road home, after filling up the tank, when a guardian angel in the form of a petrol attendant offered to check my tyre pressures.[40] The one front tyre had been inflated near to bursting point, and the other deflated. Preoccupied as I was with the problems at Table Top, I doubt I would have noticed until it was too late. It brought to mind the car accident in which Joe Mavi, the general secretary of the Black Municipal Workers' Union, was killed, and many like him since. We will never know how many of them were set up.

The week after the funeral Mr Wolthers of Premier contacted me. He urgently wanted to set up a meeting at its head office in Newtown. I thought it was about a dispute with SA Milling over membership of the pension fund, and went with Philip Mayoli from the Salt River mill. I could tell from his schoolboy state of excitement when we got there that something else was afoot. We were to be introduced to the chairman, Tony Bloom, and would get to sit in his antique chairs. It seemed he wished to convey his condolences personally.

This was one indication that Neil's death was a turning point for at least a section of the business establishment. Hogan was found guilty of treason, but she was not a trade unionist.[41] At the end of it all, the state was not even able to mount a credible show trial. Of the trade unionists charged, Fine was acquitted and the charges against Gqweta and Njikelana were withdrawn.[42] This surely went to show that it was not possible to maintain a line between industrial and political democracy. A few years later, Bloom went with the bosses of Anglo and a few others to meet with the ANC in Lusaka.

Neil's death, and the stoppage, also precipitated a crisis in the established union movement. TUCSA issued a statement by dissociating itself from the strike, and refused to blame the government for Neil's death. This amounted to a declaration of political bankruptcy, which put it to the right of many employers.[43] Two unions disaffiliated, and more were to follow later.[44]

The symbolism of a white dying in detention was also important. Neil was not the first or only white to die for the struggle, but he was the only one to die in detention, at the hands of the state, and that was important in establishing this was not a racial struggle. But these are also attempts to find meaning in death. In truth it was a terrible loss, to me personally and everyone he had worked with, especially the workers he had organised.

After the customary speeches at the graveside, the coffin was lowered into the orange earth. People were milling around while two cemetery workers in orange overalls started to shovel earth onto the coffin. We will do this, I said to one of the workers, as I borrowed his spade. It is what Neil himself would have done at my funeral. There would be no onlookers when there was physical work to be done in the new South Africa that we imagined. A group of us took turns to silently fill the grave.

More than one of the speeches at the graveside was to the effect that in death Neil had brought people and unions together. This is the sort of thing you say at funerals, and I myself said so. It was not untrue, but not altogether true either, as the fracas over the flag on the coffin had shown. Neil was in the process of leaving one circle and entering a new one when he was detained. I looked to someone from his new circle for a stand-in for Neil. Perhaps Doug Hindson was too aloof a person to be regarded as part of any circle: in any event, it was he I approached. He suggested his partner, Maria Hambridge.

Maria, he told me, was from the English working class and did not have the hang-ups South Africans had about race. She also had a no-nonsense air about her, and was breastfeeding their baby, Sean. The baby was the reason she was available at all, and the Union needed someone urgently. I was worried what Beatrice and Mildred would get up to on their own, as were the office-bearers of Johannesburg and Kempton Park. It was agreed to see what she could do on a voluntary basis.

It seemed that it was only after Neil was gone that the workers were at last able to discipline Beatrice and Mildred. The factories were not being visited and the workers' problems were not being attended to, so it came as no surprise to find money had been misappropriated. It was only Mildred who tried to put up some kind of defence. I felt sorry for her, but she had in the end succumbed to Beatrice's influence, and had to go.[45]

Maria's 'lack of hang-ups' sometimes meant she went in where someone with a better understanding of local dynamics would have been more sensitive. However, she compensated for this with her energy. What the workers especially appreciated was her refreshing disregard for the bosses' notion of how a white woman should behave in their company, by continuing to breastfeed Sean, by now quite large, in meetings around the boardroom tables. I am not sure whether it was on account of the men in suits or the men in overalls that one of the managers of EPOL, unable to contain his disgust, complained.

The line the Union had now to take, in the aftermath of Neil's death, was to press on in a spirit of renewed determination. I am not sure whether it was the EPOL Isando workers that first organised the EPOL workers of Pretoria West, or the workers of SA Milling that organised the Pretoria West flour mill, but within a short space of time ballots were being conducted at all Premier plants in Pretoria, and very high percentages, above 95, voted to be represented by the Union. By the time of the Union's conference, a Pretoria branch had been established.[46]

Unity from above or below?

Having put off taking a decision as to whether to register, there was no alternative but to carry on as before in the workplace and at branches, operating as one organisation. This was not as easy to do where there had been animosity between FCWU and AFCWU, as was the case in Paarl.[47] At Fatti's & Moni's in Bellville, by way of contrast, the committee had operated as one ever since the workers returned to work. The only animosity among workers was between those who had turned their backs on the strike and those who had stuck it out to the end, the latter led by Friday and Spasie, who were now chair and vice-chairperson respectively.

The committee had been especially effective in defending jobs. When management dismissed a worker without following the agreed procedure, workers went on an overtime ban, until eventually he was taken back.[48] No sooner had management done so than a supervisor assaulted another worker with a bailing hook. If anyone else had done this, he or she would have been dismissed on the spot, but the supervisor in question was Skilpad, who had led the breakaway from the strike. This time there was a full-blown stoppage over management's refusal to do what was required of it in the circumstances. This was what lawyers would later label a dispute of right, in which it was not permissible to strike. Skilpad was eventually dismissed, however. It was a fitting demise for a Judas.[49]

In the spirit of pressing on with renewed determination, the other vital task confronting us was to get the 'unity talks' on track again. The 'solidarity committee' in the Western Cape was not remotely representative of trade unions in the region, of course, since the vast majority of 'organised' workers belonged to TUCSA affiliates.[50] It was, however, uniquely placed to get the talks going again, because neither of the two federations, FOSATU or CUSA, had a significant presence in the Western Cape.[51] Nor did SAAWU or the wave of trade unions it had inspired, which were now characterised as 'community trade unions'. We were thus able to be an 'honest broker'. On behalf of this solidarity committee I wrote to all those who had been represented in Langa to propose another meeting. FOSATU said we would have to wait until after its congress.

What FOSATU now had to say was important. I believed what Joe Foster told me at the funeral about how solidly its affiliates had supported the work stoppage, because it tallied with our own observations. It really did have organisation in workplaces, as well as leadership structures which were able implement the call for a stoppage.[52] This could not be said of all the other organisations we now had to do with. So when FOSATU at its congress declared it was prepared to disband and form a new federation, it seemed like a bold step forward. The problem was that it wanted a 'tight' federation of industrial unions, and we did not.

On top of this, Joe Foster had presented what was described as a 'keynote address' at the congress, entitled 'The Workers' Struggle: Where Does FOSATU Stand?'[53] 'Keynote address' made it sound like it was pitched for an academic audience, which in a way it was, I think: not the academic community as such, as much as the circle of intellectuals drawn in the main from the white Left, which were one of its mainstays. This was FOSATU's response to the charge that it was politically timid. The reality is that workers do not set great store by the written word, and much of it would have been over the heads of the members.[54]

I had not read it when, at about the same time, I wrote an article for *Grassroots* newspaper, which was published in the name of FCWU.[55] In their different ways, both papers addressed the question of the relationship between trade unions and politics. Their points of departure were quite similar. Both emphasised in different ways why it was important that workers' organisations remain autonomous.[56] Both called into question the credibility of the community trade unions. The burning question, however, was how trade unions could engage in politics without compromising their autonomy, at a time when the country was increasingly being placed on a war footing as the government pressed ahead with grand apartheid schemes.[57] It was therefore not so much a question about 'where FOSATU stood in the workers' struggle' as where it stood in the political struggle playing out in the community.

FOSATU acknowledged the need for its members to be involved in the community, but the reason it gave for doing so was peculiar: 'since our members form a major part of those communities.' This was not by any stretch of the imagination true. Even in manufacturing it could not have supposed a majority of workers had been organised. In the final analysis, what mattered was not so much the coherence of the argument as the tone struck in a revealing phrase. This was the phrase 'there has emerged into our political debate an empty and misleading political category called "the community"'.[58]

For activists who relied on 'the community' as a basis to organise, this was a slap in the face. It was also not realistic to suppose that what a working-class movement such as FOSATU was advocating would be possible without their active or tacit support. The *Grassroots* article was an attempt to enlist their support, to address a common political problem: the apathy of the coloured working class in greater Cape Town. This, I implied, was due to the dominance of the middle classes in the coloured community, which community organisations were replicating.[59] It was this division in the coloured as well as Indian community on which the recommendations of the President's Council were premised.

The President's Council was to constitutional reform what the Wiehahn Commission had been to labour reform. Soon after the *Grassroots* appeared, it published the first of two reports, which in effect proposed to give coloureds and Indians a more effective say in government, while still denying the African majority the vote. Sonny Leon, the former leader of the coloured Labour Party, was among those to welcome these proposals: they were 'the fulfilment of his wildest dreams', he said.[60] This was the same person who had been guest speaker the Union's 1976 conference, at which I was appointed general secretary. It was an indication of how far the Union had travelled politically since, that the management committee meeting unanimously resolved to 'openly oppose' these proposals, and make members aware of their implications.[61]

The other important recommendation the President's Council made was to establish 'community councils' in the urban areas, where Africans with permanent rights could elect local representatives to address local issues. Even if the community was an inexact concept, it signified a political space that was bound to be fiercely contested, and the position FOSATU had taken would look increasingly like political abstentionism. However, I doubt whether many knew what this position was when the next round of unity talks took place at Wilgespruit, outside Johannesburg.

The Wilgespruit Fellowship Centre, which BJ Vorster had not so long before described as a 'den of iniquity', was the closest thing there was to a safe space in apartheid South Africa for its opponents. While we waited on the hillside looking across the pristine valley to see who would pitch up, Norushe, who had recently been released from jail and had a fine voice, led the delegates in a song that was not part of our repertoire in the Western Cape: about young people leaving home and following freedom, into places their parents did not know.[62] The immediacy of the bond established was palpable.

Principles and factions

The outcome of the Wilgespruit meeting was an agreement 'to discuss in detail plans for unity on a more permanent basis' and to hold a further meeting to discuss the detail.⁶³ Unity on a more permanent basis was code for a new federation. However, the optimism this resolution generated was misplaced. Organisations we had expected to attend were absent without explanation. The position adopted by MACWUSA, which had not been at Langa, had also cast a long shadow.⁶⁴

The basis of working together is principle, said MACWUSA's spokesperson, Government Zini, wearing a full-length black leather coat that looked as if it could have cost an ordinary worker several months' wages. He proceeded to list its 'principles'. There were five, including non-racialism.⁶⁵ I had by this point already had the opportunity to introduce myself to the leadership of MACWUSA, when we established our Port Elizabeth branch office. It did not seem they were glad to be sharing offices in the same old, double-storey building in North End.

'There could be no compromise on registration,' Government said. But there was no support from the other organisations for MACWUSA's 'principled' stance, and it was already getting dark when Government stood up to make a further flowery speech, ending with the announcement that MACWUSA was withdrawing and would not accept any invitations to any further meetings. 'The history of trade unions was littered with splits,' was my response, taking advantage of my position as chair to respond before he could stage a dramatic exit. 'A split could never be in the interests of workers, no matter in the name of what principles it was justified. Our Union was affiliated to SACTU and that had been a federation of registered and unregistered unions.'⁶⁶

'When the crunch comes,' Gatsby Mazwi of BMWU added, after the MACWUSA delegation had already left, 'it is not going to matter whether a union is registered or not registered.'⁶⁷ I liked Gatsby, because he was witty and charming. Above all, he seemed independent-minded. So you had to wonder why, at the meeting to discuss the detail of 'unity on a more permanent basis' a couple of months later, his union was part of a group that did a complete about-face on the position they had taken at Wilgespruit, and he would describe unity between registered and unregistered trade unions as a 'marriage of convenience'.⁶⁸ Not long after that, BMWU dropped 'Black' from its name and became a 'Municipal and General' workers' union.

Whoever proposed that MACWUSA be invited to the next meeting and that it be in Port Elizabeth must have realised that Government had painted his union into a corner, and wanted to offer it a way out. It was proposed and agreed to restrict the size of delegations. That meant ordinary workers would not be able to attend, which I always thought was a mistake. Workers were a restraining presence. It was also important for workers to witness what was done in their name, which is why we encouraged branches to send delegates to these meetings whenever it was possible. WPGWU did the same, but FOSATU did not even allow its affiliates to attend meetings in their own right.[69] Neither did CUSA on the occasions when it was present.

As it happened, CUSA was present at the Port Elizabeth meeting, and the CUSA delegation included the newly appointed legal officer of the newly established National Union of Mineworkers (NUM). This was my first encounter with Cyril Ramaphosa, who would soon become general secretary of NUM. The story doing the rounds was that Anglo had agreed with CUSA to grant NUM access, in order to avoid having to engage with a more radical union.[70] However, Gatsby and others who had been present at Wilgespruit were nowhere to be seen.

They arrived several hours later, as a group of seven unions, including MACWUSA, and from the moment of their arrival behaved as a faction on a mission. Representatives of each stood up to state the same position, which was the 'principled' stance MACWUSA had taken in Wilgespruit, with two more 'principles' added. As they did so, there were complicit smiles between them at each other's clever remarks. I was again chair, and in a juvenile moment dubbed them the 'Magnificent Seven'. In the movie of that name, 'the seven' came to the aid of the poor peasants.

I have no first-hand knowledge of how this about-face happened, but it was obvious there had been a meeting to which only they had been invited.[71] It was also obvious that MACWUSA itself did not command the authority to have won the others over to its 'principled' stance. Someone else had intervened politically, and one of the 'principles' added provided a clue as to who that was, and why BMWU had changed its name: Black Consciousness was a 'reactionary fallacy', it said, and called for the 'total rejection of reactionary organisations – local and international'.[72] This was Party-speak, but there was no way of telling a Party operative apart from a SACTU operative, and SACTU in exile had particular cause to be concerned about the formation of a new federation.[73] There also seems to be little doubt that the target of this political intervention was primarily FOSATU.[74] If FOSATU was not itself a 'reactionary' organisation, it was seen as keeping company with them, notably the ICFTU. Doubtless the Party would also have taken umbrage at FOSATU's advocacy of a working-class movement, now that Foster's 'keynote address' had circulated more widely. The Party, after all, regarded itself as the sole authentic representative of the working class.[75]

'The seven principles are mandated by the workers and they cannot be modified,' Donsi Khumalo pronounced with breathtaking arrogance at the end of a frustrating day.[76] Donsi was representing GWUSA, a general union established by MACWUSA a short while before. Not long after the meeting, there was a falling out between him and the leadership of MACWUSA/ GWUSA, and the next time I encountered him he was representing a union in Pretoria. 'You claim to discuss the issues with the workers,' I snapped back, 'but each union has separately pronounced the identical principles ... You are refusing to budge from a preconceived standpoint.'[77] It would have been better coming from someone who was not white.[78]

The effect of a faction driving a particular line in a meeting is to polarise positions. Ramaphosa was busy extolling the virtues of Robert Mugabe in bringing about reconciliation in Zimbabwe when 'the seven' pitched up the following day and reiterated their 'principled' stance. The other organisations then began to espouse their own 'non-negotiable' principles in which 'worker control' was the common theme. In CUSA's case, it was 'black worker control'. This was a way of saying that they did not believe 'the seven' were accountable to, or controlled by, workers.[79] So, one of the consequences of this meeting was to lend weight to a critique of the 'community union' model.

However, the momentum built up had been lost and the trust created through an inclusive process squandered. GWU wanted to press ahead anyway, on the basis that 'some unity is better than none'.[80] FOSATU then proposed that we meet with GWU and themselves. We were not amenable to this. The only meetings we would attend were meetings to which all were invited, we said. An unofficial reason was that this would have looked like the banding together of organisations with whites in leadership positions. Probably the most important political consequence of the breakdown in the unity talks was that when the second report of the President's Council came out, the trade unions were not united in their response.

A bunch of *skollies*, was what Pendlani said to me about 'the seven' straight after the meeting, while we were still in shock at what had happened. This was not of course remotely true. Their representatives were young, black, male and snappily dressed. They were a nascent elite. Sipho Pityana from MACWUSA would one day be a director general in the Department of Labour. Sydney Mufamadi would become a Cabinet Minister. The Congress movement was bound to want to recruit individuals such as these, although some of them would not make the cut. What was true was the lack of respect displayed towards someone like Pendlani. It was as if anything he might say would have made no more impression on them than if they had really been *skollies*, trying to part an ordinary worker from his wallet.

At the Union's conference that year, we resolved to intensify our engagement with the community, to explain the need for trade union unity more widely. The idea was to claim the political high ground, and counter the line taken by the seven.[81] A letter also arrived from Oscar in jail in time to be read. 'It is with great pleasure to me to have found this opportunity to send you greetings at your Annual Conference however with great regret to this miserable situation I have been subject to,' the letter read 'I have been harbouring under pretention that I shall be free before the end of July. All hopes were dashed by the refusal of my application for discharge at the end of the state's case ... I told court that allegations are fabrications. I am fully convinced that even without that Cross Roads affair, of which I have nothing to do, I could have been arrested. Our head office was raided. East London officials were detained, not once. Look at what happened in Johannesburg, shall I be wrong to say Comrade Aggett was murdered? Surely comrades you are with me on that point.'[82]

The state obviously wanted to put Oscar out of action for the remainder of his political life, and there was faint reason to trust in the independence of the judiciary when the charges were terrorism and murder. I knew from my father how it worked in practice. Ever since the Nats had come to power, the JP, as he called the Judge President of the local division of the Supreme Court, had been a party loyalist. The JP would take particular care to ensure that politically sensitive cases were entrusted to a reliable judge.[83]

There was even less reason to trust in the independence of the lower courts than the higher courts, and some months after Neil's death there was an inquest in the Johannesburg magistrate's court. I had been brought up to believe that magistrates were no more than civil servants, so to me it was always a foregone conclusion that the inquest into Neil's death would exonerate the police. However, no amount of cynicism about the legal process could have prepared me for the perversion of the truth that was the magistrate's finding, at the conclusion of the inquest at the end of 1982. If anyone was to blame for Neil's death, said the magistrate, it was Auret. Supposedly it was his responsibility to have alerted Neil's torturers to the consequences of their actions.[84]

20

Congella Road

YOU ARE PROVIDING AMMUNITION for 'the enemy', Leila Patel, the editor of *Grassroots*, told me. By talking about problems between community organisations and trade unions, you are exposing weaknesses in our ranks.[1] At least she was prepared to take the trouble to come and see me about the FCWU article I had written. The response of many others was snide remarks behind our backs.

Speaking of 'the enemy' and 'our ranks' was of course war talk, and South Africa was at war, even if the fronts on which it was being fought were across its borders. When I was exempted, military call-up seemed like little more than an initiation ritual for white youth. One of the few reported casualties had been the son of the factory manager of Moberg's. Who was the enemy, and what he was fighting for, were not questions anyone would have asked the droves of Moberg's workers that attended his funeral then. I doubt whether any would have attended such a funeral now. By dint of casualties learned about by word of mouth, it was clear that conscripts could be expected to fight. Coloured youth in Paarl, particularly, were leaving the country to join the ANC's military arm, MK.[2]

The idea that the Union could engage with the community, and win the political high ground, seems a little quixotic in this context.[3] Inevitably, perhaps, community organisation was increasingly assuming a sectarian character, as the example of the United Women's Organisation (UWO) showed. Liz Abrahams, Virginia and Athalie had helped establish UWO, and it had started out promisingly, doing what a women's organisation had to do, by taking up women's issues. Before long, however, the focus seemed to shift to recruiting women in leadership for the Congress camp.

The significance of MK in the popular imagination could not be gainsaid. The risk for unions was that this would encourage a belief that all other forms of struggle were secondary, and that their leadership would be encouraged to jettison the work of organising and representing workers in the belief that victory was around the corner. An associated risk was that strikes would be instigated with the object of creating a climate of insurrection. There was a fine line between a political strike, such as the Aggett stoppage, and what I would privately regard as using workers as cannon fodder. The strike at Langeberg in East London at the end of 1982 exemplified the risk. Months before wage negotiations were due, workers were demanding not only a wage increase, but the release of Oscar Mpetha. This was obviously not a demand management was able to fulfil, and on the second day the workers were dismissed.[4]

Athalie and I were on leave on the Transkei coast at the time. As I travelled back and forth along the back roads between the Transkei and East London to try to salvage the situation, I pondered the possibilities. Had the Union leadership come under pressure to strike over an overtly political demand, perhaps to prove a point to SAAWU? Did they believe that by so doing they would create an insurrectionary situation? In the tense climate that prevailed in East London, these were real possibilities. It was also possible there was an agent provocateur on the committee, whose brief it was to destroy the organisation. The scenario I was above all concerned to avoid was the one facing SAAWU at Wilson-Rowntree.[5]

This was a factory plum in the middle of East London, employing close to 1,000 workers. It might have become SAAWU's flagship. On one of my visits to East London, a comrade I bumped into expressed his amazement that the workers of Wilson-Rowntree had gone on strike again, for the nth time in so many weeks. A month or so later, after yet another strike, the workers had been dismissed. It did not look as if the boycott SAAWU had called was taking off. Coming after a succession of other defeats, the survival of SAAWU in East London was now at stake.

One of the things that set the Union apart from all the other unions in the unity talks was that it had survived. Survival was important, because trade unions were for the long term, and one of the reasons the Union survived, as I have explained, was that there was local control, in terms of our decentralised model of organisation. The ultimate test of the decentralised model would be making Durban, its furthest-flung branch, viable. We could not be a truly national union without a successful branch in Natal, as the province was then called.

'This is the third or fourth time I am asked to join a union: what makes this one any different from the others?' was the question workers in Durban posed to us. It was unfair to expect Peggy, the branch secretary, to have been able to answer it on her own, when she herself had no experience of what the Union could be. Following a visit to Durban by Noko and Liz Abrahams to assess the situation, we asked the East London branch to suggest a volunteer to help her. It nominated Mandla Gxanyana.[6] Mandla was from the East London intelligentsia and not a member of the Union. He had also been with 'the seven' in Port Elizabeth, although I did not know it at the time. A couple of months after he arrived in Durban, Peggy died tragically in childbirth, and he became acting branch secretary. It soon became clear how well he had absorbed what was distinctive about the way the Union operated. Although the head office would pay him a salary, this was strictly a temporary arrangement. It was up to him to secure a membership. If he was able to do so, then he would have a job, not because he had met some performance standard that head office had set, but because there were sufficient subs-paying workers to provide him with a salary.

Mandla could not have succeeded in making inroads in a large urban centre without help. It was here that the Union's openness to the community paid dividends. Someone from the community put him in contact with Bobby Marie, an activist from Merebank who came from a Black Consciousness background. Bobby's banning order had recently expired and, crucially, he had an old Volkswagen Beetle that he was willing to use to ferry Mandla to the factories. Our focus was on the south of the city centre. Durban was important symbolically, of course, because of the 1973 strikes, but ten years afterwards the food factories were almost entirely unorganised.[7] Our organisational paths had also not crossed with Sweet Food, although this was where it now had its head office. Jay Naidoo had just been elected as its general secretary.

If you took the N2 from the city in the direction of what was then Louis Botha airport, you could not miss the cement-and-steel superstructures of these food plants, including some in sectors of which the Union had no prior experience, like the sugar mill and the sorghum brewery. Interspersed among them were fig trees and tropical greenery. The first factory where the Union secured a majority was Carnation Foods, a small factory producing canned goods and sauces, but before long it had begun organising a far larger and more strategic factory, Robertsons Spices.[8] The challenge in Durban, as in the other big urban centres, was to bring the Union closer to the workers. To do so it needed an office nearer to the factories and, as always, a meeting place. Here the support of the Catholic Church was critical.

Parallel to the N2, taking the Sydney Road exit from the City, you come to Congella Road. There, in Clairwood, adjacent to a hall operated by the Young Christian Workers, the Union rented an office from the church. Further down the road, on your left, there was a stone church whose square steeple is visible from the N2. It was here that the Union met with the workers of Carnation and Robertsons and, later, other factories as well. By the time of the conference in 1982 the Union had a majority at Robertsons Spices and before the year was out we had negotiated a recognition agreement.[9] By early 1983 the Union had gone on to organise the SASKO mill in Mobeni.

This union is not like the unions that the workers knew in Durban, I recorded Zwane, the branch chairperson, telling the conference. The branch had on its own initiative collected the money to send a delegation of eight to Paarl by kombi. These other unions were like insurance schemes to which workers contributed, Zwane went on. The leaders of these unions always disappear after a time with the workers' money, without the workers seeing the benefit of it. Our Union showed that what was important was unity.[10]

At the same time as we were consolidating our organisation in Durban, the Durban-based National Union of Textile Workers (NUTW) was seeking to expand to Cape Town. Joe Foster set up a meeting between its general secretary, Johnny Copelyn, and me. NUTW wanted to employ Virginia to help them do so, and was offering more than the Union paid her.[11] Some six months after that, MAWU offered a job to Bobby, who was still working on a voluntary basis. So the emergent unions were now generating a labour market of their own, in which a second wave of paid official would find employment. Without diminishing the magnitude of the task confronting this second wave or of subsequent waves, they were not pioneers. They were inheriting structures and traditions of organisation that already existed.

Athlone and afterwards
The second report of the President's Council recommended the establishment of a so-called Tricameral Parliament, in which coloureds and Indians would be represented in separate houses, alongside whites, but the African majority were not represented at all. As outrageous as that proposal seems today, there were influential people among the coloured and Indian middle classes who saw it as 'a step in the right direction'. Dominee April, who had helped establish the Union at George, accepted a position on the President's Council.

You would not have been as likely to hear someone from the emergent African middle classes say that 'independent' homelands were a step in the right direction, yet there were still enough takers for the opportunities the homelands provided, including the opportunity to become bosses in their own right. The example I used to give was of Temba Bakery in Bophuthatswana, which we had organised, and which was a partnership between Premier and the Bophuthatswana Development Corporation.[12] The number of African and coloured members in our Union were by this point roughly equal, and we could not afford to be ambivalent about the political situation unfolding. Government was attempting to buy off the middle classes, and what the President's Council was proposing was an 'anti-worker' constitution. At a packed management committee meeting in January 1983 a resolution was adopted condemning these proposals as well as the coloured Labour Party's decision to accept them.[13]

The anti-worker constitution made the formation of a federation all the more imperative. The GWU had in the meantime had second thoughts about 'some unity being better than none at all'. It invited all the unions that had been at Port Elizabeth, including 'the seven', to a meeting whose sole object would be to discuss the possible formation of a federation.[14] GWU received a letter in response from MACWUSA accusing it of being 'a bunch of whites manipulating black workers for its own obscure ends'. Hennie du Toit commented on this letter by drawing an analogy based on his own experience of organising in the workplace.[15] 'When a worker starts acting divisively', he said, 'there is always someone pushing him from behind to do so.' Hennie was one of a new generation of coloured leaders that had emerged in Grabouw, following the strikes.

The only opposition to the Union's decision to press ahead with the federation was from Bryce*, who like Mandla was an African from the East London intelligentsia. In fact, it was in large part because of the success Mandla was having that Bryce had been appointed on the same basis to our Port Elizabeth branch. 'Some people are keeping quiet because they don't share the same opinion,' he said, after a succession of speakers at the NEC had advocated pressing ahead with the formation of the federation regardless of MACWUSA/GWUSA. He was clearly referring to himself. 'We won't be happy if some unions don't attend the meeting, because a minority can be powerful.'[16]

This was a strange thing to say in a Union meeting. The Union preached that a minority had no power. However, MACWUSA/GWUSA had its own approach to organisation, and Bryce was clearly taken with it. The demarcation between MACWUSA and GWUSA was indicative of their political ambitions. MACWUSA aimed to organise workers in motor assembly and its components, the industries on which the economy of Port Elizabeth was founded. GWUSA would organise every other workplace. The organisational logic seemed to be that two interlocking trade unions would have the capacity to control the regional economy.

This, however, presupposed that the majority in a majority of workplaces were properly organised. In the workplaces we were interested in, which were mainly branches of conglomerates we had organised elsewhere, GWUSA had a handful of members in each. What I was told was that these 'cells' were not too concerned whether their fellow workers were members or not. Instead, they operated as gatekeepers to keep the likes of us out. This seemed like exercising power over ordinary workers rather than for them. A minority can of course be powerful when it occupies key positions in the workplace hierarchy.

Bryce's task was very much easier than Mandla's, because he did not have to start from scratch. The branch already had Harvestime, a factory with a substantial membership. He had only to maintain that base, and expand it by organising new factories. It was up to him how he did so, in terms of the Union's decentralised model. Although we would have warned against making promises the Union could not keep, the only criterion by which we could monitor his performance was the number of workers paying subs.

The contrasting fortunes of Mandla in Durban and Bryce in Port Elizabeth exemplify two very different approaches to worker organisation, adopted by political activists from a similar background. The fact that workers at Durban had decided to collect money to send a delegation to a Union meeting 2,000 kilometres away showed how workers had been empowered, although empowerment was not a term we used at the time. In contrast, in Port Elizabeth workers at Harvestime seemed increasingly unwilling to pay their subs. Bryce's response was a familiar one: to blame the workers for their backwardness. These same workers, it later emerged, were afraid of him.

These very different approaches to worker organisation mirrored the division that surfaced again at the next round of unity talks in the spanking new hall that CTMWA had created in its renovated Athlone offices. 'The seven' were there, including MACWUSA/GWUSA, which claimed to have had nothing to do with the letter in its name about 'a bunch of whites'. If it was fabricated by the Special Branch, however, the SB was very likely quoting what MACWUSA's leadership had actually said. This was the SB's way of 'pushing from behind' to promote discord.[17]

'Unity should develop organically from struggles on the shop floor,' Thozamile Gqweta intoned at the start of the meeting. Gqweta was the nearest to a Biko figure that the emergent trade union movement had produced. The tragedy was that most of those present did not believe he knew what he was talking about.[18] This was not just because of what had happened at Port Elizabeth, where Gqweta had not been present, but because 'the seven' had not prospered since. It was now clearer than ever that these were not organisations that in any sense empowered ordinary workers. The political dynamic had also become clearer. If the community unions were left out of a new federation, their funding would likely dry up. Their officials would then be without a job. Inside a federation, they would be exposed for having few if any members. Either way, the formation of a new federation threatened their existence and the destruction of their political base.

At the same time, the threat to SACTU which the emergent movement posed had not dissipated. 'Mention has been made of SACTU,' remarked Gqweta in response to Liz saying that the Union had previously been affiliated to SACTU. 'SACTU participated in the adoption of a document in Kliptown in 1955 which represented the aspirations of the people. A federation should not hold views in conflict with this document and should not compete with SACTU both locally and internationally.' Sydney Mufamadi was more explicit later in the day. 'Those who are in a hurry to form a federation', he said to applause from SAAWU and others, 'have an ulterior motive which is to set back the broader struggle.' I was not going to allow him to get away with this slur unchallenged, but he refused to withdraw it.[19]

The five organisations 'in a hurry' represented the best-organised formation of workers in the country. Each had significant delegations of workers and, although FOSATU had come as one delegation, this time each of its affiliates was represented.[20] When a delegate from FOSATU formally proposed a resolution to form a new federation halfway through the second day, Hennie seconded him. These five organisations would now drive forward the process of establishing a federation and agreed to form a committee to do so. The other unions now had to jump one way or the other. Unexpectedly, and with obvious bad grace, SAAWU then agreed to join the committee. Others prevaricated. Mufamadi protested that it was GAWU's democratic right to participate in this committee even if was not prepared to commit to the federation.

'The sun is setting,' Friday pointed out, speaking in isiXhosa. Many branches had sent delegates to the meeting, and the spirit among them was high. 'There are people here whose only aim is to obstruct the meeting from reaching a conclusion.'[21] The obstruction was to continue, however. GAWU somehow managed to insert itself onto the committee despite not committing to join a federation. Others would do so later, among them CUSA. Someone suggested it be called a feasibility committee. It was an unfortunate title since it implied that the establishment of a federation was still up for debate.[22]

A centralised versus decentralised model of organisation
So the feasibility committee was divided from the outset between a core group committed to a federation and those who felt they could not afford to be left out, because this was where power was located or because they were playing for time. The wild card would be the role that self-interest played on both sides of this divide. Although it was not as obvious as it would soon become that the trade union movement would be a stepping stone to positions of power, I doubt whether anyone who was part of the process did not sense this possibility. Certainly, I did.

What the organisations that made up the core group had in common, they thought, was a belief that workers should control their own organisations. This was what the slogan 'workers' control' was about, although some would regard it as referring to control over the production process. And even though in terms of their different models of organisation there were different conceptions about how 'workers' control' could be realised – how to empower workers, if you like – there was at least consensus about the bright line to be maintained between paid officials and workers. This was what distinguished us from a union in which a paid official could be president. I was dismayed therefore to discover in researching this book that, already at this early stage, some among 'us' had breached this line in a fundamental way by entering into schemes whereby their shop stewards became 'full-time'. I will discuss this disastrous development in a later chapter, at the point at which I first became aware of it.[23]

The other thing the core group thought they had in common was that they organised industrially, as did CUSA. 'If there is already a union formed, others shouldn't go out and form new unions. Unions are springing up by the day,' Ramaphosa said, no doubt with the threat in mind of another union moving into the mining sector. 'Everyone must put their cards on the table, and say which industries they are involved in.'[24] There are different ways of organising industrially, however. FOSATU argued for 'broad-based industrial unions' rather than 'narrowly defined' industrial unions.[25] This was in fact what the Union had now become, despite having had a narrowly defined industrial base for most of its life. The reason for favouring broad-based industrial unions was that the same big conglomerates controlled the sectors in which we operated, and nowhere more so than in food manufacturing.

There was, of course, nothing new about bigger fish swallowing smaller fish in the sea of free enterprise. However, the pace at which it was happening had intensified during the recession, and one of the first of the smaller fishes to be swallowed whole was Fatti's & Moni's. After a fierce tussle with Premier, settled on the steps of the Supreme Court, it was bought by Tiger. Straightaway Tiger began taking a tougher line towards the Union.[26] Yet the way in which the Union went about organising industrially was also different from that of the FOSATU unions. FOSATU's organising strategy was to mobilise all the workers in an area to join FOSATU unions, much like a general union. It was slower to develop a strategy of targeting firms producing the same product.[27]

The more important difference between FOSATU and ourselves, however, concerned its centralised model of organisation. In a centralised model, control resides with the head office because, in the final analysis, the head office controls the money. The only monies a branch handle are the funds allocated to it. That leadership at branches should not be placed in temptation's way seemed to be their thinking. The most problematic manifestation of this approach was to recruit workers at unorganised plants by getting them to sign stop order forms, on the assumption that employers would eventually honour them. This led to workers being counted as 'paid up' who had never actually contributed a cent to the organisation, although we did not know this at the time.[28]

It was the head office, in the centralised model, that not only determined what officials were paid, but that actually paid them. It was also the head office, by all accounts, who fired them. The general secretaries of FOSATU unions were, of course, accountable to an elected worker leadership in the same way as I was, but it is very difficult for workers that are employed full-time in a factory to exercise control at a central level. The adoption of this model, so far as I could see, was driven by foreign funding. Foreign funders were clearly not going to route their assistance to branches. The question was whether foreign funders were, in a similar manner, going to determine how the federation would be financed.

The issue came up towards the close of the first meeting of the feasibility committee. 'At this stage of our development', Piroshaw Camay of CUSA pronounced, the federation needed to be financed from the outside.[29] 'We are not interested in a federation that relies on money from the outside,' was my response. 'If we want a federation controlled by workers, then the workers must pay for it. If we want a federation controlled by the ICFTU, then we must ask the ICFTU. We don't accept that you can take money from the outside "for a stage". Once you take the money, you acquire a taste for doing things in a manner which you cannot afford. It later becomes difficult to do anything any other way.'[30]

We had not aired our concerns over foreign funding before, even with the GWU, and this was as far as we were able to take it.[31] There was no prospect of weaning organisations from foreign funding when there was not even agreement that workers had to pay subs to be regarded as members.[32] This emerged when, most embarrassingly for Ramaphosa, CUSA failed to put their cards on the table and say how many paid-up members they had, and in which industries. And SAAWU and GAWU refused to do so because they did not accept paid-up membership as a gauge of a trade union's actual strength.[33] In effect, they wanted their membership claims to be accepted on trust. This was no small matter. Paid-up membership was the only basis on which to determine how many votes a trade union would have in a federation.

Soon afterwards, FOSATU wrote a letter to us proposing 'a new initiative encompassing all those unions who are serious about unity'.[34] This entailed withdrawing from the feasibility committee. In our view, this would have been a mistake. Those withdrawing would be portrayed as splitters and tarred as anti-Congress. In truth, those organisations that refused to disclose how many members they had were obviously not ready to form a federation.

A clash of political cultures
One reason SAAWU, GAWU, MACWUSA/GWUSA and others had to play for time in the unity talks was the moves in the offing to form a 'united

front' to oppose the government's constitutional proposals. The Union supported the idea of a front, but it was obvious from the outset that this would have to be a very different kind of front from the one which we had contemplated forming with trade unions, and which had now given rise to the proposal of a federation.

The problem with the 'disorderly committee' was that 'organisations which represented nobody' had been allowed into the meetings, Trevor Manuel and Cheryl Carolus told us on a visit in early 1983. They were referring to the attempt to form a front between trade unions and community organisations to oppose the Orderly Settlement of Black Persons Bill, nicknamed the Disorderly Bill. The name 'disorderly committee' was apt for other reasons as well. The solution Manuel and Carolus proposed was to exclude organisations that were not 'main line'.

This was the first of several meetings in which it gradually became clear that to qualify as 'main line' an organisation's perceived allegiance to the Congress movement was more important than how many members it had.[35] This, of course, could not be said in so many words, and a lot of their argument was based on what went on in the 1950s. I found myself becoming increasingly irritated at this harking back to the 'golden' fifties. You cannot look at a period 30 years back in isolation from what happened afterwards, anymore than I can look back at the 1980s today in isolation from what happened in the 1990s and afterwards.

For the trade union movement in the 1980s, as I saw it, the critical historical period for us to look back to was the 1960s. The legacy of the 1960s, I had by this time concluded, was a distrust of politics and political organisations that could not simply be ascribed to repression. It could also be ascribed to the fact that many leaders in SACTU had jettisoned the trade unions: some to go underground and others to advance their careers, while unions like the TWIU had drifted into the TUCSA camp.[36] If we were not to repeat the mistakes of 1960s, it was important how decisions were taken in a united front, which concerned the risks organisations would be exposed to.[37]

Carolus was from the UWO, which I have already discussed. Manuel's organisation was CAHAC, a civic organisation representing residents in the coloured areas of the Cape Flats. The centrality of local government in the President's Council's proposal underscored the importance of civic organisations. Establishing effective civic organisations was clearly a formidable task, but at root and base the task of building any form of membership-based organisation is the same: it entails identifying what the material interests of the members are, and effective ways in which to advance them. My beef with Carolus, Manuel and others like them was that they did not seem much interested in the painstaking process of building an organisation.

The Union is open to joining a front, we said, but we were not interested in forming an organisation. We spelled out why. A front is by definition not an organisation, and does not have members. It is open to all who join it. This gives rise to problems as to how decisions are reached, and whether they are binding. The unions were busy establishing their own organisation, a federation. A federation was the most appropriate way for worker organisations to engage in political issues. This was also, by the way, how it had been done in the golden fifties. So it was that deliberations about the formation of a federation and that of a united front proceeded parallel with each other. This was to have a number of adverse consequences. For one thing, it created a problem of capacity for the Union and for other unions participating in these different deliberations. It also raised a question of priorities. The priority for us had to be the federation.

When we were notified of what was styled as a 'conference' in May 1983, to decide how the different organisations would work together to oppose the constitutional proposals, I somewhat naively supposed our views, which GWU and CTMWA shared, would be taken into account.[38] But the openness of a front was precisely what the activists driving this initiative objected to. I listened in disbelief as Carolus outlined the structure of an umbrella organisation, a federation of sorts, in which we would have an equal vote with other kinds of organisations, including those that could not remotely be regarded as 'main line' if membership was the criterion. She proposed it be called the Cape United Democratic Front.

The venue was in Kensington, the middle-class coloured suburb that was Manuel's home ground, and most of the meeting was taken up with lengthy introductions of the different organisations. The unions, and specifically FCWU, were the only organisations present with a significant mass base in the Western Cape.[39] There were also two affiliates of FOSATU present: NAAWU (represented by Joe Foster) and NUTW (where Virginia now worked). It was an early indication that FOSATU was becoming more open to engaging with community leadership. It remained to be seen whether this would mean a loosening of the 'tight' federation.[40]

Evidently Manuel and Carolus had already discussed this structure with activists somewhere else. We were expected to rubber-stamp the proposal. This was how 'democratic centralism' operated. The centre would lay down the line and the branches were expected to implement it. Manuel, in a leather jacket, seemed to regard it as his designated task to secure the trade unions in the Western Cape for what would become the United Democratic Front (UDF). He responded with less and less patience to our objections as the meeting progressed, as though we simply did not get it. What he did not seem to get was that this attitude merely strengthened the resolve of the unions to stay out.

The conference ended inconclusively, with snide remarks about the unions having frustrated the formation of the front.[41] However, SAAWU and GAWU, who had all the while been frustrating the formation of a federation of trade unions, were quick to affiliate. SAAWU would soon be held up as a model trade union, while the Wilson-Rowntree boycott would become a cause célèbre of the UDF. On the ground in East London, however, SAAWU was a spent force as a workers' organisation. Like a cactus flower that opens for one night a year only, it had blossomed and then shrivelled up. All that remained was a desiccated husk.[42]

There was a follow-up meeting a month later. Most of the trade unions' objections to the structure of the proposed organisation were focused on what were envisaged as area committees.[43] Area committee meetings were needed, a representative of the UWO told us, to ensure 'a structured, disciplined campaign'.[44] Being subjected to the discipline of an undemocratically constituted committee was precisely what we objected to. She was a white intellectual from the same kind of background as I was, which is probably why she became my stereotype of the political hack that slavishly toes the party line.

'All our branches saw the need [for area committees],' she said. 'The UWO's proposal is mandated by all its branches.' Some of the trade union delegates must have snorted at her mention of mandates, because they saw her as paying lip service to notions of accountability which the trade unions had introduced into political discourse 'Organisations must have respect for ways in which other organisations get their mandates,' was her response to the snorts.[45] 'We need a central structure to coordinate because for us we need directions from worker and community organisation,' she said in defence of what Carolus had proposed. The irony of claiming to want direction from worker organisations, in the same breath as she disregarded what the worker organisations present were saying, seemed to pass her by.

I have never understood what is 'democratic' about democratic centralism. It seemed to me that once the line had been laid down, those that did not toe the line would be labelled 'reformist' or 'reactionary', either to bludgeon them into line or put them beyond the pale. The label 'workerist' was not yet commonplace but would soon be mustered for these purposes. Our role, as they conceived it, was simply to rally our members to their line. We could not see how working-class people would have a voice in their structure, let alone control.

So when the UDF was launched in Mitchell's Plain with much razzmatazz a few months later, I was ambivalent. It was a front only in the sense of being a front for the ANC. No doubt it was better having this kind of front than none at all while political organisations were banned, but it seemed to me to be overly concerned with establishing power over others rather than organising for the long haul. It rankled for some that we had not affiliated. Our messages of support on this or that occasion would mysteriously get lost. We were no longer fashionable in circles in which we had formerly been. Some would say we had betrayed the Congress tradition of which we were once part, though this accusation conveniently disregarded the state of the Union as I found it in 1976.

Because the UDF was a front for the ANC, it came as no surprise that soon after the ANC was unbanned, the UDF was dissolved.[46] All that remained of the organisations that had formed it were the student organisations and perhaps some faith-based organisations. CAHAC and the UWO had long before lapsed into irrelevance.

The long haul

Yet there was no gainsaying that the formation of the UDF changed the political landscape at the time. There was now an internal political organisation utilising the space that the government had created by its political reforms, just as the unions had utilised the space created by the labour reforms. For people who had grown up with no high expectations, it was enormously appealing. We were therefore careful, despite our fallout with its Western Cape leadership, to emphasise our support for the UDF.

On the same weekend as the Union's own conference, in Johannesburg, a national structure for the UDF was formed. The conference resolved to accept an invitation to meet with the national leadership of the UDF, as well as endorsing their declaration of opposition to the new constitution.[47] We were in effect in a front with the Front. This was the first conference in the history of the Union to be held outside the Western Cape. The venue was Wilgespruit. It was too small to accommodate the number of delegates that attended – Kempton Park alone had over 40 delegates – so we met in an open-air amphitheatre under the gum trees. The guest speaker was one of our own members, now retired: Aunt Violet, who had once carried an umbrella for Oom Joe.

This was so that the branches outside the Western Cape could hear about the founding of the Union from the horse's mouth, as it were: a coloured worker who had been at Moberg's 42 years before, when the Union was founded. The busloads of workers that had travelled from the Western Cape had heard it all before. Most would also have been familiar with Aunt Violet's rhetorical flourishes and idiosyncratic phrases. The revelation for them – and for me – was that when she started speaking, the language she used was not Afrikaans, but isiXhosa.

I have already mentioned that Aunt Violet understood isiXhosa, but I had no idea she was a isiXhosa speaker. What had made her declare this so publicly? It must have taken her something like 40 years to muster the courage. It was safe for her to do so now, because she was in an organisation in which different identities were represented in numbers, and respected. It went to show how long the haul might be, for people to give expression to their identity. Also, it showed another way in which workers were being empowered.

Mandla was at the conference, together with Inthiran Moodley, who had replaced Bobby as an organiser in Durban. However, the delegation of workers that accompanied them did most of the talking. It was clear their branch was flourishing. Port Elizabeth, on the other hand, was not present, for the first time since the branch had been re-established. There was not even an apology from Bryce.

Our deal with Bryce had been that the head office subsidise the branch for six months. That should have given him time enough to organise enough workers to generate a salary for himself. However, they had not even organised Epol, where the workers had everything to gain from belonging to the same trade union that was negotiating at a national level with six of the company's plants.[48] Either Bryce did not want to upset MACWUSA/GWUSA or he actually believed in their vanguard model.

A delegation from head office went to Port Elizabeth to assess the situation. There was a huge unpaid telephone account, but no other indication of financial irregularities. Workers were simply not paying their subs. Yet they were more than willing to do so while we were there. Bryce clearly did not inspire confidence. We told him he was not up to the job, and he did the decent thing and resigned. That, you might have thought, would be the end of the matter, but a month or so later he changed his mind. Someone must have prevailed on him to try to withdraw his resignation because it was strategically important to have someone of his political persuasion in the Union.

We held a marathon meeting of the branch executive committee to discuss his request to withdraw his resignation. It was at the house of a member in one of the informal settlements that had sprung up on the outskirts of New Brighton or Zwide, because Bryce believed the office was bugged. It was also the Special Branch's doing, he told the meeting, that the telephone account was so exorbitant. The same thing had happened with MACWUSA.[49]

I did not know whether to believe him about the telephone accounts. However, years later it emerged that the Special Branch had indeed been running up the telephone accounts of unions in Port Elizabeth, to bankrupt them. It was a form of economic sabotage.[50] The issue of the telephone account, however, was beside the point. The subtext of Bryce's request was that he should remain in the Union because he was politically a very important person. My point was there was no room in the Union for VIPs who could not or would not organise workers.

This was not something any member of the branch could have said to his face. If there had been money in the branch account to carry on paying a salary, I have no doubt they would have let him do so, not because he inspired confidence, but fear. Even Bernard Mahlakahlaka, who had resigned as chairman to further his studies at Fort Hare and had volunteered to help the branch during his vacation, was careful not to criticise Bryce to his face. This was the beauty of the decentralised model of organisation. The fact that there was no money in the branch account, a situation for which Bryce was to blame, decided the matter. There could be no expectation that the head office was going to bail the branch out. When it was reported to the members that Bryce would no longer be employed by the Union, at a general meeting the next day, there was a palpable sense of relief.

While going through the books after Bryce was gone, it was discovered that he had 'affiliated' the Union to the UDF and even paid an affiliation fee.[51] Not long afterwards, a smear pamphlet was distributed in the Port Elizabeth townships attacking trade unions that were led by 'whites and women'. I think the reference to women was on account of the fact that Deborah Komose, from East London, had stayed on in Port Elizabeth to help the branch back onto its feet. At about the same time Bryce wrote a letter claiming payment for the month after he completed his notice period. When I rejected his claim, an unsigned letter arrived addressed to 'Comrade Jan Theron' from 'East London Comrades'. The tone was threatening, and it closed with the comment, 'We also hope you will pay other comrades their wages.' Not long after that the Union started being bad-mouthed on the trains.

During all of this, Norushe, who really was an East London comrade, was in detention. After having been arrested in East London he was handed to the Ciskei, where he was kept for more than five months. Sometime in March 1984 he was released without being charged and soon afterwards disappeared. The consensus was that he had 'skipped' the country.

21

The road to Pretoria West

IT WAS AN OMINOUS DEVELOPMENT for the Union to be bad-mouthed on the trains from Mdantsane, because someone in the Congress movement must have sanctioned it.[1] You would think it made no sense to destroy the Union's most precious asset, the belief among organised workers and the working-class communities they came from that the Union was a good thing. But it did make sense if you were engaged in a hegemonic project.

It was in areas in which belief in the Union was strongest that working-class communities would most readily follow its lead on an issue such as the community council elections.[2] Where organisation had always been weak, on the other hand, our influence was limited. At Ashton, to the Union's embarrassment, two members of the branch executive committee were elected as councillors in Zolani township. The more critical elections, however, were for the House of Representatives, as the coloured chamber in Parliament would be known. We had already started to prepare our members for this at the start of 1984, before it had even been announced there would be elections.[3]

The UDF's plan, however, was to collect a million signatures in opposition to the new constitution. This was its 'Million Signature Campaign'. You do not build organisation by setting unrealistic targets, and this was not a target the UDF could by any stretch of the imagination have reached. Also, a petition was a singularly inappropriate way to engage with the working classes, many of whom did not read or had reason to be apprehensive about putting their signature to documents. The leadership of the UDF had opted for razzmatazz rather than utilising the political space that had opened up to engage with ordinary people.

We had a practical proposal as to what the UDF could do: address a series of meetings that we would arrange at our branches, and that would be open to members and non-members alike.[4] Meetings in the rural Western Cape would be especially important. These areas would have a major impact on the outcome of the elections, and the Union was more often than not the only counter to the Dominee April-type figures that held sway there. Yet with the date of the election fast approaching, and the government and the Labour Party campaigning to register coloured voters, there was no move on the part of the UDF to take up our offer.⁵ In the end the Union was left to confront the Labour Party and its backers on its own.

Political pressures like this made it a trying time to be a worker leader. Pendlani had now retired, and been replaced by Noko, who was fiercely loyal to me and popular with the coloured members, because his second language of choice was Afrikaans.⁶ Yet it had become increasingly obvious that a national union needed a president who was fluent in English. Also, despite the importance the Union attached to having a worker at its head, it was hard for a machine operator stuck away in Tulbagh to command the respect of the clever young men (and some women) of the UDF or, for that matter, the UDF unions.

'UDF unions' was a new label for the community unions, led by SAAWU and GAWU, which had affiliated to the UDF. After another round of unity talks, in March 1984, SAAWU and GAWU claimed they had been expelled from the talks. Although this was not true, as I shall presently explain, the UDF took their part, and this led to further strain in our relations with the UDF. Some in UDF ranks also believed SAAWU and GAWU were expelled because they were unregistered.⁷ A militant stance on registration resonated with the UDF's rejection of all institutions of the apartheid state, and especially those seen as integral to the government's reforms.

In this way, the issue of registration surfaced again, in 1984. However, unregistered trade unions were able to utilise the Industrial Court in the same way as registered trade unions were, as a result of another amendment to the Labour Relations Act (LRA).⁸ The court was not just integral to government's labour reforms: it was its jewel. What was surprising, given their stance on registration, was that SAAWU and other UDF unions were as quick to utilise the court as registered unions were.⁹ Even though there was some uncertainty about what kind of court it really was, it was certainly part of the judicial system, tasked with upholding 'the law'.

Attracting unregistered trade union litigants to the court was surely no less of a coup for government than to get community leaders elected onto community councils and, in the longer run, far more important. 'The law' in respect of labour relations was at this point rudimentary. It would not have been possible to develop it without the buy-in of the emergent trade unions. A 'turn to the law' by the emergent trade unions would reinforce a global trend which the Wiehahn Commission evidently wished to encourage, by expanding the reach of the judicial system at the expense of the administrative arm of government.¹⁰

I would like to think the Union's own turn to the law was not as sharp. It began with a dispute at Epol Pretoria West. This was a plant where over 95 per cent of the workforce had joined. This was not unusual for Pretoria. What would have bothered the bosses was how many workers in the middle layers chose to be represented by the Union. While most had by now accepted that it was a necessary evil that ordinary workers join trade unions, they were far from happy about workers in the middle layers doing so. But because they could not deny anyone their freedom to associate, they confined their opposition to the proposition that trade unions could only negotiate for those below a certain 'cut-off point'. This constituted what they referred to as the 'bargaining unit'.

The 'cut-off point' was one of the most hotly contested issues in recognition negotiations, and the bosses marshalled all sorts of arguments as to why it did not violate the freedom of association of workers to have one.[11] We did not find their arguments convincing. Workers joined a union precisely because they felt the need to have someone to negotiate on their behalf. Our strategy was, therefore, to win over everyone who was not part of management, although it was not without risks. Workers in the middle layers were inclined to believe they were a cut above the ordinary workers, and also to side with the bosses, as I have already observed.

The only category of workers in the middle layers which was not routinely regarded as above the cut-off point, and outside the bargaining unit, was drivers. Maybe this was because until this point drivers had not taken much interest in trade unions, or maybe it was because the bosses did not consider them part of their 'core business'. The 'core business' of Epol, the suave and unflappable MD told us, was not to feed the pets of the middle classes, as I had supposed, but to supply agribusiness, notably South Africa's burgeoning poultry industry. He was aptly named Fowler. The industry leader was Rainbow, which was known to be an exacting client. Its specifications as to what its fowls were fed were precise and inflexible, and it was the drivers who ensured timeous delivery of the feed. They were, in fact, no less key to its operation than workers engaged in production in the mill, and one of their number played a major role in organising the Pretoria West workers. This was a tall and charismatic man whom everyone referred to as Mr Mabelane. Like many men from the far north of the country, he wore a light green suit with a badge on its lapel: a silver star on a strip of green and black cloth.

I do not believe it was by chance that Mr Mabelane was only vice-chair, however. This was the workers' way of countering any wayward tendencies among the middle layers while keeping them on the same side. They elected a strong and silent worker from the mill as chair, one Joseph Mlambo. No doubt Epol thought it was being a progressive employer by appointing an African personnel officer at the Pretoria West plant, but so far as the workers were concerned he displayed wayward tendencies from the outset. They called him a snake, and the first dispute at the plant was over a demand that he be dismissed for his anti-union utterances It was only resolved after a two-day strike, by allocating him to clerical duties in the offices, where he would no longer have to do with the workers.[12]

There was another two-day strike less than a year later. This time the issue was the dismissal of five workers. The reason given was that they had played a leading role in the 'dismissal' of two of their colleagues. These two had not really been dismissed, of course. Only the bosses had the legal authority to do that. The workers had physically expelled them from the premises. It was not clear on what basis the five had been singled out for what had been the conduct of a much larger group.

It was some four years since Premier first agreed that where there was a dispute about the dismissal of an employee(s), 'such employee(s) shall be reinstated until the dispute settlement procedure has been exhausted'. This was the first occasion when the Union had invoked it. The wording of the agreement was unambiguous, and there could be no doubt what the parties intended when it was entered into. However, Epol simply refused to abide by it. Accordingly, Epol Pretoria West was the factory where I spent more time than any other in 1984. It became the testing ground for the most far-reaching of the rights the Union had been able to secure in a recognition agreement.

Pretoria West is the old industrial area of Pretoria, where the bulk of the Union's membership was located. From the city centre your point of departure is the central station, which the locals call Bosman Station. It is still an imposing building from the outside, despite the fire in the late 1990s, started by commuters angry at the trains being late and doubtless a lot else beside. Mitchell is the road that takes you from the Station into the heart of Pretoria West, past Ruto, reputedly the largest grain mill in southern Africa at that time and part of the Fedfood group.

The more scenic route is to turn left at Fountains Circle, on the outskirts of the city, and to travel through the hills that ring its southern perimeter, past the Voortrekker Monument. The road takes you to the slagheaps of what was at that time Iscor, the state-owned steel corporation. Staal Road, where the main Iscor plant is located, is off Industrial Road. Iscor at that time was still a bastion of the whites-only section of the established union movement, although there had been large-scale retrenchments there. Before the decade was out it would be converted into a private company. The Epol plant straddles the block between Staal and Bessemer, parallel to the railway line, and across from a row of red face-brick houses of the same design utilised by the SA Railways to house white workers all over the country.

We are not refusing to abide by the agreement, Fowler told us. We will pay the five workers their salary until the dispute is resolved. 'Reinstatement' does not require us to let the workers back into the workplace or put them back in their jobs. In the end we were forced to accept this position, or risk having the entire workforce dismissed. There was no court we could have approached to enforce what had been agreed. The meaning the bosses were now giving the term 'reinstatement' was precisely how the Industrial Court had interpreted it.

The turn to 'the law'

To understand how and why the Industrial Court came to this interpretation, it is necessary to explain the context in which our turn to the law began. The rapidity with which the Union had grown until this point was fuelled by the fact that in almost every industry where it was organised it had been winning real wage increases for ordinary workers. Nowhere was the increase more dramatic than at Simba-Quix, in Molteno, in the remote interior of the Eastern Cape. This was where Ouma Rusks, an iconic South African product, was made.

Negotiations took place in a makeshift venue in the factory, but this was not the standard to which we were by now accustomed.[13] At their insistence, the big conglomerates always hired hotels as the venue, precisely because they wanted to keep us away from the factory premises. But there was no hotel in Molteno, and during a lunch break we were able to wander through the factory, which must once have been the outbuildings of a farm. It was like wandering back in time, to an era of labour-intensive manufacturing, when it did not matter that workers were not fully occupied because it cost so little to employ them.

The big conglomerates had been willing to grant real wage increases until now, because the bosses perceived that wages were too low, especially the wages of African workers. Now Simba-Quix, which was part of the Fedfood group, had agreed to triple the wages. For us this was an unprecedented increase.[14] Even so, I was not particularly enthused. It was obviously not tenable for a big conglomerate to be operating as though it was still the 1930s, even in a rural backwater. It was surely inevitable that it would restructure its operation and retrench workers, even though it had not threatened to do so.

This realisation underscored how fleeting the impact of collective bargaining was in raising the living standards of the working class, and the limitations of what trade unions could achieve in the absence of a political organisation. This was more especially so for a trade union like ours, founded in labour-intensive manufacturing. The wage negotiations for the pineapple canneries a few months later made this clearer still. We thought we had taken a step forward by bringing a third cannery to the table, Deepfreezing and Preserving, only for WPP to pitch up and announce that they would no longer be taking part.

WPP were closing their factory on the West Bank, its management said, and no longer had a workforce.[15] The fact that it had not even bothered to notify the Union of this beforehand seemed to me to show how confident the bosses of East London had become since the demise of SAAWU. It was procedurally unfair not to notify the Union, the Industrial Court might have said, and this would have been in line with guidelines for retrenchments which the court did in fact develop during 1984.[16] Yet even if it had been an option for AFCWU to refer a dispute to the court at this point, I doubt whether the Union would have done so.

A finding that their retrenchment had been procedurally unfair would have been cold comfort to the workers of WPP. It would also not have got them their jobs back. 'What was the real reason for retrenchment?' was the question that had to be answered, if the workers were to have labour justice, but no court would have had the appetite to answer it until after the transition to democracy.[17] WPP was not closing for the same reason as the deciduous fruit canneries were. In fact, it was opening up a new factory at Kidd's Beach, so it could employ a workforce from the Ciskei. Labour-intensive manufacturing was low-wage manufacturing, and that was more easily done under a despotic regime.

This was the parent company of WPP – H Jones – doing what multinational companies in all parts of the world would increasingly do over the next few decades to maximise profits. I had been to Australia the year before, at the invitation of the Food Preservers' Union of Australia. However the trade union movement there was fragmented, and as luck would have it, H Jones was based in a state where another food union operated.[18] Although it was unlikely Australian workers would come out in support of the WPP workers, so long as there a prospect they might do so, we were going to see what came of it. 'Let us await the outcome of the court proceedings,' H Jones would undoubtedly have told its workers, had there been court proceedings pending. 'If the trade union in South Africa has confidence in the courts, who are we to second guess them?'[19]

Undoubtedly, as the economy deteriorated, there would be many situations in which there was no prospect of an organisational solution for a trade union. But trade unions would also be given to believe there were no organisational solutions by those who had an interest in promoting legal remedies and resort to the courts, in situations in which their legal prospects were no better. The agents of this development were a new breed of lawyers who, spotting a business opportunity of note, lobbied for an amendment to the LRA that would entrust to the court the power which the Minister of Labour had to reinstate employees pending the resolution of a dispute.[20]

This was the so-called status quo remedy, and it was tailor-made for lawyers, because it was decided entirely on the papers. The more 'skilfully' drafted the papers, the better the prospects of workers being 'reinstated'.

What these skills boiled down to was a command of legal discourse and an ability to select that version of the facts likely to be most palatable to the court. A sentence inserted or omitted might determine the outcome. However, reinstatement, as interpreted by the court, did not mean what we understood it to mean, namely putting workers back in the position from which they were dismissed.

The irony was that it was trade union lawyers who persuaded the court that all an employer had to do to 'reinstate' workers was to pay them their wages.[21] It was an interpretation that undermined not only a right the Union had secured in its agreement with Premier, but one that tens of thousands of workers had placed their jobs on line to defend (and not, of course, in our union alone). It was an early indication of how disempowering the turn to the law would be for workers and worker organisation.

Private arbitration
What was galling for me, personally, was that almost all this new breed of lawyers were my contemporaries, or younger than me, and seemed to believe that it was through their agency, and the agency of the courts, that workers had rights, as though these rights had not existed until the court pronounced their existence.[22] Whereas the lawyers the Union had worked with over the years – attorneys like Hymie Bernadt – subsidised what they did for the Union from other sources, the primary source of income for this new breed would be the new branch of law that the court opened up. And what made labour law lucrative, in the case of the lawyers that acted for trade unions, was that foreign funders were footing the bills.[23]

Were it not for the foreign funders, the turn to the law would never have been as sharp. They would surely argue that by footing the bills they were promoting the rule of law. But wittingly or unwittingly, it was a particular version of the rule of law they were promoting, which did not ask searching questions of the bosses and would facilitate the schemes of multinational companies to maximise their profits. The Union was all for the rule of law, where it had a say in what the law should be, in forums in which workers were represented. That was why we would have been better off with a tripartite tribunal, such as the Industrial Tribunal had been, than the Industrial Court.

If dismissal is an exercise of power, not every dismissal was an abuse of power. There was conduct 'we' did not defend, even though we would always plead for leniency. To this extent, 'we' and 'they' subscribed to a common morality. So we did not claim what the workers had done at Epol Pretoria West was right. It was lawless to have expelled the two, who were of course non-members. Lawlessness would be even more damaging to our standing in the working class than slander on the trains, and we told the workers so. The workers accepted the criticism in good spirit.[24] What we did claim, as grounds for leniency, was that the two had behaved provocatively, instigated no doubt by lower management – Pretoria, after all, was the stamping ground of the far Right.

Hard cases often concerned what they regarded as dismissible offences, and we did not – insubordination was a common example. Months had gone by during which the five workers remained on suspension, before Epol picked up sufficient courage to hold a disciplinary inquiry. The outcome was that the five workers whom the bosses had wanted to dismiss from the outset were now dismissed. The two the workers had sought to expel were brought back.[25] The stakes were high and the Union had to somehow prevent the workers coming out again, which would surely end with the dismissal of them all. So we resolved to challenge the dismissals on the basis that they were selective. Five had been dismissed for conduct in which a far greater number were involved.

This was obviously a hard case, and not one we would have dreamt of pursuing in the Industrial Court. There was an alternative. Although we remained resolutely opposed to compulsory arbitration, the Union did not object in principle to voluntary arbitration. Although there were cases where we would have gladly done so, we had not voluntarily referred a dispute to arbitration since the Juices case more than six years earlier. The problem with voluntary arbitration was that it was voluntary. An employer would never agree to arbitration unless it had something to gain by doing so. On the other hand, a system of compulsory arbitration could only work if there was agreement as to when it kicked in.

Underpinning the lawyers' assumption that law could be divorced from power was a belief that there was a distinct category of disputes that were amenable to a legal solution: 'disputes of right' as opposed to 'disputes of interest'. Dismissal, in this scheme, was not an exercise of power. It was a question of applying the relevant legal principles to determine whether it was fair. The Wiehahn Commission set great store by the distinction between 'disputes of right' and 'interest', as I have said, and employers routinely sought to introduce it in recognition agreements. Routinely we would oppose it. The distinction is not as clear-cut as you are making out, we would argue. In hard cases it is bound to be contentious. The Union could never accept that there were disputes that only an arbitrator or court could determine.

While the events at Pretoria West were unfolding, I had occasion to pursue this argument to extraordinary lengths at Blue Continent Cold Storage. This was not a significant plant for the Union or its parent company,[26] Barlow Rand itself. It was a significant issue, however, and much as Barlow Rand claimed to want to preserve the autonomy of plant managers, its head office was always on hand when important issues were at stake. It was represented in this instance by André Lamprecht, its head honcho in labour matters, and a mover and shaker of note, he told me, almost in so many words. His mission, it transpired, was to steamroller me into accepting the principle that all 'disputes of right' must be referred to arbitration.

The factory manager at Blue Continent turned out to be none other than Mr Kets, who had ordered me off the premises of Rainbow, precipitating my arrest. He was no longer the commanding figure he had been. The white dustjacket he wore had become a mark of his subordination to the jackets and ties. During a break, he sidled up to me, out of earshot of Lamprecht, to assure me that what had happened at Rainbow was not his doing. Mr Kets also had nothing to contribute to a battle of wills over what must have seemed to him an arcane point. Old-style factory managers like him were being displaced by the likes of Lamprecht.

The battle of wills took place over more months than any other negotiation I had been part of. In the meantime, Epol Pretoria West agreed to voluntary arbitration for the same reason we did: the stakes were too high to allow the dispute to escalate. In their case this was because of the commanding presence the Union had within the company as a whole. The arbitrator we agreed on was Peter*, a contemporary of mine whom I had known from student politics. The company's star witnesses were the two workers who had been expelled.

No one could be better placed than the victims to identify who was responsible for their expulsion, you might think, and this was my first opportunity to cross-examine in legal proceedings of any kind. Peter told me afterwards I had a future at the Bar, but the fact that each of the company's witnesses began to lie almost as soon as my cross-examination began had little to do with my forensic skills.[27] One of the workers lied so extravagantly that he reminded me of Lennie, the informer at Fatti's & Moni's, under interrogation. I must have been just another white male to him: it was terror of authority that made him lie.

It was a phenomenon I was to encounter time and again, after I had left the Union and joined the new breed of lawyers myself. No matter how carefully you have been over the evidence with a worker, some will lie under cross-examination, needlessly. This was another reason to keep workers away from courts and court-like proceedings like arbitration, but not long afterwards we were again faced with a situation in which an entire workforce faced dismissal over the conduct of a few. This was at DairyBelle's brand-new, state-of-the-art dairy processing plant at Clayville, near Tembisa. DairyBelle was part of the Imperial group, and this was also the first major dairy plant the Union had organised. The workers had come out over wages and refused to heed management's ultimatums. Now, members of the committee stood accused of intimidating non-strikers. The personnel manager at Imperial was refusing to return my calls. I swallowed my pride, and appealed to Lamprecht in person.

Lamprecht flew down for a meeting which took place in his room at the Cape Sun, the poshest new hotel in town. I do not remember who accompanied me, but Lamprecht had given me to believe he would be alone. Instead, he opened the door of an adjoining room with a flourish, and out came the Imperial personnel manager with a swagger. They had obviously decided between them the price the Union would have to pay for the workers to get their jobs back: agreement to refer the allegations against the committee members to private arbitration. Buoyed by the outcome at Epol Pretoria West, we agreed.

We asked for Peter to be the arbitrator. A similar principle was at issue: among several hundred on strike, only committee members were accused. This had to be a case of making scapegoats of the leadership, and Peter would surely exonerate them. Instead he found all but one guilty, and recommended they be dismissed. Amongst them was an engaging young man nicknamed Funky, who was one of the most promising of a new generation of worker leaders that were emerging. Forgetting whatever sense of legal protocol I had acquired during my upbringing and education, I went up to Peter, after the hearing, to ask him what the Union could do to review the decision. It was an innocent enquiry, I thought. He reacted as if I had slapped him in the face.

Realising my mistake, I left the venue by kombi, together with some of the dismissed workers. Funky was among them. The workers had accepted private arbitration on my recommendation. I felt personally responsible for the dismissals. Seeing how upset I was, Funky told me that the Union should not bother about a review. The subtext was that he had in fact done what he was accused of doing. That was, as I recall, brandishing a sjambok, that old symbol of white authority. It amounted to a threat of assault.

I toyed with the idea of apologising to Peter for having impugned the correctness of his finding, but could never bring myself to do so. Even if the workers were not always right, no one had actually been assaulted and I was also not prepared to concede it was right to dismiss the Union's leadership. However, the circumstances that had made it necessary to resort to arbitration, both in this matter and Epol Pretoria West, rendered academic the disputed provision about 'disputes of right' in the Blue Continent negotiations. At some point it was quietly dropped.

Most other unions, I suspect, had by this time long since agreed to refer 'disputes of right' to arbitration, whether by a private arbitrator or the Industrial Court. Realising the potential business opportunity, lawyers and other labour relations practitioners established the Independent Mediation

Service of South Africa (IMSSA). This was an NGO to punt the services of a panel of 'impartial' mediators and arbitrators most of whom were lawyers like Peter. Lamprecht, I seem to remember, was on its board.

Populism
The delicate line the Union maintained towards the UDF was consistent with a principle we regarded as inviolate. Membership of a trade union was open to all, and no one could be refused membership or expelled on account of their political affiliations or beliefs. But the political situation in the Eastern Cape was tenser than anywhere else in the country. The UDF's attempts to establish its hegemony over other organisations contributed to these tensions.

At about the same time as the events at Pretoria West were playing out, workers at KSM Queenstown also 'dismissed' a worker, one Joshua Ngoma, by physically expelling him from the workplace in much the same way as had happened at Epol. Ngoma, however, was not only a member of the Union, but had been chairperson of the branch. His offence was that he had attended a meeting of the Black Consciousness organisation AZAPO, and had refused to apologise for doing so.[28]

Ngoma had returned to work after being expelled, only to be expelled again by a group of workers. This happened several times. By the time Mpemvushe and I got there, eleven workers had been dismissed for their role in expelling Ngoma, and he was quietly working in the garden outside the management's offices. The remaining workers were now refusing to work, and demanding the reinstatement of the eleven. The question was whether we could avert the dismissal of them all.

The workers had been whipped up to further a political agenda. This, Mpemvushe and I both suspected, was Derek's doing. Derek* was a zealous young activist who had been appointed branch secretary, but had no experience of worker organisation. He was also our translator. The tack we took with the workers was that the grounds on which workers could be summarily dismissed were set out in the agreement we entered into with the bosses. The bosses could not dismiss on the grounds of political affiliation, and so the workers could not either.[29] This might be characterised as a bid to instil our version of the rule of law in the mill. But after we went back to East London, as we had to do, the workers refused to back down and were all dismissed.

'We do not believe in situational ethics, Mr Theron,' said Mr Ironside at a meeting with KSM management, looking at me determinedly as though, as an educated person, I ought to know what he was talking about. The phrase 'situational ethics' rang a distant bell, and much later I established why.[30] It related to a theological debate raging in the Anglican Church.

Ironside was part of the new breed of manager at Tiger. Apparently you did not get a top job in Barlow Rand unless you were devout.

I had been trying to explain the situation that gave rise to the attempt to expel Ngoma, in a bid to save jobs. But we were not in anything like the commanding position in Tiger that we had been in Epol. Also, Tiger was always more hard-line than Premier. It would only re-employ selectively, and about half the workforce of KSM lost their jobs. The re-employed workers then elected Ngoma as chairperson of the Union.

He invited me to stay overnight at his house in the township during the next round of wage negotiations, and impressed me as a thoughtful young man. During the same visit I met with the dismissed workers. They had been told Ngoma was trying to start another union, they said, and did not understand what the discord between AZAPO and the UDF was about. The younger among them might eventually have found employment elsewhere. For the older workers, this would probably have been the last job they had. Derek had in the meantime disappeared with their UIF cards, which for some unknown reason he had asked them to entrust to him. Consequently they were also not even able to claim unemployment benefits.[31] They had been used as cannon fodder, I concluded, and were understandably bitter.

There were tensions with the UDF elsewhere. These flared into open hostility in Hout Bay, sparked by criticism of the Union by a local UDF affiliate, the Hout Bay Action Committee, because members attended a meeting on the housing situation in Hout Bay at which a Labour Party representative was present.[32] The same issue attracted an uncomplimentary reference about the Union in *Grassroots* newspaper, and came to the attention of the UDF regional leadership. At their request, I brokered a meeting between our members and the Action Committee.

No one from the regional leadership attended this meeting. Perhaps they would have learned something if they had. What I observed bore out many of the Union's concerns about the way the UDF had been set up and its style of politics. The Action Committee, it seemed, was a two-person organisation, made up of Mr M and his wife. So far as the workers were concerned, their trust had repeatedly been betrayed by them. I found myself in the position of having to shield Mr M from the fury of Lilian van Neel, a local matriarch, in whose council house we were meeting, and to soften the barbs from the other workers. There had to be an explanation for their uncompromising antagonism.

Although 'populism' was not a label we ever used, it is a precise enough description of a leadership that claims to have the interests of 'the people' at heart, when in fact it was pursuing its own interests. So far as the Hout Bay workers were concerned, Mr and Mrs M were in it for themselves. Years later, after the transition to democracy, I was interested to note that

Mr and Mrs M were being accused in the Hout Bay community of enriching themselves from schemes purportedly established to benefit the people.

Fragmentation

Mpemvushe continued to assist East London branch for some time on a voluntary basis, but was not taken back when WPP opened its new factory at Kidd's Beach. So far as I know, the Union also never found its way into the new factory. Some would say that the WPP workers lost their jobs because of the Union or centralised bargaining, and it would be idle to deny that these were not factors in WPP's decision to relocate. However, the real reason they lost their jobs was that the law allowed the bosses to cut the costs of labour by moving their machines down the road to another 'country'. It nicely illustrates the role law plays in constituting an economic and political order.

Increasingly, this would be an order in which money and power were concentrated in the hands of a few large retailers and finance houses. We need to free up the markets, Raymond Ackerman of Pick n Pay would say. We can sell a standard loaf of bread for less than the big bakeries are doing. Let the consumers decide where they want to buy their bread, and what they are willing to pay. The response of the bakeries was to look at how they could restructure their operation and cut costs. The operation that was the most vulnerable to being squeezed by the retailers, however, was not the standard loaf but the confectionary departments, where mainly women worked. The first to retrench was the first bakery we organised, Attwell's.[33] Within a year all the confectionaries at all the bakeries nationally had shut down.

Other manufacturers were also restructuring operations in order to cut costs. The first retrenchment at Epol was again at the first plant we organised. Sixty workers at East London had to make way for machines.[34] The closure of its most labour-intensive plant in Cape Town was even more devastating. The Premier bosses told us it was cheaper to buy feed for their poultry plants from their competitor, Meadow, than from their own factory.[35] About 350 workers would lose their jobs.[36] It was some comfort that the Union was later able to organise Premier's poultry division, which for a few years seemed to be holding its own against Rainbow. Then it too was closed.

When Rainbow took over Epol some years after I had left the Union, it was not so much a case of a big fish swallowing a small fish – Epol was not small – as a manufacturer securing its supply lines. Maybe this was because it did not depend on the big retailers to sell its product as much as fast-food outlets. Epol Pretoria West now supplies Rainbow. The horizontal relationship between Epol and Meadow would no longer be so important, and the boundary between one industry (milling) and another (poultry) would become blurred.

We did not know at the time how far this process of industrial restructuring would go. We also did not foresee how the other important amendment to the LRA introduced in 1983, providing that a labour broker could become the employer of workers it placed with a client, would contribute to this process, and blur the boundaries between different industries even further.[37] What was becoming clear, however, was that the period in which the Union had been able to secure real increases in the wages of ordinary workers was over. It had lasted for five or perhaps six years. What capitalism had given with one hand it was now taking away with the other, by way of retrenchments and closures.

It is a pity, in retrospect, that there was no plausible alternative to industrial trade unions at this point. Alternatives might have been helpful in responding to industrial restructuring.[38] They would also be important in a context in which trade unions globally were under attack.[39] We were fixated on industrial trade unionism, however, because this was the only way we could get a federation off the ground. So in response to FOSATU's plan to withdraw from the feasibility committee meeting altogether, we held a series of bilateral meetings with each of the unions making up the core group, to dissuade them from taking this course.[40] The way forward, we argued, was that all participants commit to having one union for one industry.

This would mean breaking up some unions, such as GWU, and the merger of others. Since both CUSA and FOSATU had food affiliates, these were the meetings that mattered most for us. We specifically requested CUSA's affiliate, Food Bev, to attend our meeting with CUSA. It did not.[41] We did not need to make the same request of FOSATU. Although we still had little idea of whom Sweet Food represented, Chris Dlamini, its president, was also president of FOSATU.

The unity talks were being deliberately sabotaged, Chris said, in our meeting with FOSATU. The UDF unions were actively organising against the formation of a new federation, and SAAWU had even tried to recruit members by passing itself off as FOSATU. FOSATU wanted nothing more to do with it.[42] Chris had attended a number of meetings in the unity talks, where he was generally an enigmatic presence. I was surprised that it was he who was putting FOSATU's position, and even more so at his anger at the UDF unions. It was not a good idea to get as angry as this, I thought, since it was bound to colour relations with the UDF itself.

The proposal we put to FOSATU and the other organisations was to offer the UDF unions observer status, on the basis that they were, objectively speaking, not ready to form a federation. That way we would put them on the back foot, and avoid any accusations of breaking away. FOSATU was not persuaded but agreed to let us give it a try. Consequently it fell to me, on behalf of the Union, to put the position at a meeting in April 1984.[43]

'We commend FCWU for the diplomacy of their presentation,' said Sydney Mufamadi of GAWU in reply. 'However, at no stage have we said we are not committed to a federation. We ask FCWU to say openly whether we are expelled from this meeting.' It was true that GAWU and the others had never said they were *not* committed. The point was that, even now, they were not prepared to say they were committed to a federation. It suited the UDF unions to interpret the offer of observer status as expulsion, as they did.

There was quicker progress towards forming a federation in the months following the departure of the UDF unions than at any time in the previous three years.[44] A constitution was soon drafted, which affirmed the distinction between workers and paid officials, and that workers would be the decision-makers in its highest structures.[45] All that remained to be decided on was a name, a declaration of principles, and the amount of the affiliation fees that unions would pay.[46] But the name and the declaration of principles pertinently raised the question of the relationship to SACTU and the Congress movement. 'Has FCWU disaffiliated from SACTU?' had been SAAWU's parting barb, shortly before leaving the meeting in a huff.

So far as I was concerned, our past affiliation was past. What was of more importance was that the federation would be autonomous, as indeed SACTU had once been. That raised the question of how the federation would be financed.[47] It boiled down to whether the federation wanted to be self-sufficient or would rely on outside funding. I drafted a proposal which required trade unions to contribute five cents a week in respect of each union member.[48] This was an amount each should easily have been able to afford, and would have been enough to cover the running costs of a relatively lean federation. The other organisations balked at even this. Five cents a member was more than they could afford, they said. It soon transpired that this was also because they did not want a lean federation and had more grandiose plans.

22

Hospital way

THE GOVERNMENT OF PW BOTHA seemed to be having a smooth run towards the end of 1984, when delegates from different parts of the country descended on Grabouw for the Union's conference. The Labour Party had agreed it was not necessary to hold a referendum on the National Party's new constitution. Samora Machel had traded the uniform of a guerrilla fighter for epaulettes and gold tassels. Mozambique would no longer allow its territory to be used by the ANC or MK. Things were also not looking good for labour globally. Mrs Thatcher had sprung a trap for the miners in the UK, with the threat of closing down pits. No one expected an elected leader in a major industrialised country would go to such lengths to defeat a trade union, and the shock waves were to reverberate far beyond its borders. She was prepared not only to sacrifice the mines, it later transpired, but also the industries that had made Britain great.[1]

Much as PW Botha considered himself a soulmate of Thatcher, he could not, however, emulate what she was setting out to do to the trade unions without jeopardising his showpiece reform. Workers must have sensed that. Most of all, they saw that the Union was working for them, as it was supposed to. Organised workers were still organising unorganised workers, and even though there was hardly a branch that was not affected by plant closures or retrenchments, the Union had grown fivefold since 1976. Perhaps more significantly, for the first time the number of AFCWU members exceeded the number of FCWU members.[2]

The conference welcomed a surprise visitor, just released from jail, pending an appeal against his conviction on charges of terrorism. This was of course Oscar Mpetha, who had in the meantime lost his leg to diabetes and was now in a wheelchair. The fact that Oscar had been elected as one of three 'patrons' of the UDF did not in any way influence the debate on the UDF, which reaffirmed the position the Union had taken.[3] The conference also welcomed for the first time delegates from unlikely places such as Potchefstroom and Klerksdorp, and, unlike previous conferences, each branch was asked to give a report at the start of proceedings.

You could see from these reports how vibrant branches were empowering ordinary workers. Sometimes it was the chairperson of the branch who gave the report, but, more often than not, where a branch had one, it was a paid official who did so. In the case of Ceres, which for some reason was the first to begin, this was Susan Goliath, and her report set the tone.[4] Whether big or small, the measure of a branch's success was that it had committees in the factories that were effective in addressing workers' problems. Yet workers also had to develop a broader perspective of their problems than they could get from an exclusive focus on their own workplace. Goliath referred to the closure of Juices, the cannery, to illustrate the point. Branches with their own paid officials were able to articulate this broader perspective more clearly than the others.

An effective paid official was, as always, a paid official who was accountable to the workers, and the basis of accountability was still financial. The benefit of the policies of financial discipline and zero tolerance of corruption was now evident to all. Branches had been able to increase salaries, and some now owned their own motor vehicles. This was true for branches that relied on hand collections as much as stop orders. The proof of this was that the income of AFCWU had doubled in the space of a year.[5] The Union now had enough of a surplus to pay in cash for a building in Salt River big enough to house both the head office and Cape Town branch. This branch, now the biggest in the Union, was able to make a substantial contribution to the purchase price.

Given how important the material basis of organisation was in the Union's model, it is perhaps not surprising that the most impassioned debate was not about the UDF, or the need for us to push forward with a new federation, but about a proposal to increase the subs, which had remained at 30 cents a week for something like six years. No one objected to an increase. The issue was rather that the head office was proposing differential increases: three different rates, according to what workers earned, with the lowest-paid contributing 40 cents, and the highest-paid 60 cents.[6]

This proposal was still a far cry from a system in which subscriptions were calculated as a percentage of what workers earned, as is the case today. There was, however, talk among some of the unions involved in the unity talks that 1 per cent of the basic wage should be the yardstick of what was a reasonable amount for workers to pay. This disregarded the enormous variance in wages in the industries that we and others organised.[7]

Percentages feed differentials. That was why the Union was opposed to percentage wage increases. They widened the gaps that already existed between what workers in different occupations earned. Subs calculated as a percentage meant workers that earned more paid more. They also meant the Union would never again have to approach the workers for an increase, unless of course it wanted to change the percentage. Workers that earned more would also have to pay more in terms of the head office proposal, but not nearly to the extent they would on a percentage basis. It was, I thought, a sensible proposal. I did not anticipate there would be objections in principle to members contributing different amounts. The fear, especially among the branches whose members were predominantly lower-paid and female, was that workers that paid less subs would be treated differently.

The proposal was roundly rejected, along with a proposal that members pay 40 cents rather than 50 cents. A lean delegate from Grabouw felt so strongly about this that he insisted on standing up to berate the table, even after the matter had been resolved. The conference was in Grabouw and delegates were being accommodated in the hostels used by the seasonal workers: low-paid women. 'You need to learn a lesson from this', he said, wagging a metaphorical finger. 'The table should never have allowed such a divisive proposal to be presented.'

The lesson I had learned afresh was never to discount the insight of the ordinary workers. The fear that workers who paid less would be treated differently was not far-fetched. It affirmed that money matters, and that material conditions determine the outlook not only of individuals but of organisations. Where a section of the membership contributed more to an organisation, the organisation was bound to give greater weight to that section's concerns. By the same token, where a section of the membership contributed nothing, the organisation was likely to disregard them altogether. This was the situation of the members whose factories had closed or who were retrenched. It was also, of course, the situation of seasonal workers during the off-season, although they were still members in terms of the constitution.

In practice, however, retrenched workers would not have attended meetings without encouragement from above. The Union did not think of doing so. The working class was fragmenting before our eyes, into a section with relatively secure employment and, as a result of the closures and retrenchments, a section with none at all. It was this, together with the widening gap between those in better-paid and lower-paid jobs, which would later make possible the fragmentation among workers in employment, between those in secure and those in insecure jobs.

It was not clear how the Union should mind the gap between those in better paid and lower-paid jobs. The wages in milling and baking had increased more rapidly than in the canneries, for example, because the milling and baking bosses relied to a much greater extent on expensive machinery and employed far fewer workers. Capital-intensive manufacturing was more easily able to increase wages than labour-intensive manufacturing. Where capital-intensive manufacturing had labour-intensive operations, it was shutting them down. The confectionary departments in the bakeries were a case in point.

In the meantime, material conditions were changing my own outlook. Since the first increase after I was elected general secretary, I had insisted on accepting a smaller percentage increase than other employees. This was only partly out of white guilt. Unless I minded the gap between what I earned and others did, I could be accused of reproducing a form of white privilege in the organisation. But with inflation at 16 per cent I was now confronted with the realisation that in real terms my salary had been diminishing ever since 1977.[8] Athalie and I were only able to come out as a household because we now had two salaries rather than one. The circumstance that was to worsen our material position relative to other officials in the Union, and even workers, was rent. Rent control, a perk for the white working class, was being relaxed. Soon it would be scrapped altogether. Our landlord was quick to take advantage. At about the same time the property developers must have been putting out feelers about how to transform the cemetery across the street into yuppie homes.

Rent was also what brought to an abrupt end the smooth run the PW Botha government was having, shortly after the conference ended. The rents paid in the Vaal townships of Sebokeng and Sharpeville were reportedly the highest in the country, and, in terms of the government's new constitutional dispensation, the local authorities it had set up were now expected to raise funds from the communities they were supposed to serve. Taxing the local community was the only way they could do so.[9] The rent issue triggered a series of stay-aways. The stay-aways in turn triggered a heavy-handed response by the police. Other townships, whose residents had rent issues of their own, came out in solidarity, in Soweto, Tembisa and on the East Rand. The two-day stay-away of 5 and 6 November 1984 was provoked by the government's decision to send the army into the townships. Although it was largely confined to the Witwatersrand, it was regarded as the most serious challenge to the existing order since 1976.[10]

Unlike in 1976, however, this challenge came as a consequence of organisation. This was first and foremost a local organisation responding to a local issue, in the form of the Vaal Civic Association, together with local trade unions. The Union was among those with members in the region, but I would not have known to what extent they were involved.[11] What was most noticeable, however, was that the UDF did not appear to be alive to the issue of rent at all, and played no direct role in the events that ensued.[12]

Unions played a central role in organising the two-day stay-away. Prominent among them, in a break with its abstentionist political past, were FOSATU unions. The immediate consequence was that FOSATU's Chemical Workers' Industrial Union became embroiled in a major dispute with SASOL over the dismissal of some 6,000 workers, and a number of trade union leaders implicated in organising the stay-away were detained.[13]

No doubt because of the unions' role in organising it, students by all accounts adopted a respectful attitude towards workers during the two-day stay away. This was also in contrast with 1976.

The unionists were eventually released without anyone being charged. This was possibly due to pressure from the bosses, but reform-minded elements in the PW Botha government must also have supported their release.[14] They would have understood that negotiations with the unions had acquired a symbolic importance, whose significance extended far beyond the workplace. That would also explain why SASOL, a state-owned company, agreed to take back the workers it had dismissed. So within the space of six months, despite the adverse economic situation, or perhaps because of it, the political situation had begun to change quite rapidly.

The Union was, in the interim, having to deal with unfinished business of its own. The issue of its relationship with the other food unions, and in particular FOSATU's Sweet Food, had to be confronted. Before it could do so, the issue of its dual identity as the registered FCWU and unregistered AFCWU had finally to be resolved. The formal unification of FCWU and AFCWU could no longer be postponed. Conference had already decided the name would be FCWU. It remained for the national executive committee to decide whether the Union would be registered or not. Although it was by now a foregone conclusion that it would be registered, the problem was to contain any fallout this decision might provoke. So in March 1985, along with the other delegates from the Western Cape, I boarded a bus we had hired to transport us to Durban, where the meeting was taking place.

It was a bad decision for me to go by bus. I had been to see a medical specialist about a small sore on my penis. He had referred me to Groote Schuur Hospital for a biopsy, and I was still waiting for the results. In the interim I had only to keep my wound clean, which I was hardly able to do in the garage toilets of Beaufort West and Bloemfontein. I could easily have persuaded the office-bearers to let me fly, but I did not want to do so.

The national executive committee resolved, unopposed, that AFCWU adopt the name and constitution of FCWU with effect from 1 April 1985. This formally brought to an end the division that had existed since 1947.[15] However, I had no illusions that the issue of registration was entirely resolved. Important branches had prevaricated to the last. Moreover, the Eastern Cape branches still opposed being registered, although they had not opposed the resolution.

The fallout that I had been above all concerned to prevent, for the five years or so when registration had been a burning issue, was any kind of breakaway or split in the Union. The question was whether these branches would abide by the NEC's resolution or not. The branch I was most worried about was East London. With WPP having closed down, and Epol having retrenched, there was little to counter the dominance of Langeberg. Someone I did not trust had risen to prominence there, seemingly out of nowhere. This was Lulamile Mati. It was he who was representing the branch at the NEC.

Registration was seen in a negative light in the area, Mati told the NEC, and there were many organisations that were hostile to the Union that would use the fact that it is registered to attack it.[16] I had a sense of foreboding when he spoke of attacks, and shortly after the NEC there was the most serious attack on the Union since the breakaway at H Jones. All indications were that he had orchestrated it. However, I was not part of the delegation that went to East London to meet the workers, or at the management committee meeting that followed, where Lizzie, Mandla and Pendlani reported back on what had transpired. It was in fact the only management committee meeting of any significance I did not attend during my tenure as general secretary.

The first intimation of what was brewing was at a branch executive committee meeting held shortly after the delegation's arrival in East London. Someone from Langeberg raised a complaint against Lizzie, the member of the delegation whom the workers in East London knew better than anyone else. The complaint was bogus and a ploy to turn the Langeberg workers against her. She was someone who was capable of turning a meeting around, and it was necessary to neutralise her influence. The ploy seems to have been effective. Two days later, at a lunch-hour meeting at Langeberg, the delegation was made to sit down and listen to what 'the workers' had to say. They apparently had something to say about me. I never got to hear what it was, because our delegation regarded it as not worth repeating, but presumably it was racially inspired. That could also be attributed to Mati's influence. The delegation was pelted with empty cans as they left.[17]

It was hurtful to know that workers with whom we had a bond, like Lulamile, who had been treasurer for all the years the Union had been organised there, could allow this to happen. What was even more hurtful was to see an organisation so painstakingly constructed being so easily destroyed. The committee afterwards refused to hand over the books of the Union, saying they intended to retain the name of AFCWU. That was only Mati's opening gambit. His intention now, as it had perhaps been all along, was to establish a trade union of which he would become the secretary. That is what he proceeded to do.

I was dealing with a different kind of crisis at the time. The outcome of the biopsy was that the sore was cancerous. At the age of 35, the prognosis was death, unless the treatment the doctors were proposing was successful. Whereas the people I trusted socially were radical, the doctors I trusted were conservative, medically speaking. But on the trip to Durban my wound had become septic, and the radicals at Groote Schuur Hospital prevailed. They were proposing to amputate my penis. For my part, I was seriously considering letting the disease take its course, and taking my chances.

It never came to this. Although my material position had deteriorated, I was still able to draw on the resources available to the middle class. John Frankish, a doctor who had been associated with the General Workers' Union since its Advice Bureau days, was a friend in need. He directed me to medical journals and other resources. There were others. It was probably never on the cards anyway that someone born and bred in respectable Rondebosch would be treated as someone born into the working classes might have been. I was nevertheless given a small taste of what that might be like from the way the consultant, one of the radicals, discussed my case during his ward rounds. I might as well have been a cadaver. It made my blood boil and the registrars accompanying him blush. In the end, the conservative doctors prevailed, but for years to come I had to live with the fact that only time would tell whether the approach they had advocated had been right.

I got the fright of my life. I resolved to change the way I was living, and reduce the stresses I was under. I had achieved enough of the goals I had set myself over the past nine years, including the formal unification of FCWU and AFCWU. Others would now have to deal with the breakaway in East London. Others would have to assume a more prominent role in the unity talks, even though these were at a critical stage. The problem was that I had not yet found a way to replace myself, although I was trying.

Athalie and I believed that it was in response to the imminence of death that she conceived, in much the same way that birth rates go up after wars. We knew it would be a difficult pregnancy, because she had miscarried before. At six months she was hospitalised. So not too long after I had been a patient, we were again reliant on the public health sector, and I was once again wending my way to hospital, this time as a visitor, to try to maintain her morale, and do whatever I could to prevent a premature labour.

Mowbray Maternity Hospital occupies a three-storey building off Durban Road. This is the transport route that connects Klipfontein Road and the townships on either side of the N2. The chronic cases were on the top floor, overlooking the railway line from the city to the Southern Suburbs, and the station. The parking lot by the station served as a taxi rank even then, although unofficially. Most of the taxis were still illegal, although the government of PW Botha was shortly to legalise them.

I was permitted to visit any time of day or night over the next three months, provided I shed my shoes and put on the green cloth slippers that the hospital staff wore. Against all the odds, Athalie reached term and my son Leif was born. We had wanted a child and I felt an enormous sense of relief as well as the joy of the new-born. I kept the green cloth slippers as a memento.

About-face
The inaugural conference of the federation that we now know as COSATU had taken place in the interim. Even though I was perhaps the only person who could claim to have been at every formal step of the way up to this point, I had no qualms about not being there. My time had passed. In any event, there was a parallel political process taking place of which I was not part. Increasingly, this seemed to be where decisions were taken.

This parallel process was the politicking that went with the project of establishing the hegemony of the UDF, first of all, and then, after the UDF was banned, what was referred to as the Mass Democratic Movement (MDM). I accepted its inevitability. All the Union could do was to try to influence how it happened. It could happen in a top-down manner, where Congress activists tried to impose their will on organisations, as had occurred in Queenstown, or it could happen in a bottom-up manner, as with the Transvaal stay-away.[18]

The stay-away, and a subsequent joint call for a 'black Christmas' at the end of 1984, showed that almost all the unions participating in the feasibility committee were now in a front with the Front, and strengthened the hand of those in the UDF who supported a bottom-up approach. There was talk of establishing properly organised civic structures and street committees.[19] If these structures gave ordinary people in the community a voice, it would have been a most acceptable way to establish hegemony.

The PW Botha government put paid to any such possibility. It declared the UDF 'the enemy' and put the leadership of the UDF on trial for treason, in Pietermaritzburg and Delmas. At the same it increasingly resorted to dirty tricks. As in 1976 and subsequently, this involved exploiting divisions in the community, and specifically the African community. Although I believe that at root and base these were class divisions, they manifested as divisions between the UDF and AZAPO, and between the UDF and Inkatha.

As the dirty tricks became more blatant and more gross, my attempts to maintain a relationship with my father became more fraught. He had by now retired, but assiduously followed the events of the day as reported in the press and on SABC television. I would try to contain my irritation at his gullibility regarding the denials of state complicity in what was portrayed as 'black on black' violence. I never succeeded in doing so. Although the terror tactics that activists resorted to were not of the same order, some were complicit in barbaric acts, and the space for a more open political process was narrowing. Increasingly, the politicking was directed at the leadership of the unions and the places where they stayed.

It was the leadership living in the African townships that were most susceptible to the menacing overtones of 'we know where you stay'. Two activists from Johannesburg found out where Pendlani stayed in Mbekweni, and one of them, I gathered, must have been Monde, who had spent the night on my floor before the Langa summit. Their sole purpose seemed to be to bad-mouth me. I suppose I was seen as 'too independent'. No doubt there were others who were approached in a similar manner but did not tell me about it.[20] Perhaps I was too independent. On the bus to Durban, Liz Abrahams handed me a circular letter from John Nkadimeng, the general secretary of SACTU, not addressed to anyone in particular. I had received the same letter as we were departing.

'The aim should not be to form a federation for the sake of a federation', he intoned. The tone was patronising, to say the least. 'We have been closely following the developments in the unity talks, not as observers, but as participants in our own right.' This confirmed what I had long suspected, that at least since the Port Elizabeth meeting some participants in the unity talks were 'under instructions'. In spite of the 'good work', Nkadimeng went on, a number of questions remained. 'What is the basis of this unity?' was the first of them. It was the very question that the UDF unions had been posing all along, in a bid to delay the formation of a federation. 'Whilst we do not claim expertise on these questions,' he wrote, 'our experience has always proved us right'.[21] Stalin himself might have penned that sentence.

As a disciplined member of the Congress Alliance, Liz must have expected me to convene a meeting of an inner circle where the letter could have been discussed, as we used to do in her *voorkamer*. However, there was no longer an inner circle to convene. What needed to be discussed between meetings, I would discuss with the office-bearers. What SACTU wanted was another postponement. Nkadimeng said as much: 'there should be no deadline for the formation of the federation. The federation should not be formed hastily and for expediency.' But there was no way in which the office-bearers could have contemplated a postponement without disregarding the express mandate of its members, adopted at conference. This was to brook no further delays in forming the federation.

For me, it was an issue of autonomy: the autonomy of the Union's own decision-making processes and the autonomy of the federation we wanted to establish. SACTU, on the other hand, seemed to be hoping for some kind of reincarnation, with its own people in positions of power. Foisting on workers a leadership that had not earned their trust was a betrayal of what I had always thought SACTU had represented, and a betrayal of the tradition of the Union. I would have said as much to Ray, if I had the chance to do so.

I received a message from Ray sometime after Nkadimeng's letter and my return to work. She said she would like to meet with me again, and I arranged a holiday in Zimbabwe, thinking we could meet there. It would have been a very different meeting from the one we had in Maputo, if it had come off. SACTU's wanting to put its own people in positions of power looked very much like the kind of thing the Party would do. Ray was in the leadership of both. It was best not to say anything to my friends in Harare, I decided, but their house was so closely and so obviously monitored by a man in a motor car for the duration of my stay that it was obvious there had been a leak. I put it down to the shambolic state of ANC security and resolved to have no more clandestine meetings.

FOSATU, in the meantime, had received the same letter from SACTU that we did, and blinked. The first I knew of it was a letter from Joe Foster on my return to work in May, after an absence of more than a month. 'I have been directed by FOSATU Central Committee', it said, 'to make the following proposal to your Union.'[22] The same letter had gone to the other organisations that made up the feasibility committee, which was to meet in little more than a month's time at Ipelegeng Community Centre in Soweto. This meeting would finally approve the draft constitution of the new federation and clear the way for an inaugural conference.

The letter said FOSATU's 'proposal' was to circulate the constitution to all independent unions. It was not clear what FOSATU meant by this, because quite a number of new unions had been established since the inception of the unity talks. At a hastily arranged meeting of the feasibility committee, the night before the Ipelegeng meeting, it emerged that FOSATU meant to invite all of them to the meeting and had already done so. I think all the other members of the feasibility committee felt the same sense of betrayal and dismay that I did.[23] It was an extraordinary about-face, if you consider the respect for process that FOSATU itself had displayed throughout the unity talks. The understanding had always been that no organisation would act unilaterally in matters concerning the new federation.[24]

Betrayal, however, is too emotive a term for an organisation in disarray. Johnny Copelyn represented the one face of FOSATU. Although he was reputedly a key figure in FOSATU's old guard, I cannot remember him taking any part in the unity talks up to this point. This made his intervention all the more peculiar. If we were not happy to proceed with the next day's meeting, he said, we could all join FOSATU instead. It was an outrageous proposal. Cyril Ramaphosa spoke for us all in rejecting it with disdain.

Chris Dlamini represented FOSATU's other face. As its president, he was nominally part of FOSATU's old guard, although I doubt whether he had a significant part in determining its direction. The vehemence with which he had opposed the further participation of the UDF unions in the unity talks was fresh in my memory. Now, he was vehemently advocating their participation and the participation of others about which we knew very little. My immediate thought was that he had been recruited.

We now know Chris was in fact recruited. He is on record as saying he became a member of the Party in the 1980s, although not precisely when. My guess is that it was when he was detained. Jail was always the most fertile of recruiting grounds.[25] Be that as it may, there can be little doubt he was now under orders. The line he was taking, which FOSATU's central committee seemed to have endorsed, was SACTU's line, and also the Party's. So it is likely that others had been recruited at about the same time as he had, and a 'takeover' of the trade unions had already begun.[26]

You might argue, as some have done, that no harm resulted from this about-face, given that the inaugural conference of COSATU did take place before the end of the year. It was to all intents and purposes the constitution drafted by the feasibility committee that was ultimately adopted by COSATU. The position the feasibility committee had taken in relation to the community unions would also be vindicated, in so far as COSATU would commit itself to the principle of one union for one industry. The UDF unions and others that organised generally would have to disband if they were to be part of the federation.

My own view was that the autonomy of the federation had been compromised even before it was established. No good could come from a wholesale abandonment of past positions, with so little acknowledgement of why organisations had adopted these positions in the first place. A zigzag this extreme would be interpreted as a sign of weakness, and an indication that trade unions were ripe for the taking. The disregard for process would also set a bad precedent, as became evident the next day.

Some 40 unions were present at Ipelegeng. Some we already knew, and some we had heard of. There were others whom we knew nothing about, and of whom little was heard afterwards. These included unions affiliated to AZACTU, an organisation I, for one, had not known existed. If there was to be any progress at all toward the formation of the federation, blunt instruments would be needed. Donsi, the colourful character who had wanted to foist seven principles on us in Port Elizabeth, ought to have appreciated the irony of the blunt instrument that was selected. It was non-negotiable 'principles'.

The union Donsi represented was based in Pretoria and went by the acronym RAWU, as did a different union from Cape Town, which has a place in my story, as I will presently explain. These were the principles that had been adopted by the feasibility committee, said Ramaphosa.[27] That was news to me, I thought to myself, as I listened to John Ernstzen of CTMWA list five short points. We had agreed Ramaphosa would chair the meeting the night before. We had also agreed that he needed to confer with a smaller group about strategy. These must have been 'principles' agreed on between themselves over breakfast.

To put forward principles at a meeting of organisations from such different traditions was bound to divide it. That was why, since the Port Elizabeth meeting, our focus had been on securing agreement on a constitution. The constitution was premised on a principle of non-racialism in so far as it made no reference to race. We had been given to believe the CUSA unions could live with that, but they would never support the principle of non-racialism.[28] In all variants of the Africanist tradition, 'non-racialism' was code for support for the Congress movement. The other important principle that Ernstzen listed was 'industrial unions'. This was code for having 'one union for one industry'.

It is difficult to say what, in the final analysis, had been achieved at Ipelegeng. The organisations responsible for drafting the constitution had gone through the motions of consulting others. Yet the unions that accepted the five principles were essentially the core group formed after Athlone. All that had changed was that CUSA's most important affiliate, NUM, was committed to pressing ahead with the inaugural conference. The other CUSA unions were not. We will never know whether that outcome could have been avoided.

It was the feasibility committee in another guise that met later that month in the Johannesburg offices of CCAWUSA, where the decision was taken to press ahead with the inaugural conference.[29] It was only in retrospect that I was able to put my finger on what was different about this meeting from all those that had preceded it. For the first time since the 'unity talks' began, FOSATU was not present. It is a curious fact that I do not know what became of FOSATU. Other accounts of the period also do not say if and when it was formally dissolved.

It was agreed at this meeting that the inaugural conference would take place on the basis of the constitution as drafted. There was no further discussion of any of the unresolved issues about the new federation. Key for us was the question of how the federation would be financed and of foreign funding. If the federation was to rely on foreign funding, it would greatly weaken the hold the affiliates had over its leadership. The plan was to hold an inaugural conference in October, but a letter of invitation to unions that I had drafted was considered too hard-line, because the criteria it expected trade unions to meet were too onerous. It was left to Ramaphosa to send out a suitably softened letter.

The UDF unions were all there when the inaugural conference took place, including SAAWU, GAWU and some new recruits. Someone must have promised them a place in the leadership of COSATU, and Sydney Mufamadi became assistant general secretary. The bloc that subscribed to the Black Consciousness or Africanist tradition was absent. Within a year AZACTU and CUSA had amalgamated to form NACTU.[30] Chris Dlamini was now COSATU's deputy president and Jay Naidoo its general secretary. The other office-bearer from a former FOSATU union was Maxwell Xulu from MAWU, who became treasurer.[31]

No one wanted COSATU to be seen to be as politically timid as FOSATU had been, least of all, perhaps, the newly elected leaders who came from FOSATU unions; but once again, the zigzag was too extreme. At a rally in Durban held after the inaugural congress, the newly elected president of COSATU (who had not been from a FOSATU union) made a speech laced with populist rhetoric, in which he called Chief Buthelezi a running dog of the apartheid state.[32] Buthelezi was the kind of 'moderate black' who began to attract a surprisingly passionate following among whites at this juncture, including my father. This should not have blinded anyone to the fact that Inkatha had support, particularly among Zulu migrants.[33] It was as though nothing had been learned since 1976 or any of the subsequent occasions when 'the system' had exploited disaffected migrants.[34]

What was also worrying was that the COSATU leadership seemed to be intoxicated by its own rhetoric. It was claiming 'a giant' had arisen before the infant could walk. Manufacturing was shedding unskilled workers unchecked, and organisation in the mines had yet to be tested.[35] These two sectors accounted for more than half the membership COSATU claimed to have. In virtually all other sectors of the economy organisation was weak and fragmented. There were also serious divisions between unions that had to be overcome. The lax criteria set for the inaugural conference meant that unions were able to inflate their membership. There was also little to deter them from doing so, because the affiliation fees that trade unions were to pay COSATU were so low.[36] This was bound to exacerbate the divisions and mistrust.

It also meant there was nowhere near enough money coming in to fund the budget that COSATU's leadership had in the meantime drawn up.[37] Although I was by this time a member of the central executive committee of COSATU, to which its leadership was ostensibly accountable, I knew nothing of its decision to go overseas to secure foreign funding.[38] The outcome was that the Swedish and Norwegian trade union federations agreed to contribute what amounted to about 90 per cent of its running costs.

The response to the attack on Buthelezi was the formation of UWUSA. Simultaneously there was an escalation of 'black on black' violence. In Paarl, the same hostels where the migrant workers at Jones stayed, whom the Union had still not succeeded in organising, became the stronghold of persons who styled themselves as AZAPO or were branded as such. From there, they launched attacks on the residents of Mbekweni identified as being UDF. Lizzie Phike was in detention in Pollsmoor Prison at the time, together with an organiser of Paarl branch and others. She had to hear from the Special Branch that Abel, her only surviving son, had been stabbed to death in one of these attacks. The police set up roadblocks to prevent anyone from outside Mbekweni attending the funeral.[39]

23
Albert Road

THE UNION HAD TO GO THROUGH a process of restructuring of its own to realise its commitment to one union for the food manufacturing industry. It was untenable, to start with, for the head office of a trade union with members in all provinces to be accountable to a structure composed only of representatives from the Western Cape. The management committee would have to be replaced.[1] The argument was hard to fault, yet I had deep misgivings.[2]

The management committee meeting had become a vibrant forum, where a most vulnerable section of the working class was represented. It was an institution that had loomed large in the lives of many, over many years. Many important debates in the Union had taken place there. I tried to reassure the members that the branches would still have a forum where they could meet, at a regional level. They would also still be represented at the conference, the workers' parliament.

Sweet Food did not have branches, as we understood them. There was also nothing like our annual national conference in the FOSATU tradition. The nearest thing to a decision-making forum in which local representatives had a direct voice was the AGM of their 'Transvaal branch'. We were invited to one of these, which took place in an East Rand community hall early in 1985, a few months after the celebrated stay-away of 1984.[3] It was on a Saturday, and we were surprised to find that the formal proceedings were already over by the afternoon, when we got there. What we witnessed was more like a *jol,* dressed up as 'cultural activities'. Scantily clad maidens in bright-red Swazi-print skirts pranced across the stage, and there were copious quarts of beer in the corridors. Noko, who was still president at the time, was appalled.

Sweet Food would eventually accept our branch structure, but it would be too much to ask that branches retain autonomy over their finances. In the FOSATU tradition paid officials earned the same salaries. If each branch determined its own salaries on the basis of how much money it brought in, some of them would face a pay cut. Every model has its drawbacks as well as its benefits, I told myself and others. It looked like inconsistency that, in the Union's decentralised model, paid officials at some branches were earning less than others. After all, the Union was opposed to differentials in the workplace between workers doing equivalent work.

However, this was not in fact the reason why we were constrained to abandon our decentralised model, as I have earlier explained. Having stuck our neck out for a policy of one union for one industry, we were obliged to do what it would take to implement it. This meant going along with the centralised model everyone else had adopted, in which there was be one centre of power, at the national head office. Ironically, this was at the very point at which the bosses started moving strongly in the opposite direction. More and more of them adopted the Barlow Rand model, in which the head office treated branches as autonomous. The benefit of this model for the bosses was to be able to deny accountability for what local management said and did. In time, the scope of this denial of accountability would be widened beyond our wildest imaginings.[4]

Only a founder could have pushed through as many changes as quickly. I was not the founder of the Union, of course, but very few now had a personal recollection of Ray. For most, I was the one person at head office who had been there all along. That made me something like a founder, and also very conflicted about the role I was playing. On my say-so, a system that had enabled the Union to survive the years of repression, and that had ensured that leadership remained accountable, was being dismantled. The sad thing was that few COSATU unions had an inkling of what we were giving up.

I was also conflicted because, having got the fright of my life, I had resolved to leave the Union. I did not announce my intention to do so. Continuity of leadership is especially important in an organisation in flux. Instead, I began planning an exit strategy. Part of it was to take long leave – what would now be called a sabbatical – in terms of new, standardised conditions of employment for all the Union's staff.[5] These were very different conditions of work from what applied in the factory, Liz complained. There was even some talk, I afterwards heard, that the inclusion of a provision for maternity leave was actuated by self-interest.

This was to some extent true. Parenthood brings with it certain realisations. Athalie had resolved to spend the first year with our son, and the maternity leave provision allowed her to do so. Since it was unpaid leave, I was now the sole breadwinner and was increasingly anxious about our material circumstances. I would object to spending anything on what I saw as luxuries. I was also increasingly desperate to preserve a boundary between my work and our home life, in a misplaced attempt to cope with my anxieties about the direction in which the Union was headed. Our relationship became more fraught.

The appointment of an assistant general secretary was intended to smooth the way for my replacement, as well as the appointment of three national organisers: one for each of the three largest sectors in the Union.[6] Lizzie Phike was one. The only one who had not been a member before becoming a paid official was Mandla Gxanyana.[7] Having paid officials who had been members was intended to establish the trade union equivalent of a career path, so that there would be a natural progression from being a worker to becoming a branch official, and from being a branch official to becoming an official at a higher level. In theory, this should make it a sustainable system.

Yet it was a system that worked best at the branches, because the relationship between the paid official and workers at the branch was clear: the members knew what the paid officials were expected to do, and what to do if they failed to deliver. It was much more difficult to ensure accountability at a national level, where officials were expected to articulate the Union's position on issues such as recognition, as well as provide guidance on policy issues on which the Union had yet to formulate a position. It was too much, perhaps, to expect someone with little or no formal education like Lizzie to do so.

Even at the big urban branches it was a problem to equip workers with the skills they would need as paid officials. Cape Town by this point was more than twice as large as the entire Union had been when I started, and all of the officials it employed except Athalie had been members. However, it did not seem any would be capable of replacing her while on maternity leave, so the branch recruited someone from the white Left to do so. This was Miles Hartford. There were others from the white Left employed elsewhere in the Union.[8] Any movement from below needs an intelligentsia it can trust, and this is what the white Left had been for the trade unions.

The reason it could be trusted, you could argue, was no different from when I was recruited. However, the political context had changed. A position in a trade union such as ours signified a position of power. I had up to now made a point of trying to visit every factory the Union had organised at least once, so that workers could see their general secretary was not just another white.[9] Yet this could not change what a white man in power symbolised, at a time when white men were in Angola fighting to preserve 'the South African way of life'.[10] The days in which someone like me could occupy a position like mine were numbered.

On the other hand, I owed it to the members to keep out personally ambitious individuals who did not have the workers' interests at heart. Jay Naidoo seemed to have the workers' interests at heart, and I was planning to sound him out about his becoming general secretary during a visit to Cape Town to discuss the merger between our unions. Instead it was Jay, accompanied by Chris Dlamini, who sounded me out about his standing as general secretary of COSATU. Jay was the first effective general secretary of Sweet Food, and was like a founder, for this reason. Taking him out of the union at this point would undoubtedly complicate a merger. On the other hand, there was no one else we could have nominated as general secretary of COSATU. It had to be someone who was not white, and without too much political baggage.

This was the first visit by Sweet Food representatives to the FCWU head office, and I wanted to show it off. Everyone in the Union was extremely proud of the building. Among the affiliates of COSATU, hardly any owned buildings where they were based and ours had been bought entirely with workers' money, for cash.[11] It was a monument to our model of organisation and the value of financial self-reliance. It also made the kind of aesthetic statement such a monument should make: it was gracious yet modest. I had expected them to be bowled over.

The 'lower' main road from the city centre is called Albert, presumably after the husband of Queen Victoria. The corresponding stretch of the upper main road is named after her. It would have been during her reign, or not long afterwards, that many of the buildings near the circle at the divide between Salt River and Woodstock were built. Comforted by the proximity of royalty, the king of Norway was busy adding another, close by, to be called Community House.

Near to the circle, on the railway-side of Albert Road, there is a row of double-storey tenement buildings. The head office was the one with the curved gables, and easily accessible to workers from many factories that the Union had organised: I&J, Atwell's Bakery and SA Milling, as well as a cluster of oil and flour plants in Maitland, and Jungle Oats. The logo of Jungle Oats was of the killer cat from which the Tiger Group had taken its name, gazing at the consumer improbably from a backdrop of grain fields.

If the visit of Jay and Chris to the Union's head office building had made any impression on them, it must have been negative. Soon after the inaugural congress the leadership of COSATU took another decision without a mandate from its affiliates: to move from what seemed to me perfectly adequate offices into something much grander and more imposing, a multiple-storey office block in central Johannesburg. Self-sufficiency clearly did not come into it. There was not even a plan as to what to do with all the surplus space.[12] The European Economic Community came to COSATU's rescue. It 'offered' to donate a million rands to COSATU to buy the building. This was a vast sum at the time, but no one else seemed to find it odd that South Africa's largest trading partner would want to help out South Africa's largest trade union federation. FCWU was the only union to oppose acceptance of the donation.[13]

TOP: Friday Mabikwe of Fatti's & Moni's, Bellville, leads the congregation in prayer at the memorial service for Neil Aggett at St George's Cathedral. (Courtesy of Independent Media)

BOTTOM: St Mary's Cathedral in Johannesburg was packed when I delivered a tribute to Neil. Sipho Kubeka is translating into isiZulu. (Photograph by Paddy Donnelly)

TOP: B(e)aring the flag: Mourners assembled behind an ANC flag begin marching towards the cemetery, in perhaps the most significant display of political defiance since the 1960s. (Photograph by Paddy Donnelly)

BOTTOM: The police look on while the march proceeds down what is now Beyers Naudé drive. (Photograph by Paddy Donnelly)

OPPOSITE TOP: A marcher holding one of the posters produced to commemorate Neil. (Photograph by Paddy Donnelly)

OPPOSITE BOTTOM: For many, this gathering of trade unions and other organisations at the cemetery was the most diverse they had witnessed. (Photograph by Gavin Younge)

On the anniversary of Neil's death the Union organised meetings in different parts of the country. This photo is from a public meeting in Athlone, Cape Town. 'Msuks' Qotolo, an activist from Gugulethu, is translating. (Courtesy of Independent Media)

Athlone, 1983: The Union's delegation to the 'unity talks' in Athlone where a resolution was adopted to form a new federation. Sitting in the front row, second from the left, is Mandla Gxanyana, and next to him Athalie Crawford. Visible in the second row is Inthiran Moodly, and on the right, Phillip Mayoli and Alfred Noko. (Photograph by Merle Brown)

TOP: Chris Dlamini is seen here addressing a meeting of COSATU's Western Cape region, of which McWellington Mtiya (seated) was chair. Mtiya was one of the members expelled from the Union in 1990. (Photograph by Chris Ledochowski)

BOTTOM: John Pendlani speaking at a farewell bash for me, on the occasion of my taking long leave. Lizzie Phike is translating and Irwin Pereira is on the left. Leif is on my shoulders. (Photograph by Chris Ledochowski)

TOP: This is a photograph of a COSATU demonstration against the LRA in central Cape Town, in about 1989. Note the posters calling for Oscar Mpetha's release. (Photograph by Paul Grendon)

LEFT: Liz, Luska and Lossie: Liz Abrahams, now retired, is sitting next to Luska Ndinisa and Ellen 'Lossie' Rogers on the couch at the Gugulethu offices of the Union, on the occasion of the Union's 65th birthday. Chris Dlamini, who joined the group, is standing behind them. (Courtesy of UCT Special Collections)

TOP: The cake: The representatives of the parties to the tripartite alliance, Zwelenzima Vavi of COSATU, Jeremy Cronin of the SACP and Thabo Mbeki of the ANC, are gathered around a birthday cake. The general secretary of the Union is partially obscured behind them. (Photograph by Mark Wessels /Sunday Times)

BOTTOM: Munaadiah Moosa in front of the factory she was instrumental in organising in 1978. She's still organising workers at the time of writing. (Photograph by Jan Theron)

TOP: John Pici with the silos of Tiger Oats in the background, in about 2009. (Photograph by Jan Theron)

BOTTOM: The factory at Lambert's Bay was once the flagship of a fishing empire. It was already in decline when this photograph was taken. Now the factory is idle. (Photograph by Jan Theron)

One union for one industry

Power could be exercised to further the interests of workers and it could be harnessed for a political objective. There was no contradiction in harnessing it for the objective of establishing democratic rule, according to the ANC's (and the Party's) concept of the National Democratic Revolution (NDR). But this seemed to me to depend on what kind of democratic rule you had in mind. In terms of the Union's model, democracy began at the local level, at the branch general meeting. It was here where local leadership was elected. In terms of the Sweet Food model, local leaders were elected by the executive committee.[14]

It was an issue we had debated with Jay and Chris on their visit. The provisions of the FCWU constitution regarding general meetings were too open-ended, Jay had argued, because there was no limit to the number of members who could attend and vote. This allowed a well-organised faction to grab power, even though it only represented a minority of members. The only way to prevent a power grab, I replied, was for the majority to be well organised. In retrospect, that seems a bit glib. I would shortly be reminded that majorities are not well organised all the time, even in a well-organised trade union.[15]

Sweet Food's organisation was quite haphazard: a sugar mill here, a factory manufacturing sweets and chocolates there, much of it in industries which had no point of contact with FCWU. It had also begun organising bakeries in Durban, and had spearheaded a strike over wages. But the bakeries were controlled by the same mills we were organising elsewhere, and wage levels were determined nationally. The only way workers in Durban could have achieved their demands would have been by collaborating with the Union.[16] Collaboration between different trade unions in the same industry was also the organic way in which to merge their operations.

The one industry in which Sweet Food had a strong presence was brewing. Apart from the sorghum breweries, which were a money-spinner for the government, brewing was dominated by a single company, SA Breweries (SAB).[17] FCWU collaborated with Sweet Food in organising SAB's brewery in Newlands, Cape Town. In doing so it emerged that SAB workers were educated. Applicants for a job there needed a matriculation certificate, although this had not yet been the case when many of the older Cape Town workers were first employed. Wages at SAB were also higher. Indeed, they were among the highest in the manufacturing sector as a whole. As a result, SAB was a magnet for the upwardly mobile.

A more educated workforce meant there was a more educated worker leadership, and even officials of the Union found their facility with high-flown English intimidating. Our first formal meeting with Sweet Food suggested this was part of a larger problem: Sweet Food's worker leadership was articulate (in English) and confident to a degree that FCWU's was not. No doubt this had to do with the emphasis on formal worker education programmes in the FOSATU tradition, which was a justification for accepting foreign funding.[18] I suspected, however, that it was mainly upwardly mobile workers who benefited from these programmes. There were hints of this in one of the cultural activities I witnessed at the Sweet Food AGM. It must have been billed as a poetry recital, and involved a young man stringing together as many English words as could be uttered in the same breath, stream-of-consciousness style. The more highfalutin the word, the louder were the gasps and cheers from the audience. What was one to make, I wondered, of people applauding language that was barely comprehensible even to a first-language speaker like me?

I was also sceptical (and still am) about claims that formal education programmes produce better leaders. Leadership is essentially about integrity, and integrity cannot be taught in courses. People who run courses are also not likely to understand intuitively what kind of leadership a workers' organisation needs: a leadership that is sensitive towards its most vulnerable members, who are invariably the least educated and lesser skilled. My worst fear about the merger was that the interests of these workers, who had been the backbone of the Union during the dark years and the period of its revival, would be trampled on.

Even without a formal education programme, the Union was undoubtedly educating workers in all sorts of ways. You could see the benefits of this from the confidence with which Lizzie engaged with intellectuals and professionals with whom the Union had to do, or the lucid way in which Susan Goliath could explain the provisions of the Labour Relations Act to workers. Goliath was a modest person with capabilities of which the head office would have had no inkling, had her fellow workers not pushed her forward.

Observing her holding her own against the outrageously devious personnel manager of Growers, you would never have guessed that this was someone who had grown up on a farm, and only had the opportunity to finish Standard Three.[19] These paid officials had been educated through being able to assume responsibilities that would otherwise have been denied to them. Education through empowerment, you might call it. Yet it was one thing to hold your own against white personnel manager when you are among your own, speaking your own language. It was a different matter holding your own against a black 'on the same side' as you who spoke better English, in a less familiar context.

It was because I believed that both paid officials and workers would be better able to hold their own at a branch level that I proposed that a merger between Sweet Food and FCWU should be 'from the bottom up'.

The branches should merge first, which in most cases meant enlarging the FCWU branches to include factories organised by Sweet Food. Afterwards, there would be an inaugural conference to formally merge at the national level. This was the way FCWU and AFCWU had merged, but of course FCWU and AFCWU shared a common tradition and had collaborated in the factories. Even then, it had taken many years to overcome the divisions between them. The only way my proposal could have worked was if the national leadership shared a commitment to unity and was actively engaged in bringing it about. I was not up to doing so myself.

I could not look to Chris or Jay for help. As part of COSATU's national leadership, they had pushed through a resolution that 'required' affiliates organising in the same industry to merge within six months, and wanted Sweet Food and ourselves to lead the way. If what they envisaged was only a merger between the two unions, the time frame would not have been as unrealistic.[20] What they were wanting, however, was for the UDF unions to be absorbed in these mergers, because if they were left out, the problem that the 'one union' policy was intended to address would remain unresolved. Political considerations trumped organisational ones, in short.

Most of the affiliates of COSATU simply ignored this resolution, and with good reason. Disregarding valid organisational considerations and forcing the pace at which mergers took place was bound to have adverse consequences. But even though I thought it was a bad resolution, I was resolved to press ahead with a merger, because if a merger was to happen it would be better if it happened while I was still around. The adverse consequences would soon be evident in our case, and the 'bottom-up' approach I advocated had only made matters worse. It was really a pseudo 'bottom-up' approach, because it did not allow any time for collaboration at a local level to develop organically.

One of the adverse consequences was that many delegations to the inaugural conference were neither representative nor united. Cape Town was a case in point. The branch claimed some 40 factories, the vast majority organised by FCWU. But workers from two factories organised by the Retail and Allied Workers' Union (RAWU) formed a majority at the general meeting where the branch was supposed to merge.[21] These two factories belonged to the Imperial group, and most of the workers were from its hostel. They had organised buses to get to the meeting, whereas FCWU had not organised any transport for its members. The branch leadership was to blame for this, but the upshot was an outcome that FCWU members were bound to view as a power grab, and a divided delegation.

Soon afterwards the inaugural conference took place at the offices of CTMWA in Athlone, the scene of one of the most important meetings of the unity talks.[22] Despite what had taken place in the lead-up to the inaugural conference in Cape Town and elsewhere, FCWU could perhaps still have commanded enough votes to secure all the leadership positions. But this was never on the cards. FCWU had already decided to nominate Chris as president. This gesture was not reciprocated. Every other position was contested, including the position of vice-president, to which Peter Malepe was elected. Peter was a driver from Epol Isando and vice-president of FCWU.

Even though it was a done deal that the head office of the merged union would be that of FCWU, the position that was most sharply contested was that of general secretary. It was also contested at every election after that, until I went on long leave. Although I was voted in each time with a safe majority, you could not fail to be struck by the contrast between these elections and the elections for the national leadership of COSATU. In the case of COSATU, everything had been settled beforehand, and no positions were contested. The justification for this was presumably to preserve unity at a point at which the federation was establishing itself.

Crossing the line: full-time shop stewards and the fading away of the worker leader

Ours was the only merger that took place within the six months that COSATU had set. This was a point of pride within FAWU and COSATU, but I was not at all proud about it. I felt as if I had put my own personal needs before the interests of the workers who elected me. Because of the pseudo 'bottom-up' approach I had advocated, I had set FCWU up for a takeover at a time when the risk of a takeover had never been greater. It left me with a bad conscience. That is probably why, after the inaugural conference, I felt as fervently as I did about something as inconsequential as the logo of the merged union. I favoured a logo with a clasped-hands motif at its centre. This was a motif we had adopted at the time of the Fatti's & Moni's strike. The alternative was an obvious crib of the UDF's logo: people marching behind a waving flag, in the style of socialist realism. It was emblematic of the conflict that now surfaced between the populists and so-called workerists in the trade union movement.

There was nothing new about this conflict, except the context. If the objective of the NDR was to establish democratic rule, the trade unions needed to give one last push, together with other popular organisations. None of the affiliates of COSATU were unwilling to do so, and the decision to send a COSATU delegation to Lusaka to meet with the ANC was not controversial. Lizzie Phike was the only woman among them.[23] However, one last push would clearly not topple 'the regime'. The question for me was what form the backlash would take, and where it would leave trade unions for the long haul. Anyone who entertained these kinds of questions was liable to be labelled a 'workerist'.

It was probably inevitable, amid all the rhetoric about 'regime change', that there would be so little tolerance of dissension or debate among organisations that were expected to support the NDR. At the same time the careful approach to workplace organisation of the 1980s was being eroded by the way in which the trade union movement was expanding. Whole factories would join trade unions en masse, with no questions asked as to how they were organised. Whole unions would join COSATU, following the dissolution of TUCSA at the end of 1986.[24]

The ABI plant on the outskirts of Soweto was an example of a factory that joined en masse at the time of the merger. ABI was a subsidiary of SAB with a licence to can Coca-Cola, and a magnet for the upwardly mobile for the same reasons that SAB was. It was somehow represented at the inaugural conference without having been organised by any of the trade unions that formed FAWU, and its chairperson, Geoffrey*, was one of my most vocal opponents. The Garment and Allied Workers' Union, formed as a result of a merger between the Garment Workers' Union and an ex-TUCSA union in Natal, was the most significant of the trade unions to join COSATU with few questions asked. Evidently the Petersen dynasty in the Garment Workers' Union had outlived its usefulness to the bosses, and there had at last been a successful palace revolution.[25]

This was a far cry from the organic growth of the early 1980s and, as a consequence, developments that should have been closely scrutinised were not. The most alarming by far was a scheme Kellogg's had introduced. Someone whispered in my ear at FAWU's inaugural conference that it had created a position on its payroll for someone called a 'full-time shop steward', and the person employed to fill that position was none other than Chris Dlamini. Instead of doing the work others in Kellogg's did, it seemed he was now something like a paid official in the workplace. The difference was that a paid official was paid by the Union and operated from a Union office. This full-time shop steward, by contrast, was paid by the bosses and operated from their offices.

This was when I first learned that the bright line drawn between a paid official and a worker had been breached. If Kellogg's was not the first employer in South Africa to introduce such a scheme, it was one of the first, and, like Kellogg's, they were all American or European multinationals. Clearly, then, it was a scheme that they had already piloted elsewhere. It seems obvious why bosses with significant resources would find such a scheme attractive. It meant having an in-house 'manager of discontent' over which they had a considerable measure of control.[26] This was because, regardless of what was agreed when such a scheme was introduced, a full-time shop steward could not be removed without reference to the employer. Also, the opportunities to co-opt such an individual over time were boundless.

In my book, if a full-time shop steward could be regarded as a worker at all, he or she was not a worker like any other, and could never be regarded as independent. No good could come from the introduction of this scheme, and it was doubtful whether it was even constitutional.[27] I resolved to raise the issue at the first available opportunity. This was the first national executive committee after the merger. This meeting, which should have been an opportunity to draw the organisation together after a fractious launch, was coloured by what looked like a full frontal assault on the FCWU tradition. Ruling from the chair, Chris announced that the paid officials present would not be allowed to speak.[28]

Not only was it unheard of to make rulings from the chair, without debate, but paid officials in FCWU were listened to with respect, because they were likely to have a broader and more informed outlook than someone whose outlook was shaped by the particular workplace where he or she worked. This was another important reason why a full-time shop-steward was a bad idea. So most unhappily, no sooner had 'we' got together than it emerged that we disagreed on fundamentals. In the end I could do no more than raise the issue of full-time shop stewards, in the hope it would be addressed later. It never was. Chris also never disclosed he was one or the terms of the agreement entered into with Kellogg's. Despite numerous hints, he also never invited me to visit Kellogg's, so that I could see for myself what was going on, or any other factory on the East Rand that had been under Sweet Food. This was his stamping ground.

So far as I know, there were no fresh attempts to appoint full-time shop stewards during my tenure. But if there had been, I would not necessarily have known about it. There was something surreptitious about this scheme. This was pertinently brought home to me when I recently discovered that, unbeknown to me, a close comrade from the former Sweet Food had all along been a full-time shop steward and was not proud about it. 'There was no way I was assisting workers as a full-time shop steward,' he told me. 'I was floating.'[29] It also does not appear the scheme was ever openly discussed within FOSATU. People I have spoken to who were in a position to know if it were profess ignorance about it, although this might be due to embarrassment.

You can get away with surreptitious schemes where there is a weak tradition of financial self-sufficiency, as was the case in the emergent union movement and now COSATU; also, where there is a climate of fear. The fact that until the Marikana massacre hardly anyone among the intellectual hangers-on of the trade union movement publicly questioned the introduction of the full-time shop steward is itself extraordinary.[30] It also had to with the practice of labelling those who asked questions as 'workerists'. Certain topics were now out of bounds. Foremost among them was the topic of political and trade union leaders being cosy with the bosses. It touched on what the NDR was really about. So while it seemed there was an agreement with Anglo for the president of COSATU to spend protracted periods away from the mine where he was supposedly employed, no one on the central committee asked what it was.[31] Instead, there were discreet jokes about what he did for a living.

The bosses, from their point of view, had reason to go to some lengths to secure a relationship with COSATU. Since the dissolution of TUCSA, they were without a trade union partner for the first time since the 1920s. The wave of militancy post-1986 must have sharpened their awareness of this. Workers were clearly emboldened, and on many occasions did not even consult their trade unions before going on strike. This was the case at the Clover factory in Pietermaritzburg, shortly after the merger.[32] The strike, it later emerged, had been sparked by political tensions between factory foremen who supported Inkatha and union members. Inkatha had in the meantime established its own 'trade union', UWUSA, as a direct consequence of Barayi's needless and untimely declaration of war.[33] The entire workforce had been dismissed before the head office knew about it.

The bosses responded very differently in a series of strikes in Johannesburg where workers for the first time adopted the tactic of sleeping-in at the factory premises.[34] The first I would hear of a strike would be when some lawyer contacted me, because he (at that time it was almost invariably a 'he') was applying to the Supreme Court for an interdict. Or sometimes it would be a lawyer our local officials had contacted, wanting to oppose an interdict. I would then end up instructing someone I had never met to wage a battle according to rules only lawyers understood. It was a foregone conclusion that this was a battle the Union could never win.

It was hard to piece together what, in each instance, had given rise to these strikes. My sense was that they had less to do with the workplace than rendering the economy ungovernable, in much the same way as the townships were being rendered ungovernable. It was likely this that prompted government to propose amendments to the Labour Relations Act, facilitating employers in interdicting trade unions and suing for damages.[35] But the bosses could not have failed to notice that, even without the amendments, workers eventually heeded the orders to stop striking illegally or sleeping in once they obtained their interdicts.[36] So the clearest consequence of this period of militancy was to sharpen the turn to the law.

At the same time, in stark contrast to the militant strategies being adopted in the workplace, trade unions carried on referring hard cases to the Industrial Court. Our prospects in the case of Clover were in my view nil. However, Pietermaritzburg had been a FOSATU stamping ground. The Union could not be seen to be doing less for its members than MAWU, which had brought a similar case for its members at nearby BTR Sarmcol. The two cases were to run parallel with each other.[37] But unlike BTR Sarmcol, Clover was not a stand-alone plant. Although the local leadership did not seem to know this, it belonged to the biggest dairy group in the country, which so far as we knew was unorganised.[38] There was an organisational alternative to litigation.

I am not sure the attorney handling the case, a decent enough fellow, would have understood what a debilitating effect on the movement reliance on lawyers and legal strategies would have.[39] We inhabited different worlds. This was graphically brought home to me when, after a meeting with the Clover workers, I got a lift back to Durban with him. After playing me his recording of Maria Callas's opera arias in his BMW, which was admittedly second-hand, we sat down at a table in his back garden to discuss our prospects. I took it for a social meeting until I observed him reaching for his expensive-looking wristwatch and switching on the timing device.

It is difficult to see how this period of militancy strengthened workplace organisation in any way. Even the most notable victory of the period, the OK Bazaars strike by CCAWUSA, looks pallid in retrospect.[40] The workers won a significant increase after a marathon strike, but soon afterwards CCAWUSA split into two.[41] Without any serious attempt to get to grips with what had caused the split, COSATU's national leadership chose sides. It was later to backtrack on this decision, but it was some years before the union was able to present a united front to the bosses. Within ten years, what was left of OK Bazaars after a succession of store closures and retrenchments was sold to Shoprite, the emerging competitor of Pick n Pay, for the princely sum of R1.

The most significant strike of the period, the miners' strike of 1987, was an unmitigated defeat.[42] There was no disgrace in defeat in the economic circumstances of the time. Yet it had been a legal strike, which meant the bosses had an opportunity to plan how to ride it out. The one contingency for which NUM had to plan was how workers would sustain themselves during the strike. It did not do so. Time and again, under COSATU, up to the present day, unions would embark on protracted strikes in which workers were expected to sustain themselves on willpower alone.[43] Among those who did not get their jobs back after the strike was James Motlatsi. He carried on being president of NUM nonetheless. At some point it seems the union began paying him a salary.[44] This was of course precisely what we and other trade unions had objected to in the case of SAAWU a few years before. Again, the bright line between paid official and worker had been breached.

The strategy I adopted to counter the institution of full-time shop stewards was to promote accountability among the paid officials. To do so necessitated a crash course in financial self-sufficiency. Some from other union traditions were quick to learn. Others, having never had the responsibility of handling workers' money, were intoxicated at being able to operate a bank account and began freely spending in the name of the Union vast sums it did not have. It was boom time for the car hire business. Cars would be hired without authorisation and not returned. When they were eventually returned, they were often damaged.

Spending money without authorisation was not just reckless. It was a strategy to extract more money from the head office. The financial stability of the Union as a whole was at risk. It was also at risk because, almost as rapidly as the Union was recruiting new members, it was shedding members. The same was true of some other COSATU affiliates. The role of paid officials was now critical in holding the line. However, the slogan of 'workers' control' was time and again used to shut them up. In effect, it was a licence for despotic control in the unions, at a time when the government was trying to exercise despotic control through its declaration of a state of emergency.[45]

At the first ordinary conference after the merger, Geoffrey of ABI proposed a resolution that had obviously been discussed with others beforehand: 'we shall always learn from the heroic leadership that came before us,' the resolution went, and proposed Ray Alexander be 'elected' honorary president of FAWU 'in recognition of the past Heroic Leadership in our struggle'.[46] Honorary president was a silly title for someone who had not been a worker, I thought. But what was more ironic about the resolution was that someone like Geoffrey seemed unwilling to learn from those currently in the Union that had come before him.

Conversations about empowerment

There were several indications that this unwillingness to learn was due to a prejudice towards ordinary workers, and that a class divide among the members was opening up. It manifested over the issue of translation. People who were fluent in English became increasingly impatient at having proceedings translated, especially now that translation into at least three languages was needed. What made matters worse was that what ordinary workers had to say (or officials who had been workers) was increasingly mangled in the process. Someone like Goliath, who spoke little English, would find it increasingly difficult to make her voice heard. She and others like her, who actually were the bearers of the FCWU tradition, found themselves more and more disempowered.

They were also disempowered by the way decisions were taken. With a COSATU congress coming up, the first since its launch, somebody somewhere decided that it was time for unions to pin their flag to the

Congress mast. NUM was the first of a number of COSATU affiliates to 'adopt' the Freedom Charter. At the first opportunity after that, there was an attempt to do so at the Union's national executive committee. The Eastern Cape delegation blocked it, on the grounds that the branches had not had the opportunity to discuss the matter. Little more than a month later, at a national executive committee meeting specially brought forward to take place before the COSATU congress, a formal resolution was tabled to adopt the Freedom Charter 'as a guiding document'.[47]

Although the resolution itself was innocuous, there had still been no discussion about it at the branches. Among the ordinary workers who knew of the Freedom Charter, few would have known what it said, except in vague terms. Some would never have heard of it. The value of adopting it would have been in the debate it generated, but its sponsors had no time for this kind of process. They themselves were educated, and made no attempt to conceal that they had caucused for this resolution. When it was adopted, they burst into a struggle song, as if the 'workerists' in the Union, whoever they might be, had been vanquished.

So the adoption of the resolution marked the emergence of a faction in the Union which at its core seemed to me composed of essentially the same class of people, from the same layers in the factories, that Oom Joe and Mrs R had come from. Only they had been an old guard, wanting to cling onto the material benefits they had been able to wring out of the old order. This was a new guard, which evidently saw itself as part of a vanguard. I felt profoundly depressed, although I tried not to show it. The contrast with the vibrant debates we used to have at the management committee meeting could not have been starker.[48] It looked like my worst fears were being realised.

Murphy Morobe was a guest speaker at the 1997 COSATU congress. He was reputedly among the intellectuals in the UDF leadership, but since the UDF was now banned he spoke in the name of the Mass Democratic Movement (MDM). My guess is that he was also under instructions from the same vanguard. It was an intimidating spectacle listening to him denounce by name four white intellectuals who had been expelled from SACTU years before, in front of thousands of workers. The vast majority can have had no idea who was being denounced, or why.[49] The only reason I could see for such a gratuitous attack, so long after the event, was to assert Congress hegemony and as a warning to the white Left.

I felt vulnerable: all the more so because my material circumstances were deteriorating. Hardly a day would pass without an argument between Athalie and me about how to get by. The problem of nappies was resolved in a working-class way. We could not afford disposables so Di Cooper, an official of the GWU, organised a collection for a washing machine. To secure a space for my son against the scorpions from the graveyard across the street, I did something no working class person would have done. I spent my savings on renovating rented accommodation. When we were evicted soon afterwards, I found myself in the proverbial catch-22: I could not come out on the same salary as other organisers because I was living in a white area, and I could not ask for a higher salary because I was white.

The only form of political leadership someone like me could now provide was with regard to money matters, and in opposing any slippage on issues of corruption. If the head office building was a monument to financial self-sufficiency, for those who had no inkling of how it was acquired it was simply part of the pot of money and other assets which the head office controlled: the Union's capital, if you like. This pot was much bigger than it ever had been, even if the money was flowing out as fast as it came in. Almost all the bosses had by now come round to thinking that they would be fostering responsible trade unionism by granting stop orders. The numbers on stop orders were unprecedented. This was also true for other trade unions.

Siphoning off a portion of the pot for speculative investments was just the kind of thing I was determined to prevent the new guard from doing. It meant crossing a line I thought inviolate. So too did the idea that a trade union could invest in businesses that operated for profit. I knew, however, there were those who did not see things this way. While waiting for delegates to arrive at one of the Ipelegeng meetings, I had a conversation with Johnny Copelyn of NUTW.[50] He started telling me about the Histadrut, the Israeli trade union federation, which owned a significant section of the economy of Israel. This was his inspiration for the role of a trade union federation in South Africa, he told me. I was gobsmacked.

At about the same time as the trade unions and government were locked in conflict over the amendments to the LRA introduced in 1988, the bosses were floating all sorts of propositions that would ease the passage of trade unions across the line I thought inviolate.[51] What counted in money matters, as in the case of the turn to the law, was what trade unions did rather than what they said. One of these propositions was that trade unions endorse private pension or provident funds for their members, and nominate trustees to sit on their boards. There were two problems with this. One was that it foreclosed the possibility of a national contributory pension fund, something Neil Aggett had been passionate about. The other was that the conglomerates had designs on the capital that these funds would raise from their workers. This was evident from the zeal with which the Union was now courted.

We were invited to a flurry of presentations to determine which financial institution was to be appointed administrator of which fund. Although different financial institutions tendered for these appointments, the outcomes were predictable. In the case of Tiger's provident fund, the institution with the lowest tender was Old Mutual. Old Mutual happened to be the major shareholder in Tiger. With Premier it was Liberty. Liberty happened to be a major shareholder in Premier, as well as SAB. I would not be around to observe how the Union's involvement panned out, but all indications are that worker trustees were increasingly drawn into a conversation about investments and returns in which the lines between the interests of those they represented and the company were subtly blurred.

At the same time as this was going on, one of the directors of Premier came to Cape Town to put to us a proposition in which the Union's willingness to cross lines was tested in a different way. The migrant labour system was being phased out, he said, and it was no longer appropriate for Premier to maintain a hostel. Would the Union be interested in leasing the complex or buying it? Our members stayed in the hostel, and we felt obliged to consider this proposition. But since we made it clear from the outset that we would only consider a 'purely business arrangement', our discussions soon petered out. At any market-related price, the complex would be far more than the Union could afford. In any event the Union had no use for another building.[52]

The Union also had more pressing concerns. Foremost among these was the negotiation of a national recognition agreement with Premier. This was, I believe, the first such agreement to be negotiated by any trade union in South Africa, and unprecedented in its scope. It covered all the different operations of Premier nationally.[53] It was to be followed by the negotiation of a national recognition with SAB covering all its plants, including its new flagship at Pietersburg (or Polokwane, as it is now called). It was not coincidental that the first national recognition agreements were with these two companies. In terms of a sensational deal with far-reaching repercussions for the entire private sector, Anglo and Liberty Life had bought out Premier's British shareholder. Part of the deal was that Premier in turn would acquire a major shareholding in SAB.

Utilising cash the deal had generated, no doubt, Premier announced soon afterwards it was moving its head office from Newtown, which it must have regarded as altogether too downmarket for its new status. It began constructing a palatial new building in Killarney, on Jan Smuts Drive, sparing no expense. The Italian marble finishes and Renaissance artworks were handpicked, the gossip went, by Tony Bloom himself. Not long afterwards Bloom emigrated to the UK, to be replaced by Peter Wrighton. I invited Chris to the Union's first meeting with Wrighton to sign the national recognition agreement. Since the new head office was not yet completed, it was at an Epol facility at Halfway House.

We are trying, Wrighton told us, at one point during the meeting, to appoint more blacks in management positions, but we are encountering problems with the attitude of workers when we do so. What can we do about this? If he had directed the question at me, rather than Chris, I would have told him what workers at Epol Pretoria West had said: that the black managers were trying to outdo their white counterparts in arrogance and high-handedness. 'The problem is the Group Areas Act' was what Chris said, because it prevented black managers from living in the same suburbs as white managers. Wrighton, unsurprisingly, immediately agreed with him. This was someone we can do business with, I imagine him saying to his aides afterwards.

It set me thinking. The promotion of trade union leaders into managerial positions was an unresolved problem for the Union. Hennie du Toit at Kromco was a case in point.[54] He had been one of the most impressive coloured men to come to the fore in my time but had just accepted a management position at Kromco. Although he was not compelled to resign, he lost interest in the Union not long afterwards. It was to be expected: there was nothing the Union could offer him now. What, then, was in it for the new guard? In retrospect it seems obvious: it was to secure the material benefits to which it believed it was entitled, under a new order. The 'me-first' version of empowerment that trade unions went for in the 1990s, I believe, can also be traced back to these conversations.[55]

A code of conduct
There is a fine line between corruption, in the sense of eating the money, and transforming a workers' organisation into an organisation that serves the interests of its leaders; and any membership-based organisation is by its nature eminently corruptible. That is why it is never a good idea to talk too soon about 'heroic' trade union leaders. 'Do not trust me because of who I am, or what I say, but look at what I do in practice', is what I would say. 'The bosses are always looking for leaders that they can buy.' I had no proof of the bosses buying off trade union leaders by doling out money, although I have no doubt this happened. On the other hand, the bosses did not need to 'buy off' leaders in such a literal way.

As part of my exit strategy, I had a modest plan to combat corruption as well as factionalism. I wanted the Union to adopt a Leadership Code, which spelled out, in simple language, the values that leadership in a workers' organisation were expected to subscribe to. Above all else, this was honesty in dealing with the money and assets of the organisation, which was why someone involved in dishonesty and corruption was 'not fit to be a leader', as mentioned in the opening chapter. A draft was circulated to the branches and discussed at the national executive committee.[56] It was to be submitted to the conference for approval. This was to be in Durban.

Mordecai Mabasa, the treasurer, and I visited Durban shortly before the conference to review the arrangements, and spent the night at the Moon Hotel in Clairwood. It was near the hall where the conference would take place, and we knew what sort of accommodation it provided when we made a booking for delegates to stay there. It was not of a high standard. For my part, admittedly, I was trying to make a point. For the first time since I had become general secretary of FCWU, the Union's expenditure exceeded our income. Drastic steps were needed to curtail costs. FCWU delegates were, after all, used to sleeping on mattresses on the floor.

As it turned, it was delegates from the Transvaal branches that were booked into the Moon Hotel. They arrived late at night in two buses. Geoffrey led the revolt. They would under no circumstances stay there, he said, and instructed the drivers of the buses to take the delegates to the Durban beachfront where, without reference to anyone else, he was on the point of booking the entire delegation into a Southern Sun hotel with lots of stars. Funky, who had become an organiser in Kempton Park since his dismissal from DairyBelle, stopped him.

Proceedings got off to a late start the next day, and as a consequence of the fiasco over accommodation, the foot soldiers of the new guard were fired up. Even while Sydney Mufamadi was speaking on behalf of COSATU, they made it clear that they had no intention of heeding his appeals for unity.[57] The issue on which they had evidently resolved to take a stand was the Leadership Code and the annual report. Their objection to the annual report, I believe, was that it referred to the existence of factions in the Union and what was contributing to them, and provided the motivation for the Code.[58] By preventing me from presenting the report, they could prevent the adoption of the Code.

The stage was set for a meeting which, for me, was no less extraordinary than that which resulted in Vormat's dismissal years before, and no less hopeful. The new guard engaged in what would be called a filibuster in the United States. It went on all day and into the early hours of the next morning, firstly to prevent me presenting my report and, when that failed, by proposing that it be referred to the national executive committee for its consideration. It would not only have been a blow for me personally, had this proposal been accepted, but would have undermined the status of the conference as the workers' parliament. It was decisively rejected, and not just with the backing of the former FCWU members.[59]

24

The Transkei road

A LEADERSHIP CODE WAS NOT GOING to change the way anyone operated, especially when its adoption had been so fiercely resisted, and the events that would give rise to a second showdown with the new guard in the Union were already unfolding. I was not looking for a showdown. My plans to take long leave were far advanced, and I was trying to keep my distance from the day-to-day goings-on. In the end, however, I had no choice but to get involved. I had after all written out the cheques for strike pay that were at issue, and handed them to an official of Cape Town branch.

The branch now had something like 60 factories. Most had never been organised before, but Spekenam was the same factory I had tried to organise ten years earlier, when the memory of defeat in 1956 was still strong. The fact that FCWU had a strong base in the industrial sprawl of the Northern Suburbs must have allayed workers' fears that what had happened before would happen again. Some had also been members of FCWU elsewhere, notably their charismatic chairperson, George Xashimba, whom I had first encountered on the West Coast.[1]

A month or so before the Durban conference I got a message from a lawyer I had never heard of to come to Spekenam. The workers were on strike and his clients wanted to dismiss them. He was trying to talk them out of it. I had never before come across a lawyer who was willing to go out on a limb to avert a mass dismissal, and that day he was successful. However, Richard*, the organiser responsible for Spekenam, was obviously out of his depth. Not long afterwards the workers came out again over the same issue. Some 600 workers were dismissed.[2]

The Union had no organisation to speak of in Vleissentraal, the group to which Spekenam belonged, or in the red meat industry.[3] All we could do was to keep the dismissed workers together, in the hope that Spekenam could not get going without them. That meant providing strike pay. The cheques I had written out were entrusted to Richard, who was to keep a register of the workers receiving strike pay and return any money that was over. The only other avenue open to the Union was to apply to the Industrial Court for a status quo order, as it eventually did.

It was not supposed to matter after the merger who had belonged to which union, but it mattered in Cape Town because of the power grab at the time of the merger. It also mattered to Richard, who had been from RAWU, and preferred to rely on what must have been a RAWU inner circle for guidance. This inner circle was well connected to the UDF, and looked to its UDF contacts to devise a strategy to mobilise support for the strike. This was a UDF-style initiative: a glitzy concert by the singer Johnny Clegg, which took place while I was at the Durban conference. Unbeknown to anyone else in the branch or the Union, Richard and his circle freely incurred expenses in the Union's name. On my return, rumours were circulating that the concert had been a flop.

It was a flop, not because no tickets were sold, but because of the extravagant expenditure. This included flying Brenda Fassie and her entourage to Cape Town, and hiring an aeroplane to fly over Cape Town with a banner advertising the concert. To make matters worse, on the assumption the concert would make money workers had been promised additional strike pay. I decided to go to the hall where the workers were meeting to see for myself what was happening there. The first thing I noticed was how few workers there were: far fewer than the number in respect of which I was issuing cheques on Richard's say-so. Many, I suppose, had already been re-employed at Spekenam on condition that they had nothing further to do with the Union.

I myself took charge of the pay-out that week, and there was a substantial amount of money over. That raised the question what had happened on the previous pay-outs. It turned out that each time there had been money over, but no record of what had happened to it. I called for statements. When these were not forthcoming, I became more insistent. Then it transpired that signatures on the register of strikers receiving pay had been forged. I will spare the reader the sordid details. Nowadays South Africans have become inured to corruption: so much so that I find myself glossing over the details of the latest outrage with a sense of weariness. It was not like that then. Hardly anyone in the movement seemed to think it remotely possible that their leadership might be corrupt.[4]

Then the Industrial Court handed down its judgment. Labour lawyers hailed it for stating that it was 'unfair for an employer not to negotiate bona fide with the representative union'.[5] This was a nice illustration of legal discourse obfuscating what legal outcomes mean in practice. Predictably, the court had not allowed the 'human element', which it said had caused it 'much anguish', to outweigh the fact that the workers had disregarded an ultimatum to return to work.[6] The workers did not turn against Richard because they had lost the case, but because he was nowhere to be found when they wanted to know what had happened. Later, when I reported to the branch that strike pay had been misappropriated, Xashimba and the leaders at Spekenam were among the loudest voices calling for his dismissal.

This was something that Richard's circle was determined to prevent. He was the second of three former RAWU officials in the branch to be implicated in corruption. But whereas his predecessor had a somewhat shifty demeanour, Richard seemed such a straight arrow. First, the circle tried to fudge the extent to which Richard was accountable. The second line of defence was to concede that, while he might have made mistakes, this was due to inexperience. He had not benefited personally and, in any event, the general secretary was gunning for him. The implication was that I had an ulterior motive. This is what nowadays might be referred to as the Zuma defence.

This was how the Spekenam dispute started. To counter any suggestion of an ulterior motive, I asked the Union's auditors to produce a report. These were the same auditors who had failed to pick up that Johnny M and Stafford were paying themselves a double salary, and they were no worse than any other: auditing the books of a membership-based organisation would always be more trouble than it was worth, compared with auditing a commercial client. But they made a mistake, which could be seen as a deliberate attempt to inflate the amount misappropriated. By the time I discovered it, the damage was done. We had to fire the auditors and appoint others. Not that it actually made any difference what the precise amount misappropriated was.

It was not necessary to prove someone had benefited personally to justify their being removed from a position of leadership in the Union. It was sufficient that it had happened on his or her watch, and he or she was culpable. This was the standard Leadership Code was meant to reaffirm, but more important than the words on paper was what happened in practice. This would now be put to the test at the Branch. Since meetings of its executive had always been open for all who wished to attend, it proved easy enough for a 'well-organised minority' to prevent the branch from taking a decision about Richard. In the end it was the Western Cape region which dismissed Richard (on the same basis as the management committee had dismissed Juffrou H). Richard then appealed to the national executive committee.

The NEC meeting that followed was the lowest point of my career as a trade unionist. From the outset it was clear that the new guard, led by Chris Dlamini, had decided to claim Richard as their own, either because they believed I was gunning for Richard or simply to turn the tables on me. The delegations from Natal and Transvaal were obviously caucused beforehand, and the region that now held the balance of power, even though it had hardly any members, was the Northern Cape–Free State. This was the only place outside Cape Town where RAWU had organised. So even before I had opened my mouth, it was clear that the majority was against me.

This was my first taste of what it must be like appearing before a kangaroo court in which the presiding officer, by a none-too-subtle nod and a wink, makes no attempt to conceal his bias, by indulging one party and undermining the other. Emboldened, Richard began to embroider and backtrack. It was as if there had been nothing untoward in his handling of the strike pay. It is tedious, so long after the event, to have to explain why these attempts to avoid accountability should not have been entertained.[7]

The Western Cape was bound to support its own decision, of course. For me, the region to emerge with the most credit from the meeting was the Eastern Cape. They could see that an important standard had been violated. They did not support the decision of the Western Cape because they belonged to the same camp or had been caucused, but because it was right. One of their delegates sharply rebuked the assistant general secretary for openly undermining me in the meeting. 'I have information that certain people were working for the CIA and intelligence services,' was his response. I think he was referring to the delegate that criticised him.[8]

Confronted with the prospect of an undemocratic outcome, there was no alternative but to organise. Since my president and assistant general secretary were now openly subverting me, I had no scruples about caucusing with delegates myself, something I had never done before. A situation in which the national executive committee had been hijacked, and sought to impose itself on the organisation, was foreseen. The constitution provided that a minority of delegates to the national executive committee could requisition a special national conference. In the centralised model of organisation we had adopted, this was the one way in which the branches could hold the centre to account.

The NEC meeting was postponed without arriving at a decision, although it was clear what that decision would be. When it reconvened a few weeks later and decided to reinstate Richard, the requisition for a special conference was already drawn up and signed. The Spekenam dispute had ceased to be a dispute with the bosses about dismissed workers and recognition. It had become an internal dispute for the soul of the Union. It was unlike any other dispute in the trade union movement at that time.

Nothing focuses the mind as much as the prospect of being called to account before the same people who elected you. That is why, I suppose, Chris agreed to a statement I had drafted in the name of the national office-bearers, conceding that there had indeed been corruption, which was circulated to all the branches before the special conference.[9] It was because of similar tactical considerations that Chris and I accompanied a delegation from the Union to Lusaka, soon after the three-day stay-away in June 1988, to meet with SACTU over the issue. I felt sure where Ray, at least, would stand.[10]

As the date of the Union's special conference approached, there was a flurry of attempts to postpone or cancel it. These included a telex sent from the offices of one of the Imperial factories initially organised by RAWU, warning of dire consequences if it went ahead.[11] When it became clear the meeting was nevertheless going ahead, Richard resigned. He must have been prevailed on to do so, in a bid to pre-empt an outcome that was seen as inevitable. The conference did in fact decide overwhelmingly to reverse the NEC's decision.[12] It was fitting that the venue was again Community House, where the by-now infamous NEC had taken place.

Chris left the special conference at the end of the first day to attend a 'very important' meeting in Germany, and was not present when Sithembele Kawa, the chairperson of the Eastern Cape region, took the floor. The statement I had drafted and circulated to the branches was obviously a bid to save face, he argued. Even though the NEC's decision had been reversed, we could not leave the matter there. A leader who condones corruption is himself corrupt. The majority, however, did want to open up an argument over leadership's role in the Spekenam dispute, and I agreed with them. It was an issue that could be resolved in the elections at the annual national conference.

At the annual national conference a few months later, Chris was nominated for re-election and Peter Malepe declined to oppose him. Mandla was the elected general secretary to replace me.[13] I was pleased that Mandla had been elected. The only credible alternative had been Elliot Nduzulwana, the regional secretary of the Eastern Cape, but I was not sure he would try to preserve what autonomy the branches still retained after the merger. Having secured funding to write a book, I bade the Union farewell in the belief that an important precedent on the issues of accountability and corruption had been established.

My long leave was as much about reconstructing my relationship with Athalie as the book I had in mind, and the Transkei coast had been for both of us a kind of paradise to which we could escape while we worked in the Union. It was also the nearest thing to a foreign country we could find without actually leaving South Africa. According to the South African government, of course, it really was a foreign country, yet unlike the other homelands it was now headed by someone whose commitment to the 'ideal' of independence was doubtful: an army general called Bantu Holomisa.

That, paradoxically, made it even more like a foreign country. No one knew how this situation would pan out, and there was an attempt to oust Holomisa while we were there. It was as though we were living in a different time zone, in which there was no telephone connecting us to our past life. The newspapers arrived days after they were printed, if at all, and there would be electricity blackouts for days on end. The one road leading out of it was hazardous and long. This was the road to Umtata. So I was not able to follow the ups and downs in the negotiations between COSATU and NACTU, on the one hand, and SACCOLA, a federation of employers revived for this purpose, on the other.

I also knew nothing of the proposal of a 'workers' summit'. This would be a meeting of members of the unions affiliated to the two federations where, free from the supposedly pernicious influence of their paid officials, they would arrive at a common position. I thought it was a preposterous idea when I first heard about it, and even more so knowing that one or more of the 'workers' at this summit were full-time shop stewards. However, the enthusiasm with which the COSATU leadership embraced the workers' summit suggested there was another agenda: most likely to take over the NACTU unions, which in all industries except chemical were smaller than their COSATU counterparts.[14]

At a time when suspicion abounded, there was therefore reason to be suspicious about the workers' summit. I would nevertheless have been more circumspect than the Eastern Cape region in issuing a strongly worded press statement criticising the summit. But these were heady times. The response to the press statement was unprecedented and savage. First, the national office-bearers suspended the entire Eastern Cape region. Then they dismissed Nduzulwana.[15] Clearly they were not entitled to do any of this. The national office-bearers had never had decision-making powers. But when legal proceedings were instituted challenging what they had done, they expelled Kawa, which they were also not entitled to do.

The only dismissal that could be described as political in all the years I had been in the Union was Juffrou H's, and the circumstances there could hardly be compared with this. Throughout the bitter struggle with the Vakbond, the Union had also never thought to expel a member. The ostensible reason for expelling Kawa was that he had supported a legal challenge to the conduct of the national leadership. Presumably this was because he had deposed to the founding affidavit. It also looked like revenge against an independent-minded individual.

I had not yet heard what had happened in the Eastern Cape when, a couple of months after the event, I made the long journey to Umtata by minibus taxi, and from there to Cape Town via East London. A letter had arrived at the post office, dictated by John Pendlani to Lesley London, a doctor working for the medical fund in Paarl. The house he stayed in was being sold in terms of a government scheme to turn township dwellers into owners of what was, in his case, a one-room box, with a corrugated iron shack in the backyard. He was now facing eviction. 'I hope that God spares you till we see each other again,' the letter ended.[16] Actually, the question was whether God would spare him. He had cancer of the larynx.

Things had changed in the Union, Pendlani told me when we met, with the sense of urgency of someone who knew he did not have long to live. What could have changed so quickly? I remember thinking. He was not talking about what had happened in the Eastern Cape, about which he knew nothing, but about the state of the Union in the Western Cape. He did not mean things had changed for the better. The leadership was arrogant, and spent more time in aeroplanes than attending to problems on the ground. It was a refrain to which I was soon to become accustomed. I hear it still, and it does not make me glad to hear it: but I am glad I heard it first from Pendlani.

It was with a sense of trepidation that I took to the Umtata road again a couple of months later. Nearly a year had elapsed since I had taken long leave, and my project to write a book was floundering. It had been a mistake not to stick with pen and paper, and instead try to master a new technology on an early version of a laptop. My more profound difficulty was to master what was happening in the Union, and the trade unions more generally. Events were unfolding so rapidly that anything I had to say was bound to seem irrelevant and dated. I had got no reply to my request to extend my long leave, so I decided to attend the annual national conference in person. It was in Mandla's home town of East London.

I had always said I might need more than a year to write a book, but no one seemed to remember that now. If I wanted more time, the Conference decided, I would have to reapply for my job. I was being held to terms, and those who wanted to see the back of me were banking on my rejecting them. What was even more unsettling than this was the atmosphere. People I knew very well seemed afraid to be talking too long to me. From others I got a whiff of complicity in developments that I had no business to be asking about. Among coloured workers I encountered a sense of bewilderment at proceedings that were for the most part not being translated.

Negotiations about negotiations between the government and the ANC were taking place at the time, although this was not public knowledge.[17] The ANC had been left with no choice but to negotiate, conference delegates were told.[18] A few months later I took to the Umtata road again. This time it was to meet Oscar, who was being presented to 'the people' at a rally in Umtata together with other released ANC leaders. It was over by the time Athalie and I got there with three-year-old Leif, and someone told us we would find the leaders in the Holiday Inn, perched on the hill above the town centre. It had been built after the Transkei became independent, so that there would be a place where VIPs visiting the Matanzimas could be put up in a style to which they were accustomed.

We found Oscar on his own in a wheelchair, with the leftovers of the spread that had been laid on the tables in front of him. Walter Sisulu,

Raymond Mhlaba and others were some distance away, talking to each other. Oscar was in an exuberant mood, but the only one of the other leaders who took any interest in his white guests was Mhlaba. This was to be the last time I saw Oscar. Although I spoke to him on the phone, I did not get round to visiting him on my return to Cape Town. I gathered he felt forgotten and abandoned by the movement in which he had played a prominent part.

Reinventing tradition
I could see how quickly some things had changed from paging through a copy of the 1989 Annual Report that I brought back with me from the East London Conference. This was a very different document from any brought out in my time or before it. It was in a sophisticated new format, with voluminous annexures. It even had photographs. Additional staff had been employed to do this. So barely one year after my departure, the head office was expanding. This could only mean there would be fewer resources available for the branches.

The conference was taking place at 'a momentous time of our struggle', the report said in its opening paragraph. This was undoubtedly true. The negotiations about a new constitutional dispensation in Namibia would enable SWAPO to come to power in Namibia. Given the negotiations about negotiations taking place between the ANC and the government, it was surely a foregone conclusion that the ANC and probably the PAC would be unbanned. However, another statement in the opening paragraph was undoubtedly wrong: that this was 'a time when the mission of the working class was high on the agenda nationally and internationally'.[19]

The Party, of course, claimed to be party of the working class, and it was not a foregone conclusion that it would be unbanned. The report made a bold statement in this regard. It harked back to the first half of the twentieth century when the Communist International was established, and capitalism had got the fright of its life. For those who did not know what the Communist International was, there was a box to explain.[20] Its analysis of global developments since might have been borrowed from a Party manual, if such a manual existed. We are a Party trade union, the report seemed to be saying. When did 'we' become a Party union, I wondered, and what did this mean?

All the cracks that had appeared during the Spekenam dispute had been sealed, the report said. The cement was unswerving loyalty to the Congress movement. It seemed the Union had now taken a political position that went beyond adopting the Freedom Charter as a guiding document. This must have been a position taken at the national executive committee, as there had been no conference since the last one I attended. However, it was described as the 'political position' of the workers, and meant accepting the so-called two-stage theory of revolution. This, of course, was an article of faith in the Party.

The report was also laced with references to historical events I knew nothing about, which seemed to suggest a seamless transition from the golden fifties to the present. 'Our union was the first union to be addressed be [sic] the President of the African National Congress, Chief Albert Luthuli,' it said. In 1953, it went on, Frank Marquard had attended the conference of the WFTU.[21] This was by way of background to explaining why the national executive committee had decided to disaffiliate from the International Union of Foodworkers (IUF) a few months before. In the global trade union movement, the WFTU represented the other camp, and SACTU was affiliated to it.

There could be no conceivable benefit in disaffiliating from the IUF in 1989. The Soviet Union was among those that had 'forced' the ANC to negotiate, and this was a sure sign the Cold War was coming to an end. Within two months the Berlin Wall would fall. More pertinently, the decision to affiliate to the IUF had been taken after the merger, after extensive debate at the conference and branches.[22] To reverse this decision without bothering to defer to the conference was remarkable, and another indication of how power was being centralised, despite the outcome of the Spekenam dispute. A radical about-turn was now presented as though the Union were merely reconnecting with its past.[23]

The version of the past it was seeking to evoke was both sentimental and selective. It bore little relation to what the Union had been in 1976. It was also not relevant to the challenges the Union, and the trade union movement, faced on the ground. These included 'privatisation' and 'deregulation', which had become central planks of the policies that the neo-liberal project had inspired. There was an urgent need to discuss how to respond to these policies, but there was no mention of them in the report.[24] There was also a need to consider the Union's position if discussions under way about forming a trade union for teachers were fruitful and it applied to affiliate to COSATU.

Reinventing a tradition is something I had also done, but then it had to do with principles to which the Union had once subscribed. This reinvention was designed to bolster the authority of the leadership in the aftermath of what had been a coup. The conference was not being asked for direction. It was being told what had already been decided. In truth, the Union had not sealed the cracks at all. 'Officials must carry the political position of our members,' it said threateningly. There was a shrill undertaking 'to commit itself in fighting counter revolutionaries by organising and launching a campaign, firstly in the Eastern Cape.'[25] 'Counter-revolutionary' had always been Party-speak for people who did not toe its line.

Having the leadership laying down the line to the conference was fundamentally at odds with the model of democracy that I thought the Union had always represented, but perhaps I had been wrong about this. A vanguard party lays down the line, and Ray was a member of such a party. There were happy snaps of her with the national office-bearers in the report. Only she could have known about the obscure events in the golden fifties that it mentioned. She must also have had a hand in the decision to disaffiliate from the IUF. SACTU (or the Party) must have seen some benefit in a symbolic statement from an erstwhile affiliate, at a point at which its global reach was waning.[26] That is how out of touch Ray and the SACTU leadership were.

It tallied with my impression of SACTU on our visit to Lusaka during the Spekenam dispute. At the end of a long day, my conclusion was that the SACTU representatives present, including John Nkadimeng, had a very limited or distant experience of building a workers' organisation. Their outlook was also shaped by the exigencies of exile. The night before I had a glimpse of what might have coloured the exile experience. A contingent from the ANC security's department visited the hotel where we were staying, with questionnaires. They sought to interview each of us in our rooms. The sole purpose was to get us to name names of suspected informers. The man who interviewed me was not happy. I told him I did not know of any. Others, I heard, were effusive.

I was also disappointed with Ray's contribution at the end of the meeting. In the final analysis, the issue at Spekenam was whether you apply the same standard to your own as you applied to others. This, it seems to me, is the ultimate test for any leadership opposed to corruption. It was the leadership of the Congress and the Party that was now being tested, because Richard must surely have been a Party member. That would explain the ease with which the new guard had adopted him. Instead of addressing this issue, she concluded a rambling exposition about what the Union had achieved over the years by saying, 'How could you put all this at risk for a Richard' or something to that effect. The part I am sure of was her 'for a Richard', because she said it so emphatically, and this is what I latched onto as an endorsement of my position. But it was not an endorsement in terms I welcomed. It was as if she was cutting him loose because, in the bigger scheme of things, individuals were expendable.

It seems obvious, in retrospect, that there was always a potential conflict between a project of building democracy from the bottom up and maintaining accountability at the local level, on the one hand, and building a vanguard party on the other. What made this potential conflict overt was the proximity to power that negotiations presaged. One way to resolve it was to take over the organisation. Maybe this was already happening or had already happened.[27] Yet even now, it seems, power was not centralised enough for the leadership's taste.

Among the amendments to the constitution proposed at the 1989 conference was one that sought to dilute the representation of branches at the conference. Another proposed giving the national office-bearers the power to institute disciplinary proceedings against both members and officials. This was of course a power that they had already appropriated. The most shocking for me was a proposal by the East Rand branch, which was still Chris's power base. The annual national conference was now to be held every two years. A special conference could only be held if the national executive committee decided to hold one, or a majority of delegates signed a requisition calling one. In this way the one mechanism by which the branches were able to hold the centre to account was nullified, and the national executive committee would become a law unto itself.[28]

It was obvious where the campaign against 'counter-revolutionaries' would go next. The pretext would be a letter of support to Nduzulwana, written by one of the bright young coloured men who had come to the fore in Cape Town.[29] The branch leadership were frustrating a discussion of his treacherous conduct, it was suggested.

The conference mandated the region to organise a special meeting to remedy this.[30] The region took the hint, and before the year was out it had suspended the branch, ostensibly because it was unable to function. Things were coming full circle. It seemed the factory that had precipitated the demise of FCWU in Cape Town in the 1950s was now precipitating its demise again.

So far as I knew, the only time a branch had ever been suspended was after the breakaway at Jones. That was at a stage when Paarl branch had less than ten members left and really was unable to function. The only sense in which Cape Town branch was unable to function was that it was unable to resolve its political disagreements. This had everything to do with the political position the Union had adopted, in terms of which everyone had to toe the same line. There could be no dispute about this.[31] It was a complete reversal of the position taken in the Leadership Code, which promoted tolerance for different political views. Maybe it was this that the new guard most objected to about it.

A resolution apparently adopted by members at Epic Oil goes to show how particular this line was, although I doubt whether the ordinary workers can have had anything to say about it. 'We are entering a very exciting time in our struggle,' it stated, but 'some elements within our midst are in disagreement with the course the struggle is taking.' These elements are 'exploiting the openness of the Union to further their own political agenda'. Only someone from the Party could have framed the solution proposed.

This was to adopt 'the analysis of the South African situation' as being 'a colony of the special type' and the two-stage theory.[32]

Those who were unwilling to toe the line in Cape Town were removed by another 'well-organised minority' at a general meeting of questionable validity, presided over by Peter Malepe, as vice-president of the Union.[33] Among those removed were MacWellington Mtiya, the chairperson, and Miles, who was dismissed as branch secretary.[34] Three of the branch organisers were also suspended. One of them was George Xashimba, the former chairperson at Spekenham, who had since been employed by the Union. There had never before been such a radical intervention in the affairs of a branch. I cannot see how any aspect of it could be reconciled with any tradition of the Union I recognised.

The hardest thing for me to accept was the role that people I knew and liked had played in all this, and that the well-organised minority had included workers from factories I had been involved with. Epic Oil, for example, was the last of the Premier factories in Cape Town to be organised by FCWU. Its chairperson was a thoughtful and quiet-spoken person, Ernest Theron, who was to assume an increasingly prominent role in the Union. This also reflected the increasing dominance of upwardly mobile workers in its structures. Ernest was the first qualified boiler-maker to join the Union, whereas Mtiya had become closely identified with ordinary workers like Friday and John Pici during the Spekenam dispute, despite being from SAB.[35]

The ousting of an elected leadership could have led to a breakaway. It did not, because the ousted leaders and their supporters decided to try to reverse their ousting from within, by launching a 'Campaign for democracy in FAWU'. They also decided to challenge their removal in court. Legally, so far as I could see, they had an open-and-shut case (I was not consulted). All they needed was a suitable lawyer, who could claim the political and organisational high ground. Instead, they appointed someone with a thriving practice interdicting workers and unions. This blunder proved disastrous. The Union, for its part, appointed Bulelani Ngcuka, who in the not too distant future would become National Director of Public Prosecutions.[36]

It was always going to be a tall order to try to reverse a decision from within in an organisation as centralised as the Union now was. On realising this, it seems that the Union's public position was simply to ignore the Campaign and all it said.[37] Pamphlets putting the 'official' version of the Union were unsigned. Their killer point was that the 'campaign for democracy' was led by people who did not themselves accept the democratic outcome of a meeting, and were now splitting hairs.[38] The tone was also nastier and more personal than anything I had seen in an internal dispute. Much of this bile was directed at Miles, probably because for the upwardly mobile the white Left were seen as a class enemy.[39] Perhaps that was nothing to complain about at a time when people were being assassinated for their political affiliations in Natal.

The Union had a lot to say about being taken to court, however. They were 'taking their own union to court', a pamphlet issued by the Union screamed at Mtiya and the other applicants, whom it named. 'What are these courts they are taking their own union to? These are the apartheid courts ... The same courts that sentence our people to death. The same courts that sentenced Mandela, Sisulu and our other leaders.'[40] This represented another reinvention of an aspect of the FCWU tradition. Do not look to the courts to resolve what can only be solved through organisation, had been our position. Do not expect lawyers to do what you can do for yourselves. The FCWU position had never been that you do not resort to the courts when your rights were flagrantly violated.

While the struggle for the allegiance of the members of Cape Town branch was unfolding, Ray and her husband returned from exile. They were apparently the first of the exiles based in Lusaka to do so. Someone tipped me off when she was due to arrive, and I went to the airport to meet her. On my way into the airport buildings, to find out what was going on, I stumbled into the national office-bearers of the Union, clad in the khaki uniforms of 1950s volunteers. Led by Chris, they walked determinedly past me, without greeting or acknowledging my presence. Khaki uniforms were now to become de rigueur for shop stewards and officials alike.

Outside the building a queue was formed from the entrance of the airport to the parking lot. In front of me were some township 'mamas' whom Ray may or may not have known. She greeted them no more or less warmly than she did me. I had hoped, vainly, to exchange a brief word about what going on in the Union. I was trying to distil my meandering attempts to write a book into an article for the *Labour Bulletin*, and there could be no going back for me once it was in print.

As it turned out, there could be no going back for another reason as well. Under pressure from its own supporters, and on the basis of some kind of agreement brokered by COSATU's regional leadership, the ousted leadership of Cape Town branch withdrew their court challenge. Once it had been withdrawn, it transpired the agreement was not in writing and the Union did not consider itself bound by whatever was discussed. Instead, as had happened in the Eastern Cape, the Union brought disciplinary charges against workers.

There were ten of them in all, and the charge was 'taking the union FAWU to court or supporting people who have taken the union to court'.[41] The venue for the hearing was the complex of building where the Premier hostel was located. The Union had moved its head office there only days before, and the workers who pitched up for the hearing were reprimanded. They had been misled, it was said. Five workers, including Mtiya, declined to attend. A decision was taken to expel them from the Union. 'What kind of union expels worker leaders?' I wrote on my copy of a pamphlet announcing the expulsions.

Grounds of divorce

'Take us back to the Food and Canning Workers' Union' was the heart-rending message I received not long afterwards from Friday. Friday was not one of the five, but had played a prominent role in the Spekenam dispute. He was also still chairperson at Fatti's & Moni's, the factory that more than any other had helped shape what the Union had now become.

There were bosses in the Union, Friday explained, when we met. People in the offices took decisions. What the workers had to say no longer counted. It brought home to me more forcefully than anything else how far the national leadership had gone to destroy what it must have seen as an alternative centre of power. The national executive committee in the meantime had elected a delegation to meet with me in a belated response to my objections to the Union's refusal to extend my long leave. I asked David Lewis, who had resigned from the GWU some years before, to accompany me as my representative. It was a painful meeting, he said to me afterwards. I suppose that was because he did not think I should be burning my boats.

Somewhat to my surprise, the NEC delegation that came to the meeting was made up of people with whom I had always got on, rather than new-guard zealots. Sebei Motsoeneng from the Vaal was there, and Isaac Mahlangu from Kempton Park. Sebei would within a few years move from the mill to being a mayor. Isaac Mahlangu would become mayor of Khayalami, the metropolitan area where the brewery where he had worked was located. These were indeed 'exciting times' for talented individuals in leadership positions in the trade unions.

The NEC delegation also tried to be conciliatory, whereas I had asked the expelled workers to accompany me to the meeting. I could not go back to the Union while their expulsion stood – an injury to one was after all an injury to all – but I also had a family to consider and I needed somehow to claw my way back into the middle classes. I was looking for a divorce from the Union. The grounds for divorce were in the article which I had by now submitted to the *Labour Bulletin*. I handed Mandla a copy after the meeting. It had not yet appeared in print.[42]

The messiest part of a divorce usually concerns money matters, and this is the part of the irretrievable breakdown that I still have to explain. As well as the amendments gutting branch autonomy, there was a proposal that the subs be more than doubled. There were also to be different rates for different categories of worker, the very proposal that had been so emphatically rejected at the Union's conference in Grabouw. The justification for increasing the subs, according to the powers that be in the Union, was to 'upgrade' it. There were promises of telexes and faxes that would be bought for branches, to improve the service to members. In fact, however, the focus would be on the head office.

More departments were to be established, meaning yet more staff, which would in turn need more space. This was to be provided by buying a new building. It seemed as though the decision had already been taken that this would be the Premier hostel and the complex of which it was part. The plans for its renovations were attached to the Annual Report, and an indication of the price to be paid and the cost of renovations.[43] Even at this bargain-basement price, it was many times more than FCWU had paid for the Albert Road building. The financial position of the Union now was much worse than it had been then.[44] I doubt whether the Union could have mustered enough for a deposit.

My conclusion was that the complex was a gift. I did not say so in so many words in the article, but that was the implication.[45] This went well beyond the symbolism of not accepting a cup of tea from the bosses. If it had happened in a trade union known for its commitment to financial self-sufficiency, the same kinds of things were very likely happening in other trade unions, particularly as they adopted ever-more centralised structures and their bureaucracies became more entrenched. It was very likely happening in the broader political movement that had coalesced around the ANC. The warning lights had been flashing for a long time. I did not really expect that by blowing the whistle it would be possible to reverse this trend, but I did hope it would generate awareness of the danger of 'under the table' deals.

The one development I did think could be halted was the system of having full-time shop stewards in the workplace. Surely the Party, at least, would see it as a problem having shop stewards that were paid for by the bosses. The Party was, after all, founded on the proposition that material conditions determine how people behave. Ray, I thought, was in a position to do so. I went to see her twice after our airport encounter, and gave her a copy of my article. I also asked her to take up the matter of the expelled workers, but I don't think she shared my sense of outrage.

PART 5

From the nineties till now

*'Wisdom is sold in the desolate market where none come to buy,
And in the withered field where the farmer ploughs for bread in vain.'*
– WILLIAM BLAKE, 'ENION'S COMPLAINT'

25

The desolate market

'Uzundikhumbule wakulungelwa'
('Remember me, when everything comes right for you')
– Motto of the Ex-Natal Coal & Gold Mine
Workers Co-operative Congress

At the point at which my story about the Union began, the gap between the classes had been narrowing in developed countries – what used to be called the First World.[1] Full employment in a standard job – working full-time, on a permanent basis – seemed an attainable goal. You could understand why, in this context, people might believe that the rising tide of capitalist development would float all boats. Yet there were alternative economic models, even if not all were credible. Most importantly, there were alternative centres of power. For this reason capitalism was fettered in various ways.

In the 1980s the trade union movement – an important alternative centre of power in these countries – was decisively rolled back. Although it was doubtless an already weakened trade union movement, the state took the lead in this, particularly in the United States and the United Kingdom. It was these two countries that would inspire the neo-liberal project that held sway globally, after the fall of the Berlin Wall and the implosion of the Soviet Union. The 1990s was the decade when capitalism cast off its fetters. Instead of the First World, we began to talk of the Global North.

Aptly for me, the nineties was the decade when I began clawing my way back into the middle classes. The only way I could do so, as I saw it, was to swallow my criticism of labour lawyers and become one. So I opted to accept an offer from the Union's old firm of attorneys, who thought I would be able to bring in a slice of the business of representing trade unions. There I was fortunate to meet someone with a passion for justice, Sandy Liebenberg, who took me under her professional wing. Together we looked out over the city from the 14th floor of a high-rise building, at yet another chaotic demonstration in the tumultuous period preceding South Africa's first democratic elections.

COSATU and NACTU had in the meantime struck a deal with SACCOLA and government over the Labour Relations Act (LRA). This was the Laboria Accord.[2] It marked the point at which the emergent trade unions became established, and COSATU assumed the position TUCSA once occupied, as the pre-eminent trade union federation. The Accord also gave shape to thinking that was in the air at the time, about organised labour 'negotiating' laws and policies with organised business and government as 'social partners'. I will refer to this as tripartism. Given the elevated status COSATU now enjoyed, the funders could no longer see why they should be paying its legal bills. As a result, the business of lawyers representing trade unions was already diminishing, although the business of representing employers would continue to thrive.

The establishment of a 'powerful' trade union movement at the start of the 1990s helped foster the belief that South Africa was exceptional. It was a belief many found comforting, since apartheid had been such a global embarrassment. In truth, however, South Africa was not exceptional at all. The extent to which its trade union movement was able to buck global trends would depend on how it responded to the industrial restructuring already under way. As a consequence of restructuring in the 1980s, there was already a widening divide between workers in jobs and the unemployed.

COSATU made a half-baked attempt to bridge this divide by setting up a national trade union for the unemployed, as well as domestic workers. When this initiative failed, it seemed to abandon 'bottom-up' attempts to reach out to the unemployed. Instead, it focused on influencing legislation and policy through tripartite negotiations. This would in turn strengthen its predilection for centralised forms of organisation and legalism – an undue reliance on legislation and the courts. This 'top-down' approach also contributed towards trade unions neglecting their base. What was going on in the Union was indicative.

From my vantage point on the 14th floor, it looked as if it was unravelling. In the Eastern Cape, as a consequence of the coup by the national leadership, Kawa and Nduzulwana established a new union.[3] This was the first of a series of breakaways. The most tragic for me came about as a consequence of the coup in Cape Town. 'I am leaving a union in which there are bosses, for a union in which the workers are the bosses,' Friday, the chairperson at Fatti's & Moni's, told me. At about the same time Spasie, the vice-chairperson, succumbed to tuberculosis, the killer of the Cape Flats. She must have still been in her thirties.

I could not fault Friday's logic. Not only were individuals in leadership behaving like bosses, but they were accepting positions in management that the bosses were offering. Unlike those who had crossed over to the bosses before, these were positions in HR rather than production. Nosey Pieterse, who had been part of the RAWU inner circle during the Spekenam dispute, was an early example. He accepted an HR position in the factory where he had been a shop steward. What Friday wanted, on the other hand, was what he had learned a trade union could be. Time would tell whether his expectations of the new union would be fulfilled. I had my doubts.

Not everyone at Fatti's & Moni's followed him into this new union. Those that stayed with FAWU, I gathered, were mostly a younger generation who had not actually been involved in establishing the Union, as well as the upwardly mobile and ambitious. The appeal the Union would have for them was its proximity to power, by virtue of its affiliation to COSATU. In the meantime, COSATU had displaced SACTU in the Congress alliance, and negotiations for a new national constitution were getting off the ground. The issue of the day was whether the outcome would be power-sharing or a transfer of power.[4] Whatever the case might be, the closer you were to power, the better placed you would be to take advantage of the 'exciting opportunities' that would open up.

I was too preoccupied with learning the new tricks of my profession to offer Friday advice. Anyway, offering advice meant assuming responsibility for those who acted on it. It was not clear that there was anything to be done but wait and see how things panned out. So ten years after winning a historic victory, the workers at Fatti's & Moni's were again divided. They were still divided when the Union commemorated its anniversary, as described at the beginning of this book. Friday had in the meantime retired, but his son was now working in the factory and belonged to the same breakaway union his father helped form in 1990.[5]

Elsewhere in Cape Town, workers were divided between those who backed the 'campaign for democracy' and those who supported the powers that be. I was not sure where John Pici stood in all this. Probably he was keeping his head below the parapet, as were many others. On the other hand, Premier workers had openly resisted the Union's takeover of the hostel complex in Gugulethu. John was staying there at the time. The cell phone had by now become a popular means of communication, and every now and again a worker would phone me from elsewhere in the country – be it Durban, Pretoria, Klerksdorp or Worcester – to pass on some snippet of news or, more usually, to bemoan the state of the Union.

The women and a handful of men who had kept the Union going until 1976 were by now all gone. Had I heard that Oom Joe died? He had a heart attack, Lizzie told me, not long before Baba, his one-time lover, became one of the many casualties of the tumultuous transition to democracy. She had been beheaded by an angry mob in Mfuleni. I never heard what became of people like Mrs R, who turned bad, or those who remained true, like Mrs Amon, Hester Adams and Polly Solomons. I attended Van Graan's funeral in Pniel. There cannot have been many workers who, having attained retirement age, lived the biblical threescore and ten. Noko had retired, Lizzie's friend Luska told me at the anniversary celebrations. A short while afterwards, he was called back to SAPCo to 'help out' during the season, and dropped dead on the job.

SAPCo was still canning fruit, but Gants had closed, and would in time be transformed into a retail plaza. What capacity to can fruit still remained at Jones had been shifted to Ashton. Was it coincidence that this was also where the Union had been weakest? RFF was still operating, but most of the work it provided in Paarl was in a new division producing packaged meals for well-to-do shoppers at Woolworths. In time, the workers would join another trade union: further evidence of a trend towards small, local unions.[6] The cannery would reopen in Swaziland, where wages were low and trade union organisation was repressed.

Langeberg had in the meantime consolidated all its factories into a separate company, which was taken over by Tiger Oats soon afterwards.[7] Some years after that, the Ashton factory would link up with Ashton Canning, which presumably by then had second thoughts about its policy of refusing to employ African workers. This relentless process of consolidation would lead to ever fewer, ever larger conglomerates. As a result the Union's base in food manufacturing was shrinking, as was the base of the other trade unions in the manufacturing sector.

The 'reforms' that government was expected to adopt in order to be integrated into the global economy exacerbated this situation. Trade liberalisation required among other things that tariffs on imported goods be lowered. Egged on by its advisers, the government went much further in lowering tariffs than it was obliged to do. It also scrapped the system for the marketing of agricultural products and price controls. This in turn gave further impetus to industrial restructuring in food manufacture. Although only a government in waiting, the ANC went along with these 'reforms'. It seemed keen to show it was a quick learner.

The lines along which the working class was fragmenting were now becoming clearer. The workers that COSATU trade unions actually represented were still overwhelmingly in manufacturing and mining.[7] But since most of those who lost their jobs as a result of consolidation and restructuring were 'unskilled', the profile of the membership was changing. More and more of them had skills and education, and were in relatively secure jobs. The workers who now lost their jobs, on the other hand, were even less likely to find jobs as good as they had than in the 1980s.

For George Xashimba, the former chairperson of Spekenam, self-employment was an alternative. But like many self-employed people in a country in which the norm was a wage, he hankered for a 'real' job. When a multinational company engaged him as a contractor on a 'community-based project' to electrify Khayelitsha, I argued on his behalf – and on behalf of other township contractors – that the company was really their employer. When that project folded, he was left high and dry until he found a job as a translator.[8] He must have been in his early forties when he died.

For many, the only prospect of employment was in temporary jobs. So for the migrants among them – whose jobs had, of course, been temporary prior to the 1980s – the benefits of 'permanency' had been brief. Some stayed on in hostels, because they had nowhere else to go. I heard from John about those in the Premier hostel who got by on the charity of their comrades in jobs, all the while the Union was revamping the section of the complex where the head office was now located. It was not surprising in the circumstances that hostels became once again the launching pads for attacks on township residents, in Boipatong and elsewhere. The object this time was to derail the negotiation of a new constitution. At Langeberg Boksburg, the plant whose management had been so relentlessly hostile towards Neil Aggett, men wearing Inkatha T-shirts were allowed onto the premises in August 1991, bearing placards stating 'Down with shop stewards', 'Down with FAWU Union', 'Voetsak ANC' and the like. Later there would be bodies on the factory floor.[9]

There was nothing new about temporary employment, of course. There had always been seasonal workers in the canneries and fish processing plants, and casuals hired at the factory gates. Labour broking had been introduced by way of a 1983 amendment to the LRA, as I have already mentioned, but it appears it was primarily utilised to provide skilled workers for large projects.[10] The challenge to organised labour was nevertheless clear. Just as there was a widening divide between employed and unemployed, there was a widening divide between workers in what used to be called a 'permanent' job – which I call standard employment – and workers in non-standard jobs.

Unless unions were able to reach out to these workers, they would increasingly represent a relatively privileged section of the workforce.[11] It was much easier, however, for COSATU to expand its membership in the public sector, now that there was the prospect of an ANC government. You might think that the Party, by virtue of its claim to be the authentic representative of the working class, would be alert to the dangers of a federation representing sectional interests. This had of course been the problem with TUCSA. In fact, the Party was more concerned with establishing power over workers and stifling alternative voices within the union movement, if events in the Union were anything to go by. Maybe it would have adopted a different approach if it had had a 'plan B' when the Soviet Union imploded. Old habits die hard.

'Build democratic organisation and fight for the transfer of power' was the slogan adopted in the general secretary's biannual report. What the Union was really building, however, had more to do with centralism than democracy. The conference, for example, was now taking place every second year (which is why it was a biannual rather than annual report). Later it would become a triennial conference.[12] Having a conference every second or third year was bound to strengthen the hand of the national leadership and mute the voice of the branches. It was also accompanied by a series of other measures which were calculated to strengthen control from the top down, and which would undermine and ultimately destroy what autonomy the branches still had.

Predictably, the primary instrument of control was money. The pot of money the head office controlled was already very much larger, as a result of doubling the subs. However, branch officials complained about not having enough money to pay the telephone accounts. 'We don't know how the Union makes its decisions, or how it spends our money,' Susan Goliath told me on a visit to Ceres. 'We don't know what to report to the workers because we don't know what is going on ourselves.'[13] The office there exemplified how effective branches could enable the Union to reach out to unorganised workers as well as the unemployed. It had acquired the status in the community of a place where anyone could go for help, and on a Saturday morning farm workers would stream to town, knowing it would be open. If the Union had built on what already existed, a project to organise farm workers might well have succeeded. Instead, it was a flop for much the same reasons as 'top-down' initiatives to organise the unemployed and domestics had been.[14]

I did some sums to see whether the branches were indeed being starved of resources. In the decentralised model of FCWU, branches retained at least 70 per cent of the income they collected, and the rest went to head office.[15] This proportion was now reversed. The head office retained 70 per cent of what the members of the branch had generated, and few benefits flowed to them in exchange for this generosity.[16] Branches no longer received minutes of meetings or financial statements. It was also impossible to get answers from anyone at head office. When Goliath contacted Eric*, a diligent and hard-working official in my time, he was to be found playing card games on the computer.[17]

Branch autonomy was also undermined by vastly expanding the area each branch covered. Ceres, for example, became part of a Boland branch, which stretched as far as Paarl, an hour away by road. This rendered the branch general meeting obsolete. Along with it went the accountability of an elected official towards the members she or he served. It would in any event not be possible to service such a vast area without transport. This would fuel loud complaints, also in other COSATU trade unions, about lax officials and poor service. This in turn created fertile ground for false solutions, notably full-time shop stewards.

No one knows how many full-time shop stewards there were at this point, either in the Union or COSATU, because of the surreptitious nature of the scheme. However, there certainly were some in areas where FOSATU had operated, such as the Reef. I have been told that they were stoking complaints against paid officials at branches. It was certainly in their interests to do so. The only way the scheme could be reconciled with a commitment to 'worker control' was if that was understood to mean control *over* workers rather than *by* them. Paid officials, on the other hand, were outside their control and, accordingly, a threat. It is also easy to understand why a head office scrimping on spending money at branches might favour this scheme. It cost them nothing, courtesy of the bosses. Paid officials, on the other hand, were expensive.

What is certain is that at some point there was talk within the Union about 'dead wood' and a number of paid officials with a passion for justice were removed. Mandla negotiated a termination package with Lizzie Phike, who had sometimes used her own money to get to the factories. Munaadiah Moosa (formerly Baxter), who had done more than anyone to organise unorganised workers in Worcester, was fired. Although an arbitrator later found her dismissal was unfair, she was not reinstated, and went on to found another small, local trade union. Israel Mogoatlhe was one of the founding members of the branch at Kempton Park and its first secretary. He was fired and reinstated twice, before being dismissed a third time. The ordinary workers were well able to recognise a developed sense of justice, but the head office could not. It was not something it could measure.

In the meantime, confronted with a centralised organisation with incomparably more resources, the 'Campaign for Democracy' began petering out. This did not mean the Union was growing from strength to strength. 'Our union does not exist at all,' was the startling statement Mandla made in his biannual report to the 1993 conference. 'We are back to the early days of mergers when we were divided and were having groups.'[18] The elections were for the first time conducted by an outside organisation, IMMSA, and Chris Dlamini was defeated for the post of president by Ernest Theron. His supporters did not accept this and for some reason set Mandla's car alight.[19] Subsequently I heard that a section of the membership in Natal had blocked the payment of their subs to head office.

Blocking payment of subs might have been a legitimate response to an unresponsive head office, but not to the outcome of an election. It also raised questions about the role the Party was playing in the Union.[20] Chris was after all on its central committee, and most of the leadership were on one Party structure or another. Indeed, this was how the leadership thought the trade union ought to engage in politics. 'We need to join and encourage our members to join the Communist Party,' they were told.[21] The conclusion seems inescapable: the Union was caught in a spiral of conflict, which the Party was unable or unwilling to prevent, or of which it was actually the cause.

The problem that Chris now posed for the Union was resolved after the first democratic elections in 1994. One of the 'exciting opportunities' that the elections opened up was that a COSATU contingent would be 'deployed' to Parliament as part of the ANC slate. He was part of it, along with a few individuals who were not known to be ANC supporters. This went to show, I suppose, that the ANC was clinging to its identity as a movement rather than a political party. I had in the meantime renewed my own membership, and attended the first general meeting of the ANC branch in the area in which I stayed. Reggie September was the guest speaker and, I think, noticed me. He went on to speak of people present who had joined while the organisation was still underground, but I did not approach him afterwards.

I doubt whether any workers voted for the ANC because there would be a COSATU contingent going to Parliament, or because the Party told them to. In an isolated rural community like Ceres, the coloured working class voted overwhelmingly for the ANC in 1994 because of the promise of fundamental change that it signified. Were it not for the Union, they would not have believed in that promise. Yet even in 1994, the seeds of disaffection with the ANC were being sown. ANC bigwigs would walk into a general meeting of the Union uninvited, as if they had the right to do so without asking. 'Workers do not want to be pushed in any direction,' Goliath told me. The same year she resigned as branch secretary.[22]

Oscar was able to fulfil his slogan of 'freedom in my lifetime' by a matter of months. I wonder what he would have made of the freedom he had fought for. The most vivid thing about his funeral was not the desultory speeches, but the motorcade: a succession of glistening black limousines wending their way through shacks that had mushroomed in the open spaces of Nyanga in the years since he had been taken from it. The residents looked on bemusedly. If any of them knew who was being buried, they did not show it. There could not have been a starker contrast with the funeral of Karl, Oscar's eldest son. We had marched on foot through the township, and there were impassioned speeches at the graveside, even though the regime had nothing to do with Karl's death.

There were leaders elsewhere in Africa who had renounced the glistening limousines and all that goes with them, but the post-1994 political leadership in South Africa had no intention of following their example.[23]

It was payback time for the years of struggle. As I heard one rising star put it: 'we are no longer interested in wage labour.' A more ominous indication of where this attitude would take us was the expulsion from the ANC of Bantu Holomisa, the former head of the Transkei government, for telling the truth to the Truth and Reconciliation Commission. He had overthrown his predecessor, Stella Sigcau, for accepting a bribe, he said. Although this was a well-known fact, Sigcau was a cabinet minister at the time.[24]

This is not to say that some kind of payback was not called for. But the question was always how many would benefit from it, and how evenly the benefits would be spread. For the political leadership, it would be 'me first' in money matters, starting with the bloated salaries they paid themselves in political office and in senior managerial positions in the public sector. A trade union movement with a stronger commitment to financial self-sufficiency and solidarity would surely have rejected the bogus basis for the bloated salaries, namely that government had to compete with the private sector for the best available talent. This at the same time as executive pay in the private sector began to skyrocket. As it was, the managerial salaries did much to shape a 'me-first' mindset among the newly recruited members of the public sector trade unions.

For all there was to celebrate about post-1994 South Africa, the lack of restraint of the political leadership in money matters was most disappointing. Yet people from my background, who were never going to work in the private sector anyway, benefited in all sorts of ways from government largesse. So even though the business of representing trade unions was diminishing, I still had work. This was also payback of a kind, although I was never asked to produce a membership card. The offer of a part-time job at the university from which I had graduated was also payback of a kind. I was glad to accept it, since it enabled me to research some of the questions that were troubling me. Gradually I was able to claw my way back into modest middle-class circumstances.

My first marriage had by this point ended, and I was beginning a new life with Sandy. We are now happily married and have a son, Sam.

Externalisation and the first wave of 'black economic empowerment'

One of the most important achievements of the unions in the 1980s, I thought, was to establish the principle that trade unions should be regarded as autonomous. It was to preserve their autonomy that the unions had opted to engage in politics through an alliance. It fundamentally undermined this principle for a member of this alliance to seek to control individual unions, as the Party had done. This was not to be the only way in which trade union autonomy was undermined, however. An organisation that is not financially self-sufficient cannot claim to be autonomous, as I have often enough said.

For a workers' organisation, that means being funded by its members.

Trade unions were surely financially sufficient by 1995, technically speaking. This was the year tripartism was institutionalised, with the establishment of the National Economic Development and Labour Council (NEDLAC) and the adoption of the new LRA.[25] The new LRA created organisational rights for trade unions, including a right to stop orders. But these rights were redundant for the established union movement. Even recalcitrant employers were by and large willing to implement stop orders for them. On the other hand, being technically self-sufficient is a far cry from practising values of self-sufficiency and solidarity. Now that subs were a percentage of what workers earned, the pot of money the Union controlled was even larger still. It was as if the Union, and other trade unions, did not know what to do with all their money.

This was curious, given the crying need of the branches and the numbers of workers losing their jobs in manufacturing and mining. These included jobs shed as a consequence of 'reforms' that the government could clearly do something about. A case in point was the decision by Trevor Manuel, now Minister of Finance, to do for the clothing and textile industry what had been done for food manufacture, by lowering tariffs further than government was required to do.[26] The effect was to further the process of deindustrialisation already under way.

It was alarming to think that the ANC government could take a decision as detrimental to the working class as this. You might therefore think it underscored the need for trade unions to consolidate their base. There were lots of ways even a movement as centralised as this could have done so. Most obviously, it could have established strike funds. It could also have funded advice offices providing assistance to their unemployed members, among others, along the lines of the Advice Bureau, or promoted alternative forms of worker organisation, such as worker co-operatives, or education programmes. The leadership had others ideas, however. Somewhere, somehow, it decided to use workers' money to set up 'trade union investment companies' with the sole objective of making more money.

So it was that workers' money went towards funding what has been characterised as the first wave of 'black economic empowerment', and the inviolate line between representing workers and engaging in businesses for profit was crossed. Before I say more about who was actually empowered as a consequence, it is necessary to explain another way in which jobs were being shed as a result of restructuring. SAB had shown the way. Effective distribution was a key to the success of its business model (as was also true of its subsidiary, ABI). For much the same reasons as Premier had created a separate security company, SAB separated distribution from production at the Newlands brewery and established a new distribution depot in

Bellville.[27] The pioneering part of its scheme was to offer the drivers, who until this point had been employees, the 'opportunity' to own the trucks they drove. So they became owner-drivers, and would themselves employ the van assistants they needed to offload the massive trucks.

These trucks would still sport the same SAB insignia as before, and SAB would control distribution just as tightly. The owner-drivers still worked for SAB, after all. However, the legal instrument in terms of which the bosses exercised control was no longer a contract of employment. Instead, there was now a suite of commercial contracts, specifying the financial obligations of the owner-drivers to pay for their new acquisition. This was a critical shift. Owner-drivers could not go on strike: against whom would they be striking? Unless they were prepared to mount a legal challenge to this scheme, owner-drivers could also not utilise the LRA. They were not 'employees', and the legislation did not apply to them. Unsurprisingly, none among them was prepared to do so. Their livelihood was at stake.

This was the bosses' masterstroke. Employment had been externalised, meaning that the bosses were no longer accountable in terms of labour legislation for those who worked for them. The routine justification for externalisation would be that these were 'non-core' operations. The converse was true in the case of SAB (as it had been at Premier). All indications were that the scheme was motivated by the fear of a strike (as it had been at Premier). The expected strike did in fact happen while I was in the Transkei. From afar, it seemed like a relatively privileged section of workers pursuing their own sectional interests.[28]

Externalisation at SAB looked like 'empowerment', and that is also how SAB packaged its scheme. But it was also a scheme to remove the van assistants from its payroll. What the drivers paid their van assistants – or if they were paid at all – was not SAB's concern. It had bigger fish to fry. Several years later the government would allow SAB to relocate its head office to London, where it would very successfully launch itself as a multinational of note.[29] Not long after that, it introduced new requirements for the skills that its workers were expected to have, and began retrenching those who did not have them. It justified doing so in terms of a commitment the Union had made in its national recognition agreement to 'work together to make the company a successful world-class manufacturer'. The Newlands brewery was especially hard hit.[30]

All this went to show the lengths to which the bosses would go to shed the 'burden' of employing ordinary workers, led by multinationals like SAB. The unstated reason why employing ordinary workers was burdensome was the risks associated with their being organised. At the same time, new technologies were making it easier to replace them. It was through this combination of job shedding as a result of technology and externalisation that the supposedly highly regulated labour market in South Africa was in fact deregulated. Deregulation was also a 'reform' that government is regarded as having endorsed in terms of its own Growth, Employment and Redistribution policy (GEAR). However, it does not seem coincidental that GEAR's adoption coincided with the first wave of black economic empowerment.

If, then, GEAR represented a betrayal of the workers' interests, as COSATU would claim, it seems to me that the establishment of trade union investment companies was no less so. All indications were that large amounts of money passed hands to facilitate deals about which members were not consulted. Those who punted the deals would say they were safeguarding trade unions by securing a separate income stream for them, or that they were helping change the structure of the economy. But securing a separate income stream could only lessen what control workers retained in the organisation, and enhance the capacity of the leadership to dispense patronage. Of course, the structure of the economy did have to change, but for individuals in leadership to become inexplicably rich overnight was not a welcome change – or for workers' money to be invested in casinos rather than the productive economy.

The Union's investment company was called FAWU Investments (Pty) Ltd. Its first investment, or one of its first investments, was to acquire a major shareholding in the Oceana group. The other shareholder was a company called Real Africa, of which Johnny Copelyn was a director. Copelyn has defended his role in establishing Real Africa, which he describes as 'frontier work'; he is now counted as one of the richest persons in the country.[31] It is difficult to see how the working-class communities of the West Coast benefited from this particular change of ownership. In my book, what might be considered 'frontier work' would have been to genuinely empower fishers in these communities through forming worker co-operatives.[32] Fishing was, after all, an industry that government still regulated, and Oceana held quotas. But if this possibility was considered, no serious steps were taken to realise it. Most of the Oceana plants are now closed, among them Lambert's Bay where the company first started out.

The fallout within the Union that you would expect from the Oceana deal would have centred on the fact that it had placed itself in a hopelessly conflicted position by investing in a business in which its members were employed. Instead, the issue that drew most attention was whether the general secretary was entitled to take this decision on his own, and whether its shares represented value for money. The transaction that generated even more heat, however, was the purchase of Krugerrands. Again, the issue was not that it was obscene for workers' money to be invested in gold coins, which was what the bosses routinely did as a hedge against economic collapse. Rather, it was the commission earned by the broker.[33] Kickbacks and bribes could, of course, be expected in these circumstances. Yet the effects on worker organisation are arguably no less debilitating when those who control the income stream are able to 'reward' their foot soldiers and allies in lawful ways.

It was not surprising, then, that the Union continued to descend into a spiral of conflict, punctuated by upheavals in the national leadership and dismissals on allegedly trumped-up grounds. What was surprising, perhaps, was the alacrity with which the national leadership resorted to the courts to defend their stance. This is what workers had been expelled for doing not so long before. The net result of the Union's forays into the world of corporate finance was surely inevitable. Organisation on the ground would regress. A shocking series of events at a factory I knew all too well went to show how far this regression had proceeded. As in Boksburg, there was blood shed on the factory floor, only this time it was spilt in a fight between affiliates of COSATU.

By this time the new dispute resolution body which the LRA established, the Commission for Conciliation, Mediation and Arbitration (CCMA), was up and running. It was modelled on IMSSA, and, like IMSSA, its bread and butter would be unfair dismissal disputes, which it would try to resolve through mediation, failing which it would arbitrate. Like IMSSA, it would draw on a panel of practitioners, mainly lawyers, to do so. However, although ostensibly independent and accountable to a board composed on tripartite lines, in reality the CCMA was wholly dependent on state funds.[34] It was simply the state in a different guise. I had in the meantime opened my own attorney's practice in Observatory. Since the CCMA was now 'the only show in town' – as a lawyer colleague put it – I thought it prudent to become one of its 'part-time commissioners'.

I was not much interested in resolving unfair dismissal disputes. The potential of the CCMA, I supposed, lay in its power to determine disputes regarding organisational rights and thereby help the unorganised become organised. One such dispute I was called on to determine was referred by SACCAWU, the successor of CCAWUSA and the affiliate organising retail workers, in terms of COSATU's policy of one union for one industry. I learned to my amazement that for the previous few years it had been organising at I&J Woodstock. Now it had enough members to ask for organisational rights there. Although there had been nothing about it in the media, a FAWU shop steward had also been assassinated.[35] Evidently COSATU's policy of one union for industry was already dead, and this represented several nails in its coffin.

I&J obviously expected me to take a dim view of workers leaving the Union on whose behalf I had signed a recognition agreement, and I was indeed shocked that it had come to this, at a factory which had taken painstakingly long to organise. Yet workers had the right to do so if they believed the Union was in a sweetheart relationship with their bosses. When I&J asked for the hearing to be postponed, I decided not to allow the company to drag its feet in responding to the workers' demands. I&J then did what disaffected employers with money regularly do: they dragged their feet inordinately, by taking my ruling on review to the newly established Labour Court.[36]

What became of Premier
The biggest deal during the 'first wave' of empowerment took place in 1996. Anglo sold its subsidiary Johnnic to a consortium headed by Cyril Ramaphosa, who is also now counted as one of the richest persons in the country.[37] The Union's investment company, along with other trade union investment companies, was part of this consortium. Since Johnnic was part-owner of Premier (and SAB), this meant that the Union had a stake in Premier and SAB as well. It once again raised the question of a conflict of interests. At the same time, now that South Africa was part of the global economy, the same bosses who had been concerned about the low wages of African workers in the 1980s had become most concerned about the 'high' wages paid to 'unskilled' labour in relation to their global competitors.

Premier retrenched some 2,000 workers about a year later. The deregulation of milling and baking may have been a factor in its decision to retrench: bread would not be subsidised in the new South Africa. But the decision to close Blue Ribbon Bakery in Isando was remarkable. This was a brand-new plant and its flagship bakery. The sole reason for closing it, I was later told, was that the MD was determined to teach the Union a lesson, following a series of wild-cat strikes.[38] Since the Union had obviously not been in control of the situation at the bakery, the only lesson that could be drawn was for the Union to exercise tighter control over workers. The conclusion I draw from this is that the bosses had been emboldened to do something they would never previously have contemplated, before having trade union investment companies and the politically connected represented on their board.[39]

There was talk in Premier of further retrenchments, prompted by a fear that Tiger would take it over. This fear was well grounded. Deregulation was calculated to promote competition among the milling and baking bosses, and wage costs were a factor. Tiger was winning the competition with Premier hands down, since the Union was not as well organised in its plants – the consequence of Tiger's decentralised management model, as I have explained.[40] Premier, by contrast, had its national recognition agreement with the Union. If the Union pushed Premier too hard, it might fall over the edge. It did not seem that workers were aware of the danger.

The unorganised workers in supposedly organised plants made the militancy of the Union look even more misplaced. These were not workers able to join who chose not to, but workers who were not able to exercise their freedom to associate because they were in such an inherently insecure position. Whether because organised labour had the wool pulled over its eyes or was simply sleeping, it had, during the tripartite negotiations over the new LRA, let the bosses retain essentially the same provision regarding labour broking as had been introduced in 1983.[41] The consequences of this blunder were evident as soon as the law came into force.

At the Premier plants down the road from my Observatory offices – the Salt River mill and Attwell's Bakery – the bosses stopped hiring 'casuals' at the gate and engaged a labour broker instead. Instead of each workplace having its own workforce, working for the same employer, they became places where several workforces worked, each with its own employer. Only one employer called the shots, however. In the case of the bakery, this was obviously Attwell's. It became my favoured case study of the effects of externalisation, thanks to the understanding I had reached with John Pici to tell me what was going on there in exchange for advice.[42]

What was most disturbing was how quickly the sense of solidarity that had existed between workers in the same workplace gave way to a sense of hierarchy: between those employed by Attwell's and those employed by the labour broker. This was despite the fact that the workforce of Workforce (which was the broker's name) included some who had been retrenched by the bakery; they were doing the same work they had done previously, but for about half the wages.[43] Now that he was not legally accountable for them, Mr F, the bakery manager, felt free to become abusive and arbitrary. Every now and then he would instruct Workforce to remove someone from the premises who had offended him in some trivial way. The worker would not be seen again.

John would shake his head in impotent disbelief at Mr F's latest excesses when we met. But as supervisor of the cleaning department, he could not afford to raise his voice too loudly. There were bosses who had spotted the business opportunity of marketing cleaning as a service, called 'contract cleaning'. Doubtless inspired by the 'successful' introduction of a labour broker, Attwell's was contemplating retrenching the workers in John's department and engaging an external contractor to clean the machines. In that event, even though it would be too expensive for the bosses to retrench John because of his long service, all the other workers in his department would be out of a job.

The general manager, Mr T, had joined Premier as part of the new breed of HR and labour relations specialists. With employment increasingly externalised, there was less need for HR and labour relations managers. Some firms went so far as to externalise the HR and labour relations function itself. Premier was among those that realised this new breed would fare well in managing the production process because of their grasp of the post-1994 political environment. By the time Mr T approached John, he already had secured the endorsement of the Union for an alternative proposition.

'I don't know where it will take me … maybe it will destroy me,' John said about Mr T's proposition. This was that John become a contractor and supply cleaning services to the bakery himself. I tried to interest John in an alternative which I believed could be viable if it had the support of the Union. This was to form a co-operative together with the workers, instead of employing them. However, the fee he was offered must have been too alluring to consider alternatives. Towards the end of 1998, against my advice, the person who had brought the Union to the bakery ceased being an employee and a member, and became an employer of 14 workers.

Once again, externalisation had gone hand in glove with 'empowerment'. Shortly afterwards, the takeover Premier had feared took place. It was not by Tiger, after all, but by a comparatively unknown quantity which adopted the name General Food Industries, or Gen Food for short. The Union may have pushed Premier over the edge after all, since Gen Food bought Premier for a song. An express condition of the sale was that the national recognition agreement with the Union would be terminated forthwith.[44] Almost immediately, the company embarked on a further wave of retrenchments. Among the workers who now lost their jobs was Peter Malepe, who had become president of the Union, and 58 workers at the Salt River mill, straight after a new wage agreement was concluded.[45]

The Union was in disarray at the time. The spiral of conflict was continuing, and had resulted in another split, amid more claims of corruption, and the establishment of a breakaway union on the East Rand. This time it was reported in the press.[46] Mandla, having survived the Krugerrand debacle, was dismissed not long afterwards on some or other ground.[47] There was another breakaway in Ceres a year later. A carload of workers from Growers came all the way to my Observatory office to ask me to endorse the new union they were planning to establish. Even though I could not say what else they should do to resolve their problems that they had not already attempted, I was not prepared to endorse it.[48] Mandla had in the meantime instituted unfair dismissal proceedings in the CCMA.

By the end of the 1990s, you could say that the reversal of the gains of the 1980s was complete, and the implications of the phrase 'flexibility and mobility of labour', which the bosses bandied about, had been as completely realised as it was possible to do. In the case of Attwell's, about half the workforce on the bakery premises were not employed by it, including the workers employed by John and those employed by the labour broker. All indications were that this was the norm in manufacturing.[49] In public, COSATU was opposed to different tiers in the labour market. In practice, there were different tiers in the workplaces of its own affiliates. In the top tier were Union members in standard jobs. Those at the bottom of the hierarchy – the workers employed by labour brokers – earned about half of what workers in the top tier earned, doing equivalent work.

The neo-liberal ideal is an unregulated labour market, and the competition that ensued between different labour brokers, as more and more firms utilised them, was how its proponents thought things should be. Since this competition was more often than not driven by price, and price was determined by what workers were paid, it would drive wage levels down even further. What happened at both Attwell's and the mill next door was typical of what lawyers euphemistically labelled 'cascading outsourcing'. Workforce was replaced by Staffgro, which described itself as a 'recruiting specialist'. Staffgro provided the replacement labour for the 56 retrenched workers at the mill.[50]

John's workforce was not quite at the bottom of this post-apartheid workplace hierarchy, because the competition between the contract cleaners was not as intense. This was thanks to the minimum wage established in terms of a sectoral determination and to the bosses' organisation that had pushed for it in the first place.[51] This organisation was dominated by national firms that feared being undercut by 'new entrants' like John. The question was whether John could deliver the same service, and whether his contract would be renewed. John's position became even more precarious when, to his obvious discomfort, some of his workers joined a trade union, a splinter from SACCAWU.[52]

Dereck Cele had in the meantime replaced Mandla as general secretary of the Union. Dereck had been part of Chris' circle and, I later learned, had secured a position for Chris as manager of FAWU's investment company at a controversial salary.[53] It was during Dereck's tenure that I was first invited to attend a conference of the Union. In contrast to the vibrant parliament of the workers that conferences had once been, I observed a succession of government bigwigs addressing a largely passive membership, uniformly attired in the free T-shirts that the Union dished out, together with a 'conference pack'.

The other extraordinary thing I learned was how many officials had been dismissed by the Union since the previous conference: 24 in the space of two years.[54] This, to me, was a sign that the Union had never broken out of the spiral of conflict, and that dismissals were being used to settle scores. However, Ray Alexander, in a speech to the conference that must have been read on her behalf, said it 'shows the union leadership is developing a culture of accountability and commitment'.[55] At the birthday celebrations, I learned that Dereck had now been dismissed and was in the process of referring a dispute to the CCMA, as Mandla had done before him.[56]

What became of John

The new LRA was supposed to apply to all workers in an employment relationship. In practice, it did not. The 'labour situation' as it was conceived in the 1970s had changed. Then, workers laboured for bosses who were their employers. Now, increasingly, workers labour for bosses who are not their employers.[57] These were not just workers like John who were once in standard jobs, but also workers whose employment was externalised from inception, as well as those who worked for their own account under much the same conditions. All of these workers should be able to belong to an organisation that could bargain on their behalf. But the LRA did not enable them to do so, as the workers of John's enterprise were to discover when their trade union demanded access to the workplace.

It should go without saying that the workplace to which these workers wanted their union to have access was the premises of the bakery. But this was not their workplace in law. In fact, John did not have one. The real bosses, the bosses of the bakery, could of course have granted access, but it would have defeated the object of the exercise. Instead the bakery management crafted the letter John sent to the union, fobbing it off. This nicely illustrated the true relationship between John's enterprise (and others like it) and the bakery. They were satellites of a core business. The organisational rights that the LRA provided were of no help to the growing number of workers employed by satellite enterprises in all sectors of the economy.[58]

Very few realised how radically the workplace had been transformed as a result of externalisation in the early years of the twenty-first century, and how many ordinary workers in the sectors that had been the base of the trade union movement were now without a voice. The prevailing belief was that the labour situation was 'sorted', on account of NEDLAC and the LRA, as well as the suite of 'progressive' labour laws which the LRA had ushered in. Perhaps the most glaring example of how disconnected this belief was from reality was a law purporting to address 'employment equity' that could not even recognise an inequity as blatant as that existing at Attwell's Bakery and other workplaces like it.[59]

There were various reasons why the belief that labour was 'sorted' persisted. The intellectuals and activists of previous decades, including the rump of the white Left, were otherwise engaged. The newspapers no longer employed labour reporters. Labour studies were no longer seen as relevant at the universities, because there was not the same market for graduates in labour relations and HR. Now that trade unions in Poland, South Africa and elsewhere had exhausted their usefulness, the attention of the big institutional funders had shifted to more fashionable topics. In addition, there was a range of institutions and individuals who had an interest in carrying on as if nothing had changed, including the profession of which I was part.

A decision of the Labour Court regarding the surreptitious scheme of the full-time shop steward illustrates how 'the law' can be used to paper over the cracks. You might have expected this scheme to collapse under the weight its own contradictions. Instead, more and more trade unions adopted it, including those in the public sector, for the reasons I have suggested. When local government workers on the East Rand went on strike, and their employer applied the principle of 'no work, no pay', three full-time shop stewards cried foul: because *they* were not going to be paid rather than the members they represented. This, you might think, was quintessentially a question of public policy, to which the answer must be a resounding 'no'. However, the judge upheld their claim. They had not withheld their labour, he said.[60]

Trade unions, of course, had an especial interest in carrying on as if nothing had changed, because they had (with very few exceptions) not attempted to organise workers whose employment had been externalised. They had therefore never had to come to grips with the fact that these workers had only a veneer of rights. So it could not be expected that when COSATU appointed a commission to look into the 'new challenges' trade unions faced, it would tackle any sacred cows. This was the September Commission. Consistent with what it called 'social unionism', it advocated that trade unions engage 'strategically' with employers about restructuring, which it saw not just as a threat, but as opening up opportunities as well. Among the most puzzling of its propositions was that restructuring opened up opportunities for the 'democratisation of the workplace'.[61]

As the Commission conceived it, the prospects for 'social unionism' were best if there was 'massive' economic growth. This was the best of three possible scenarios for South Africa. In the worst-case scenario, which it termed 'the desert', economic growth was stagnant, and there was heightened conflict between labour and the bosses. It expected the ANC to take the bosses' part in this conflict. 'Powerful organisations of the unemployed, the youth and communities emerge ... Government leaders promise to look into the people's legitimate grievances but warn against false prophets.' Moreover, in this scenario, the Party splits, with half its members 'joining an alliance of Left organisations for building a workers' party'.[62]

The economic growth that did take place during the early years of the twenty-first century could not be described as 'massive'. It was nevertheless as good as it would get for a long time to come. Although jobs were 'created', these were mostly in satellite enterprises, working for labour brokers, contract cleaners and security services. Statisticians conveniently coded these as 'financial' services, which created the impression that they were new. In the case of John's enterprise and many others, these were simply the same jobs performed for the same bosses. All that had changed was the identity of the employer.

The failure of the economy to generate jobs exposed the failure of neo-liberalism generally and GEAR in particular. A scapegoat was needed. It was almost inevitable that this would be 'labour regulation'.[63] Neo-liberal think tanks and global financial institutions had long argued that labour regulation introduced 'rigidity' in the labour market and was an obstacle to job creation. It was especially an obstacle to small business (another conveniently amorphous category). Small business represented their answer to the indisputable fact that big business does not create jobs, as the loss of jobs that had already taken place in food manufacture shows.[64]

The 'rigidity' argument had some validity with respect to job security in the public sector. Jobs in the public sector had always been secure, not because of any legislative protection, but because their employer was not for profit. Now employees accused of misconduct – especially those in senior positions – could hope to drag out their case for months, if not years, all the while drawing a salary. In the private sector, job security existed mainly for workers in standard jobs, with a trade union to litigate on their behalf. The 58 workers 'retrenched' following the conclusion of the wage agreement at the Salt River mill were a case in point. The case the Union brought took four years to wind its way to the Labour Appeal Court, only for the judges to uphold Gen Food's right to shed jobs in order to boost profits.

The situation in which John's workers were shortly to find themselves illustrated how flexible the labour market was for workers who were not in standard jobs. It also illustrated how unrealistic the expectations were that small businesses were capable of creating anything like the number of jobs that the country needed. Some small businesses might create some jobs, but in an economy dominated by big business the question would always be whether they were sustainable. The rate of attrition among them was bound to be high. Although Mr T was doubtless sincere in wanting to convert John to an entrepreneur, it was a tall order to suppose that someone accustomed to the discipline of a weekly wage would truly become one.

What would a factory worker know about making provision for depreciation or even provisional tax? Where would John have acquired the discipline, after having paid his workers their wages, not to spend the balance of his fee on his family? I suspect most of it went on revamping his Transkei homestead as a way of providing for his retirement. Heavily in debt to the Receiver of Revenue, he appointed a consultant with numerous qualifications and a rampant lion on his letterhead. The consultant proceeded to eat the lump-sum payments intended to settle his debt with the Receiver.

When John's contract expired and the bakery did not renew it, I threatened legal proceedings on his behalf. That he had really been an employee all along would have been the argument. It was the same argument I had previously made on behalf of George Xashimba, except the LRA had since been amended. This amendment was supposed to help workers in this situation, but in fact went to show the impotence of legislation to do so in the absence of organisation.[65] I was relieved when Mr T, who must have been disappointed at his failure to convert John, prevailed upon the new cleaning service to employ him at the salary of a supervisor. This new cleaning service now operated at all the big bakeries in Cape Town.

Upon becoming an employee once more, John wasted no time in becoming a trade union member again. He seems to have accepted that it could not be FAWU. Instead, it was another COSATU affiliate: the SA Transport and Allied Workers' Union (SATAWU). This was one of the exceptional instances in which a COSATU trade union organised workers whose employment had been externalised. But the union did not seem to question why they should be regarded as belonging to a separate 'contract cleaning' sector, rather than the sector in which their clients were located; or why workers dusting pot plants should be in the same union as those cleaning machines. When the contract cleaners went on strike in 2005, it put no public pressure on the clients for whom its members actually worked, or on the clients of the security workers who went on an even more protracted strike a year later. All this put more nails into the coffin of the policy of one union for one industry.

No one thought to ask how COSATU or its affiliates expected workers to sustain protracted strikes like these without strike pay. So, increasingly desperate workers resorted to increasingly desperate measures, often to compensate for the weakness of their own bargaining position. Strikes became increasingly violent – the security workers' strike was the most bloody since 1922 – and were punctuated by gratuitous attacks on potential allies of the working class, such as street traders working for their own account. In this manner COSATU trade unions in particular abandoned the moral high ground, and created the political environment in which the Constitutional Court did what the controversial 1988 amendments to the LRA had set out to do: hold a trade union (SATAWU) accountable for damages wrought by its members (or maybe just criminals exploiting a disorderly Cape Town gathering for their own ends).[66]

Given SATAWU's acceptance of contract cleaning as a separate sector, the union would not have seen it as its members' concern when the bakery workers belonging to FAWU went on a national strike in 2007. John, however, was not as calculating. One Sunday he attended a meeting of the striking workers, who still regarded him as their leader. His only contribution, he told me, was when he was asked to stand up at the end of the meeting, and wished them well. Maybe the metaphor he had used to describe the strike was too graphic for the bosses' taste – two bulls in a kraal, their horns locked in conflict. The boss of the cleaning service called him in the next day. He had been instructed by his client to remove John from the site forthwith, he said. The instruction came 'from the top'.

'My history at Blue Ribbon [as Attwell's was now known] is unforgettable,' John wrote to me about a relationship spanning 30-something years which had been terminated on a day's notice.[67] Yet he had no obvious remedy against the bosses. Even if he had been dismissed, as generally happened when a client did not want the worker of a satellite enterprise on its premises, there was little he (or the CCMA) could have done. The person who could secure his job was not his employer. The cleaning service served John with a notice of a disciplinary hearing for 'getting our company [i.e. the cleaning service] or workers involved in customer's strike, dispute and strike action', and told him in the meantime to report for duty at Sasko's Claremont Bakery, where it also had a contract.

The charges were later quietly dropped. Evidently John's boss had no stomach for a hearing. Maybe the fact that John was well known among Sasko workers helped. Their workplace had also been restructured along the same lines as at Attwell's. What was telling, was the set-up with the cloakrooms. Separate facilities were provided for the workers of contractors and labour brokers, to discourage them from mingling with the workers in standard jobs: the walls and floor were bare cement, in contrast with the tiled cloakrooms for workers in standard jobs. No one challenged this form of workplace apartheid while he was there. It was as if it was now accepted that the Union represented only a section of the workforce.

In 2008 John told me that cleaning workers in the bakeries were 'sick and tired' of SATAWU and had voted with their feet, by joining a small, non-affiliated trade union. It also seemed disaffection with other COSATU trade unions was mounting. The exception was NUMSA, the one COSATU affiliate that workers still saw as strong, according to John. It might have been to counter this mounting disaffection that COSATU began calling for a 'ban' of labour broking. Mainly it was because its political clout was growing in the lead-up to the ANC's Polokwane conference. It was through political clout rather than organisation that COSATU sought to arrest its decline in the productive sectors of the economy, coupled with an increased reliance on legislation and the courts.

It was never made clear how COSATU thought a 'ban' on labour broking would work, but it would certainly need a state with the capacity to enforce it. As it was, the Department of Labour struggled to enforce its own legislation, as a result of 'reforms' which had left it with a depleted and demoralised inspectorate. The call for a ban also begged an important question: why had it taken COSATU and its affiliates more than ten years to pick up on the issue? Part of the explanation may be that the federation was increasingly dominated by public sector unions, with little grasp of what was going on in the private sector.

The problem with having public sector unions in COSATU is that their members are supposed to serve the public interest. There is a potential conflict between teachers, for example, serving the public interest and pursuing their own sectional interests. In this context a policy like workers' control – which arguably should have no application in the public sector – can be abused, to prevent management from managing in the public interest.[68] Just as a recurrent demand of public sector unions would be to replace this or that director general with someone the members regarded as more palatable, COSATU believed the way to advance its members' interests was to replace President Mbeki with someone less overtly hostile to it.

'Msholozi will deliver,' a NUM shop steward assured me, referring to Jacob Zuma by his clan name, when I asked him why he thought a ban on labour broking would work.[69] The fact that Zuma was implicated in corruption was no obstacle to helping elect him as president of the ANC at the Polokwane Conference. COSATU also saw nothing wrong with the Zuma defence, that he was being targeted by his political enemies – and that he should be regarded as innocent until proven guilty, in the organisation he belonged to as much as in the courts. Evidently COSATU's leadership had learned nothing from the Spekenam dispute.

Zuma then resorted to every legal manoeuvre possible to avoid a case of corruption ever coming to court, as is well known. Eventually the matter came before a judge who had been thick with the new breed of lawyers that emerged in the 1980s. It should have been obvious that his judgment was seriously flawed, and I found myself fully in agreement with the judgment of an apartheid-era judge reversing it on appeal. By this time, however, President Mbeki had been 'recalled', as the ANC put it. This was tantamount to a coup, and the immediate consequence was a breakaway from the ANC. The longer-term consequences for the country are with us still.

I visited Lizzie Phike soon afterwards, at her modest Newtown home, outside Mbekweni. The walls of the *voorkamer* had been stripped of ANC paraphernalia, except a dog-eared poster headed 'President Nelson Rohihlala Mandela'. On another wall was a framed poster that I don't remember being there before, headed 'Dear Lord ...' The television and sound system were encased in metal burglar bars. It looked as though she was guarding it against her neighbours. A few weeks after that, the other Liz died. Lizzie arrived at her memorial service in Paarl Town Hall at the same time as I did, wearing a brown suit. It would have been green, black and gold before Mbeki's recall. We sat next to each other, until I received a message that the Abrahams family wanted to see me. It turned out to be a ploy to separate us. Lizzie was now a member of the Congress of the People (COPE). There were rumours that Liz Abrahams also supported COPE, but I knew better than to enquire. The ANC was in charge of her funeral.

The ANC also took charge of Lizzie Phike's funeral three years later, and it followed the same stifling formula. There would be a series of speakers from the organisations that were now hegemonic: the ANC, with or without its Women's League and Youth League, the Party, COSATU and the Union, in that order. Those that did not know her were not shy to speak. Several who did know her were not asked. Lizzie's big mistake had been to leave the ANC, one comrade said. Perhaps it had been. It was too soon after the transition to suppose that a majority would break faith with the organisation that was seen as having brought them where they then were.

Certainly, Mbekweni seemed a better place now than it had been during all the years I was a regular visitor. The funeral service was in an entirely new section of the township, in a state-of-the-art indoor sports centre. The roads were tarred and there were many newish, decent-looking houses. All this, I am sure, was made possible because of teachers, prison warders and others in government jobs. The other conspicuous difference was the spaza shops, hairdresser stalls and the like that now spilled out onto the streets. You could argue that the growth of the informal economy represented some kind of a gain, in that it provided affordable services for the community and jobs for some. But these were not 'real' jobs, working for an employer for a wage. To see what had happened to the 'real' jobs that Mbekweni residents once had, you had only to go a couple of kilometres to the south.

Almost all the plants in the industrial sprawl of Dal Josaphat where Moberg's was located had long been standing idle. All that H Jones still produced, on the other side of Paarl, was jam. Jones did not even employ seasonal workers any longer. Seasonal employment was not flexible enough for the bosses, so it utilised a labour broker. Most of the workers which the labour broker placed had worked either at Jones or Moberg's before, and now earned less than half of their counterparts in standard jobs.[70] The remainder of the premises that Jones once occupied had been converted into a sprawling shopping mall, occupied by the same retail chains that had put the squeeze on the manufacturers to cut their prices. Did they consider, in doing so, whether the workers would be able to patronise their stores? At the time of writing, the highest-paid chief executive officer in South Africa, a retailer, earned 725 times more than its lowest-paid worker.[71]

Not long after Lizzie's funeral, at the end of 2012, there was a strike at Lonmin's platinum mine in Marikana. The miners, it seemed, wanted an organisation in which the workers were the bosses. Instead they had NUM, perhaps the most centralised of the COSATU affiliates. NUM had also implemented a system of full-time shop stewards. There was a large complement of them at Lonmin, where they were by all accounts treated like managers, and regarded as such by miners.[72] Lonmin's human capital manager (a new title for HR, evidently) was also expected to have an 'individual development programme' for each full-time shop steward in case he or she was 'prejudiced in terms of the career opportunities'. It even laid on cars and cell phones for NUM to use.[73] No doubt all this was done precisely to avoid the kind of labour relations disaster that in fact ensued.

After Marikana
The Marikana massacre made the 'worst-case' scenario envisaged by the September Commission look benign. Not only had the ANC declined to take the workers' part, but so had COSATU. It is safe to say that the section of the working class that COSATU represented at this point was workers in standard jobs. The majority were in public sector trade unions, which were increasingly embroiled in the departmental and institutional power struggles that have been a feature of the government under Zuma.[74] As if to justify its failure to organise the unorganised, COSATU leadership began to speak of the 'poor and working class'. The subtext was that the working class comprised only those who were employed and in standard jobs.

At the time of the massacre NUM was still COSATU's largest affiliate. The fact that Lonmin had on its board its former general secretary and a political notable, in the person of Cyril Ramaphosa, appears to have emboldened others to take a hard line, as it had done with Premier. The ANC has since exonerated him politically by electing him as Deputy President of the country. It also made a tawdry attempt to paint what had happened as a 'tragedy' rather than a massacre. The commission of inquiry into the massacre was chaired by the person who at one time was Oscar's advocate (and mine). The consensus seems to be that despite teams of expensive lawyers, or because of them, we are left with too many unanswered questions

After the massacre, the stream of mine workers voting with their feet to join another trade union became a flood. You might think that this would have prompted profound reflection on the part of COSATU.[75] Instead it exacerbated existing conflicts between affiliates and between the national leadership. The issues included the circumstances in which COSATU had sold its head office, and the direction in which the Zuma government was taking the country. NUMSA, in particular, was increasingly outspoken about the Zuma government, which it accused of trying to turn the federation into the 'labour desk' of the ANC. The conflict culminated in the suspension of Zwelinzima Vavi, who was regarded as an ally of NUMSA.

Disaffected affiliates of COSATU, at the suggestion of the Union, then decided to requisition a special congress of COSATU. Such a congress, in terms of COSATU's constitution, fulfilled the same function as a special conference had in the Union's constitution: to hold to account an unrepresentative executive. In this instance it was the central executive committee, under the presidency of a different Dlamini. However, the central committee evaded accountability by the simple expedient of dragging its feet inordinately, and then expelling NUMSA, at this point its largest affiliate. The ostensible reason was NUMSA's failure to comply with the policy of one union for one industry, by now long buried. This was no more excusable than for the Union to expel members for challenging its leadership all those years before.

COSATU is now split, rather than the Party – although the reason why the Party has not split might be said to be due to the number of dissidents it has expelled. It looks like a split between trade unions in the sectors in which COSATU was once based and the public sector trade unions. It is very difficult to see how COSATU will be able to restore the degree of autonomy it once had, let alone its integrity. A federation which should have been the one organisation capable of curbing corruption has itself been corrupted instead. An alternative centre of power has been severely weakened, if not destroyed. It is already evident who benefits.

The Union, I am glad to say, fought against the expulsion of NUMSA. NUMSA for its part is promoting the establishment of a united front, and has mooted the formation of a new federation. In this context, there is also a lot of harking back to the late seventies and eighties. However, the full story of that period has not been told. COSATU did not turn to the 'dark side' after 2012, as some have suggested, or even in 1996. The causes of its present predicament can be traced back to its beginnings. Unless the trade union movement can come to terms with its own history, a new federation is surely premature. There is also no prospect of COSATU reinventing itself.

Coming to terms with history of course raises questions about the present. I was interested to see that among the principles proposed for a new federation was financial self-sufficiency. But what can financial self-sufficiency mean for such centralised organisations? Will it be possible to wean them from trade union investment companies and full-time shop stewards? Another of the proposed principles is a commitment to socialism. Alternative economic models are certainly needed. But if the socialist tradition is to contribute toward developing such models, it will need to be reinvented, taking into account how much has changed since the 1990s.

One of the ironies of this period must surely be that the 'success story' of capitalism should be a country ostensibly committed to socialism. But the growth of China and a few other 'low-wage' economies has been at the expense of manufacturing jobs and deindustrialisation elsewhere. It proceeds apace in South Africa, despite policies to the contrary. At the time of writing we stand to lose what capacity we still retain to manufacture steel. The supposition that China represents a model that can be replicated across the globe, or in South Africa, belongs in the realm of cloud cuckoo land. So too do the projections of the government's National Development Plan (NDP) regarding the number of jobs which it would have us believe the country can create.

Agriculture and food manufacture, in particular, are routinely held up as having the potential to create jobs. Yet in the same year in which the NDP was adopted, another food manufacturing plant closed down. It was Table Top in George, some years after Harvestime had also closed. You can now buy imported frozen vegetables from your local supermarket. I was struck by a news report that when Table Top closed, it employed only 60 workers. This was less than a tenth of the workforce employed when the Union first organised it, and it was the largest employer in town. Jobs on farms that supplied the factory would also have been lost: some belonging to smallholders, in a locality where smallholder farming could be a way of genuinely empowering people.

The hard truth is that capitalism, and in particular the unfettered form of capitalism that neo-liberalism has promoted, is not capable of providing employment for all who need it, in South Africa and elsewhere. SAB, South Africa's most successful multinational by far, is about to be swallowed in one of the biggest corporate takeovers of all time. The first workers to lose their jobs will be in Melbourne, Australia, and Haryana, India. I am able to raise my voice in protest on an online petition organised by the IUF, together with 8,626 others. This will hardly cause the world's largest brewing company (as it will become) to quake in its boots. Online petitions and social media are not a substitute for proper organisation.

The prospects would not be as bleak, of course, if South Africans could have confidence in the way in which their scarce resources are allocated. Where corruption is pervasive, however, even the most capable of our political leaders lack credibility. The leadership vacuum in South Africa is also a global phenomenon. Although you might be excused for thinking that the global financial crisis of 2007 and 2008 had decisively discredited the neo-liberal project, there is no simple way to reverse the changes it has wrought. It is especially difficult to reverse the fragmentation of the working class. A fragmented working class in turn creates fertile conditions for the emergence of new forms of identity politics, and a reversion to old forms, including rampant racism. Old establishments are reasserting themselves: also in the rural heartland of the Union.

All the while, the world is becoming hotter and the sea continues to rise. The prospect that it will eventually roll back over the Cape Flats is no longer the stuff of a childhood imagination. The same people who sold us the neo-liberal project deny that this is happening, because it gives the lie to all they have stood for over the last 30 years. These are the true barbarians of the twenty-first century, and they are organised.[76] If there is an alternative to unfettered capitalism, as there must be, it will need to be better organised than they are. But as economies become increasingly deindustrialised, trade unionism is itself at the crossroads. The road more travelled would be for trade unions to carry on representing a diminishing number of workers in standard jobs and abandon the rest to their own devices, in the desolate market in which we are landed. The road less travelled will be to ally themselves with other forms of worker organisation and other organisations that prioritise human solidarity over profits.

I have no doubt what road John would have taken if the choice was open to him: not because he was exceptional in any way, but because he had experienced how organisation was able to restrain the arbitrary exercise of power. If the story of the Union has any value, it is to persuade the reader that this is what a membership-based organisation can do. This book was nearly complete when John asked to see me. He was no longer coping, and wanted to take early retirement. Like Friday and many other workers who had come to the city as migrants, his aspiration was to go back to where he came from. Barely a month later, he travelled back to the Transkei in a coffin. He had died of renal failure.

He was the last of the workers I knew from my time in the Union still working in a factory. But his eldest son works in the same bakery where his father's working life began. I don't think he knows much about the role his father played in building a workers' organisation. I also don't think the students at the universities now protesting about a lack of transformation in society know much about the movement that emerged during their parents' generation. The working class still exists, and it is as much in need of organisation and allies as ever before.

Endnotes

Chapter 1: A birthday of sorts
1. Research conducted by my colleagues and myself on labour broking suggests that the adoption of the Labour Relations Act in 1995 spurred an exponential growth in the number of labour brokers (or temporary employment services, as they are referred to in the Act).
2. The label 'neo-liberal' is disputed, in so far as the proponents of such policies no longer accept it. This seems to be a ploy, to present neo-liberal thought as representing the true state of affairs, rather than an ideological intervention by a 'thought collective' (P Mirowski, *Never Let a Serious Crisis Go to Waste*, Verso, 2014) whose origins can be traced back to the formation of the Mont Pelerin Society (MPS) in 1948. See also David Harvey, *A Brief History of Neoliberalism*, Oxford University Press, 2005. The Free Market Foundation in South Africa is an affiliate of the MPS.
3. Gold Fields, the most resolutely anti-union of the mining houses, hosted the clandestine negotiations in a stately mansion in the English countryside.
4. The bosses had recently formed an umbrella organisation, the SA Consultative Committee on Labour Affairs (SACCOLA). The trade union federation NACTU was a junior party in the negotiations.
5. See J Theron, 'Workers' control and democracy: The case of FAWU', *SA Labour Bulletin*, 15, 3 (1990), 39–64.
6. This is a development that has not been adequately documented, and is also not particular to FAWU, although I have no first-hand knowledge about other trade unions.
7. So far as I have been able to establish, Vuyisile Mini had no involvement with FCWU or its African counterpart, AFCWU, although both FCWU and AFCWU were organised in Port Elizabeth where he was based. He was executed in 1963. See Ken Luckhardt and Brenda Wall, *Organise ... or Starve: The History of the South African Congress of Trade Unions*, Lawrence and Wishart, 1980, pp. 210, 431.
8. It is convenient at this point to explain the approach adopted in this book toward the usage of racial terms. In my view there is no satisfactory alternative to the term 'African' to describe people of Bantu origin, who speak a Bantu language, or for that matter to 'Indian' to describe people of Indian origin. I have not capitalised the terms 'coloured' and 'white', or the term 'black', which is used here to denote people who are not white, i.e. coloured, Indian and Africans.
9. While I was writing this book, there was a further and depressing report of the deterioration of race relations in Grabouw, an area in which FCWU had indeed played that crucial role. See *Cape Times*, 20 March 2012, 'Race war in Boland town'.

10 Johnny Gomas was an SACP activist who, according to Cronin, played an important role in organising in the rural areas of the Western Cape before FCWU was established.
11 FAWU Leadership Code. The Code was adopted at the 1987 annual national conference (ANC), and was also incorporated in the FAWU Organisers' Manual. See chapter 23.
12 The literal translation of *baas* is boss, but the use of the Afrikaans term has white supremacist connotations that the English term does not.
13 *Umlungu* is the polite term for a white person in isiXhosa or isiZulu.

Chapter 2: Private road
1 My mother's father, Livingstone Moffat, was the son of John Moffat, and grandson of Robert Moffat, the founder of the Kuruman mission. The explorer David Livingstone married Robert Moffat's daughter, and was thus Livingstone Moffat's uncle.
2 According to my mother, her father had been put on a boat at the age of seven to go to a 'public school' in England (as private schools for the elite are known). Presumably he was introduced to the English class system there. He only returned to South Africa as a young man.
3 Livingstone Moffat was a member of Parliament, representing the Dominion Party. This party has been characterised as 'jingoistic', i.e. it fervently believed in the British Empire.
4 The entrenched provisions in the constitution could only be changed by a two-thirds majority of both houses of Parliament (the House of Assembly and Senate). Those that could vote were obliged to vote for a white representative. It was nevertheless an opportunity to voice opposition to white rule, and effective use was made of this opportunity by representatives of the Liberal Party, and increasingly by representatives of the Communist Party. Ray Alexander had been one of their last representatives.
5 This was in 1963. Vorster was Minister of Justice at the time.
6 My father told me that Vorster had said this to him, presumably telephonically, when he informed him of his appointment.
7 N Mandela, *Long Walk to Freedom*, Macdonald Purnell, 1994, pp. 447–8. I became aware my father was deeply unhappy about something Mandela had said in his book, and decided to find out at the time of my father's death. I then issued press statements reflecting his role. Mandela also got his judges wrong, and ascribes to Judge Michael Corbett, who as Chief Justice presided at his inauguration as President, a leading role in the exchange. Corbett was my father's friend, but junior to him at the time in terms of the judicial hierarchy. Corbett spoke at my father's funeral and affirmed the correctness of the press statements I had issued.
8 These were not terms that were current at the time.
9 *Betoger* means demonstrator in Afrikaans.
10 I have borrowed the concept of a pure politician from Javier Cercas, *The Anatomy of a Moment*, Random House, 2009.
11 According to Muriel Horrell, SASO's position on the role of white liberals was to fight for their own freedom, educate their 'white brothers' and 'serve as lubricating material'. See M Horrell, *A Survey of Race Relations in South Africa*, SA Institute of Race Relations, 1970, pp. 245–7.

12 He later preferred to be called Onkgopotse Tiro, and became president of SASO. After going into exile, he was blown up by a parcel bomb on 1 February 1974. The Truth and Reconciliation Commission did not investigate his assassination. See www.sahistory.org.za/people/abram-ramathobi-onkgopotse-tiro.
13 Foszia Fischer and Harold Nxasana, 'The labour situation in South Africa' (n.d.). I believe that the authors were close associates of Rick Turner, who was banned at the time.
14 See Andrew Glyn, *Capitalism Unleashed*, Oxford University Press, 2006, pp. 1–23.
15 Tony Cliff, *Portugal at the Crossroads*, n.p., 1975.
16 My fears were realistic. Solly Smith of the ANC's London office was later revealed to be a spy.
17 This was Liz Abrahams. See chapter 8.

Chapter 3: Corporation Street
1 Rex Close, *New Life*, FCWU, 1950, p. 47. I was later to discover that Botes was still employed at H. Jones and Co. at the time as a foreman. If he retained any sympathy for the Union at all, he was never to indicate it in the period I was employed as GS. I never met him.
2 Close, *New Life*, p. 26.
3 Close, *New Life*, p. 30.
4 The African Food and Canning Workers' Union (AFCWU) was established in 1947, as discussed in chapter 5 of this book. One of the strange aspects about *New Life* was that it made no attempt to explain what AFCWU was, or why a separate union was formed. It has a photograph of Oscar Mpetha in it, in which he is described as the vice-president of AFCWU. At the time *New Life* was published, he would have been AFCWU's first general secretary. See Close, *New Life*, pp. 48–52.
5 See definition of 'employee', section 1, Industrial Conciliation Act 28 of 1956. In order to understand the 'registration debate(s)' discussed in later chapters, it is important to note that the first step in becoming a registered trade union was for the trade union concerned to adopt a constitution that specified the race of the employees who were eligible for members. Coloureds, for this purpose, included persons of Indian descent, and were regarded as comprising one race, and whites another. There was a procedure whereby 'mixed' unions (i.e. with white and coloured membership) could be registered in exceptional circumstances.
6 According to the 1976 *Survey of Race Relations*, in 31 December 1975 there were 382,525 whites, 91,995 coloureds and 179,174 Asian workers who were members of registered unions, and 23,000 Africans who were members of unregistered unions. See M Horrell, *Survey of Race Relations in South Africa*, SAIRR, 1976. However, the membership figures for unregistered unions need to be taken with a large pinch of salt. Data about the number of unregistered unions is also unreliable.
7 In 1976 the Boilermakers' Union and National Union of Furniture and Allied Workers disaffiliated from TUCSA on the issue of African workers moving into skilled jobs. Said the general secretary of the Boilermakers, 'We have thousands of coloured and Asian members whose jobs we have to protect. We can't do this at the same time as putting Africans into their union.' See Horrell, *Survey of Race Relations in South Africa*, 1976.

8   See Charles Simkins and Doug Hindson, 'The division of labour in South Africa, 1969–1977', *Social Dynamics*, 5, 2 (1979), 1–12. According to this study, the proportion of Africans in skilled positions had increased between 1969 and 1977 from 9.3 to 23.2 per cent. Over the same period, the period of Africans in semi-skilled positions had increased from 54.7 to 66.7 per cent. The proportion of Africans in unskilled jobs decreased slightly over the period, to 88.4 per cent. About 8 per cent of unskilled workers were coloured and a negligible number white.

Chapter 4: Klein Drakenstein Road

1   By trade union convention, I refer to a practice that was universally accepted among trade unions or appeared to be. The distinction between an office-bearer and an official was that an official was a full-time employee of the trade union and an office-bearer was not.
2   So, for example, parents of mixed race might opt to register a child as coloured, under an Afrikaans surname. I know a person from this area with the family name Mthimkulu, who adopted its Afrikaans equivalent, Grootboom.
3   In terms of its returns to the Department of Labour, the membership of FCWU in 1976 was about 4,000. However, it certainly did not have records to validate this figure.
4   Minutes, MCM, 6 June 1976.
5   Minutes, MCM, 11 July 1976.
6   The references are to the 1976 Annual Report and the minutes of the annual national conference of 21 and 22 August 1976.
7   In Soweto and other Witwatersrand townships there were clashes between Zulu migrants living in the hostels and students and township residents. These may have been supporters of Inkatha, the Zulu cultural movement which was later to become a political party, but at around the same time there were similar clashes in Cape Town, where Inkatha had absolutely no influence.
8   This was the Coloured Representative Council (CRC).
9   She also claimed that the workers had said, 'If a European comes as general secretary they're not going to pay any more subs.' Minutes, Paarl branch executive committee meeting, 26 August 1976
10 Minutes, Paarl branch executive committee, 26 August 1976.
11 The company is now Picardi Rebel Liquors (Pty) Ltd.
12 Liaison committees and works committees were first introduced in terms of the Bantu Labour Relations Regulation Act. Whereas the liaison committee consisted of representatives of both management and workers, the works committee was only composed of workers, and thus offered workers a degree of independence. The committees were supposed to be elected under the auspices of the Department of Manpower.
13 Minutes, Paarl branch executive committee meeting, 26 August 1976.

Chapter 5: The road to Mbekweni

1   *Cape Times*, 9 September 1976.
2   *Cape Times*, 14 September 1976.
3   *Cape Times*, 16 September 1976.
4   These are the Afrikaans words for Mrs, Miss and Mr respectively.
5   According to the minutes there were 29 people present, all in all. See minutes, MCM, 26 September 1976.

6 The majority of the workers, women working on the lowest grade, were earning R16 a week with effect from October 1976.
7 If there had ever been a tradition of shop stewards taking up complaints with management, it had died out. *New Life* (p. 54) quoted a circular letter Miss Ray had written advising shop stewards that they 'must always be in close touch with the workers and if they have any complaints they must let you know'. It did not say what the shop stewards were expected to do about these complaints.
8 Section 35, Act 28 of 1956.
9 There was a provision in the 1956 Act whereby a registered trade union could seek to compel an employer to grant stop order facilities. However, this required the union concerned to first establish it represented the majority in a particular industry in the magisterial district where the employer concerned was located, i.e. it replicated the procedure to be followed for registration. It did not appear this cumbersome procedure was utilised by trade unions to any extent, if at all, and was certainly not utilised by FCWU. See section 78(1A) and (1B), Act 28 of 1956.
10 At Ashton, for example, where less than 20 members out of a potential membership of several thousand were paying subs, the secretary was claiming payment going back a year. See minutes, MCM, 26 September 1976.
11 Minutes, MCM, 26 September 1976.
12 The Anglo American Corporation.
13 Watchmen then were employed by the same employer who employed everyone else in the workplace, as opposed to an external service provider, as is the case nowadays. But their hours of work were longer. Instead of a nine-and-a-half-hour day for five days a week, watchmen worked a twelve-hour day, for much the same wages as an ordinary worker in the factory.
14 Minutes, MCM, 24 October 1976.
15 Minutes, Paarl branch executive committee meeting, 28 October 1976.
16 House of Assembly Debates, 1955, cited in Dudley Horner, 'Labour preference, influx control and squatters: Cape Town entering the 1980s', SALDRU Working Paper 50, South African Labour and Development Research Unit, University of Cape Town, 1983. Dr Verwoerd was Minister of Native Affairs at the time.
17 See Horner, 'Labour preference'. The line demarcating the 'Preference Area' from what is now the Eastern Cape province became known as the Eiselen line, after the name of the Secretary of the Department of Native Affairs.
18 Interview, Selina Claasen, Mbekweni, 28 January 2008.
19 Berg River Textiles was the only other factory in the area that employed a significant number of women. So there was probably an oversupply of women, both coloured and African, in the area.
20 Port Elizabeth was commonly referred to by the acronym PE.
21 D Oakes, *Reader's Digest Illustrated History of South Africa*, Reader's Digest, 1988, p. 411.
22 The autobiography of Ray hardly deals with the reasons for the split, and not at all with the opposition to it. RA Simons, *All My Life and All My Strength*, STE Publishers, 2004.
23 As in the case of a general secretary, a branch secretary who was employed by the Union became an official as well as being an office-bearer. These were thus the only office-bearers who did not have to be workers and members in terms of the constitution.

24 The paid officials generally earned a wage comparable to what semi-skilled workers earned in the factory. It was only Lilian, among the paid officials, who earned what the women on the lowest grade were earning.

Chapter 6: The road to Ashton
1   Minutes, MCM, 24 October 1976. Gants was established by a former manager of H Jones in Paarl, after whom the company was named.
2   The cannery in Grabouw was Highlands Canning. It was part of the same group as Oakglen Canning.
3   The term multinational company (or enterprise) refers to a business which has its headquarters in one country and subsidiaries in other countries. An alternative term is transnational corporation.
4   The number of the permanent workforce at the time was about 80 workers, most of whom were coloured men. There was no African township in Tulbagh, and the African workforce was minuscule.

Chapter 7: West Coast road
1   The inshore industry was so called because it operated closer to shore, and the boats used were relatively small with wooden hulls, as compared with the deep-sea trawlers operated by I&J and Sea Harvest.
2   Wage determinations for specific industries were set by the Minister of Labour on the recommendation of a tripartite board, the Wage Board, in terms of the Wage Act (5 of 1957). These were generally industries not covered by Industrial Council agreements, so the system of wage determinations complemented the collective bargaining system. Wage determinations remained in force until after the adoption of the Basic Conditions of Employment Act (75 of 1997), which introduced an equivalent system of sectoral determinations.
3   Minutes, MCM, 12 December 1976.
4   Minutes, MCM, 12 December 1976.
5   The banning orders were issued on 10 November 1976. A few days before, the Paarl branch formally withdrew its opposition to my appointment. 'We were against apartheid and still are,' Nellie Kilowan told the meeting. 'If we refuse a European now, then we accept apartheid.' See minutes, Paarl branch executive committee, 28 October 1976.
6   Wiehahn was appointed as adviser to the Minister in September 1976. See S Friedman, *Building Tomorrow Today: African Workers in Trade Unions, 1970–1984*, Ravan Press, 1987, p. 149.
7   The workers would be stoned by 'the youth' (as they were commonly referred to) for going to work during a stay-away.
8   Friedman agrees that the trade unions posed no threat to government at this point, and mentions a theory that provides some support to my analysis, namely that the bannings were at the instance of TUCSA. See Friedman, *Building Tomorrow Today*, p. 120.

Chapter 8: Waterkant Street
1   Becky Lan, the young white intellectual who became general secretary after Miss Ray was banned in 1952, but was reputed to have adopted a low profile because the political pressure had become too much for her. She was, however, still general secretary when she herself was banned.

2   The bokmakierie is a bird whose name is intended to replicate its call, and whose colours replicate those of the ANC.
3   Amato Textile Mills in Benoni was an example of an employer who was prepared to negotiate with an unregistered trade union in the 1950s, namely the African Textile Workers' Industrial Union, a SACTU affiliate. After a succession of strikes, the workforce was dismissed. It does not appear the union recovered. For an uncritical account, see Luckhardt and Wall, *Organize ... or Starve*, 1980, pp. 286–8.
4   Baskin describes the Food and Canning Workers' Union (which he presumably means to include the African Food and Canning Workers' Union) as being the 'jewel in SACTU's crown'. See J Baskin, *Striking Back: A History of COSATU*, Ravan Press, 1991, pp. 23, 13. The other significant affiliate, by most accounts, was the Textile Workers Industrial Union (TWIU).
5   SACTU was established following a split in the trade union movement of the time over the question of allowing African trade unions to become members of the same federation as trade unions representing white and coloured workers. The unions that formed TUCSA (initially called South African Trade Union Council (SATUC)) adopted a constitution that excluded unions representing African workers. The unions that formed SACTU had opposed this move. See Luckhardt and Wall, *Organize ... or Starve*, 1980, p. 86.
6   Luckhardt and Wall, *Organize ... or Starve*, 1980, p. 420. The photograph is reproduced at p. 160.
7   Liz Abrahams replaced Becky Lan when she was banned. This was, according to Ray Alexander, at her insistence, in preference to another white intellectual. See Simons, *All My Life and Strength*, p. 281. There was a suggestion that the Jones circle had played some dark role in the banning of Liz Abrahams, and it was probably no coincidence that it was after this banning that Moberg's workers voted out Juffrou H.
8   See minutes, MCM, 24 April 1977.
9   The first attempt to intensify production was at Ceres Fruit Juices. The Agreement stipulated that workers should not be required to start work before 6 a.m. This was an important protection for women who, in terms of traditional gender roles, had to see that children went to school. With the connivance of the Department, and despite the Union's objections, workers were steamrollered into starting work at 5 a.m. at Ceres. See minutes, MCM, 24 April 1977.
10 By creating three eight-hour shifts the bosses were able to keep production going for 24 hours a day without incurring any expenditure on overtime. The normal shift was nine and a quarter hours a day, Monday to Thursday, and nine hours on a Friday. The maximum number of ordinary hours a worker could work in a week was then 46. It is now 45.
11 At that time the minimum for a rest interval (tea break) was ten minutes and overtime was paid at time and a third.
12 When, for example, I telephoned Van den Bergh to complain about the shifts Jones was trying to force through, he was not at all concerned it would be in breach of the Agreement. The 'solution' he proposed was that I address the workers after Tredoux had done. It was a gamble that I accepted, and it paid off. All but a handful of workers voted by hand against the three shifts. Tredoux, a man of florid complexion, turned a few shades redder.
13 See minutes, MCM, 12 December 1976.

14 I surmised at the time, and later confirmed with him, that like almost all the literate contract workers with a command of English I encountered, Wellington had been educated at a mission school. The fact that someone like him had not been integrated into the chain of command at his factory suggests that, in the Western Cape at least, it was exclusively the preserve of whites and coloureds.
15 See minutes, MCM, 22 May 1977.
16 Outdoor gatherings were illegal in terms of the Riotous Assemblies Act (17 of 1956). Five or more people constituted a gathering.
17 The 'stalwart' in question was Oom Willem, the stepfather of F, who had blown the whistle on Johnny M and the typist. He was regularly in attendance at management committee meetings, where he reported that African workers in Worcester were refusing to pay subs. All the while he was 'eating the money'. If the head office administration itself had not been corrupt, he would have been found out sooner.
18 I had been able to gauge from the Association's application for a Conciliation Board how far the Union was from being representative. Of the seven employer groups that were members of the Association, six were companies and one was a co-operative, namely Langeberg. Altogether, they employed some 7,000 employees, of whom about half were Africans, at a date in the off-season. Employment in season would have been at least twice that.
19 This was the so-called minimum living level (MLL) and somewhat higher supplemented living level (SLL) as determined by the Bureau of Market Research at the University of South Africa. These measures are no longer used.
20 In all such agreements, there were also always jobs listed separately from the grades, such as the job of supervisor, various kinds of drivers and watchmen. These jobs were thus regarded as ancillary to the core activity to which the Agreement applied. Nowadays they might be regarded as services.
21 Agreement was reached to reduce the existing five grades to four, and that women on the lowest grade, earning R16, would get an increase of R4 as against R3 for the men. The gap between them was thus reduced by R1 to R3 a week. However, as a consequence of escalation clauses it would have grown to R3.81 again by the time the agreement expired.

Chapter 9: Side streets
1 'LKB se benadering was uit die staanspoor een van gesamentlikheid. Vir ons is die ooreenkoms die uitvloeisel van gesamentlike bedinging, dog die FCWU het 'n wen/verloor-benadering.' Minutes of a meeting at Langeberg Dal Josaphat held on 23 September 1977, quoted in MCM of 20 November 1977.
2 Ray Alexander told me that the original Fatti's & Moni's manufactured fruit juices, which is why it had been organised, and was located in central Cape Town. The juices that are still sold today under the Monis label are produced by an altogether different company.
3 The Union's certificate of registration still covered meat processing in the magisterial district of Bellville, which is where Spekenam's Stikland factory was located, despite the strike that had been lost. One of the benefits of the highly bureaucratic system of registration at the time was that for another trade union to be registered, it would have to prove it represented a majority of workers in respect of the same industry and magisterial district.
4 *What Is to Be Done?* was first published in 1902. In it, Lenin famously argued

for the necessity for a party of professional revolutionaries to enable workers to develop class consciousness, as opposed to merely 'trade union consciousness'. Nowadays it is accessible on the internet. See www.marxists.org/archive/lenin/works/download.

5   Means 'In our time'.
6   I recorded in the minutes that supervisors and foremen on the Union committee created a confusion of roles. 'The role of a representative of the Union is to find out what the workers think ... The role of the supervisor or foreman is to carry out the instruction of the boss about how work is to be performed.' This was of course a reference to Oom Joe. Minutes, MCM, 20 November 1997.
7   The Commission had been appointed in May 1977. The Union had made a written submission, in the form of a 12-point resolution, but decided not to give oral evidence. The resolution included a call for 'full political rights' for all men and woman and 'an end to puppet governments and separate representation'. See minutes, ANC, 24 and 25 September 1977.
8   The Minister of Labour was on record as wanting to strengthen the liaison committee system, which would be extended to all races. *Financial Mail*, 7 April 1978, cited in FCWU Annual Report, 1978.
9   The attempts to foist liaison committees on workers during this period goes some way to explain their resistance to the institution of 'workplace forums' in the LRA of 1995, although this requires that a workplace forum must be triggered by a trade union. Today there is a pressing need for a statutory form of workplace representation as a consequence of the externalisation of employment, which has resulted in a situation in which the workplace in the manufacturing and mining sectors, among others, has become a place where a multiplicity of employers operate, each with its own workforce. However, 'workplace forums' as defined in the LRA do not cater for this situation, because of their outdated definition of a 'workplace'. See chapter 25 in this regard.
10 The recognition agreement was intended to formalise relations between employers and trade unions and is discussed more fully later.
11 Minutes, MCM, 18 June 1978.
12 These industries were declared to be 'essential services' in terms of the Industrial Conciliation Act of 1956, although they were in no sense essential.
13 See minutes, MCM, 24 April 1977.
14 The introduction of an attendance bonus was at that time a popular strategy of employers, in response to wage demands. Workers forfeited it if for any reason they were absent from work. See minutes, ANC, 24 and 25 September 1977.
15 Minutes, MCM, 23 October 1977.
16 I refer both to the Industrial Court and the system that succeeded it, the CCMA and the Labour Court. There was no system of private arbitration for labour disputes. This was a later development, as I will presently explain.
17 In fact my fears proved groundless. The trade union representative was Morris Kagan, secretary of the National Union of Distributive Workers.
18 Minutes, MCM, 19 February 1978.
19 SAD, like Langeberg, was a farmers' co-operative, and had its head office in Wellington. The factory at Wellington was one of those that were supposed to be organised when I became general secretary, but in fact the Union had never succeeded in securing a majority there, and the workforce lived in fear of a liaison committee composed of supervisors and foremen. The burning issue for

the Wellington workers was that they were compelled to contribute to a savings scheme (the '*spaarklub*'), administered and controlled by a liaison committee. It was only after the workers at Wolseley and Worcester were organised that the Wellington workers (and Annie Adams) were confident enough to break the sway this committee had over them.

20 Although there were indications that SAD was willing to negotiate with the Union, it applied for the establishment of a Conciliation Board covering the three factories. When the Conciliation Board eventually sat, the validity of the comparison between canning workers and dried fruit workers was accepted, and the employers agreed to bring their wages in line with the canning industry. For Wolseley and Worcester workers, this was a significant increase. Minutes, MCM, 17 September 1978.

21 This difficulty was resolved by my encounter with a Swedish labour historian who confirmed that it was in fact the Swedish LO who had visited, and provided me with a copy of the report by one of the delegation, Kristina Persson. She is at the time of writing a minister in the government of Sweden.

22 The question of increasing subs was tabled at the conference of 1977, but it was decided to retain subs as they were. Minutes, ANC, 24 and 25 September 1977.

23 MCM, 19 June 1977.

24 The minutes record that the matter was reported to the management committee as follows: 'There were 3 possible organisers. J Mentoor had approached the President and asked him for work. The other two were young people.' The names are not mentioned in the minutes. See MCM minutes of 23 October 1977. The salary was reported to the management committee after the event. See MCM of 20 November 1977.

25 The Minister of Labour had the power to extend Conciliation Board agreements by way of a notice in the *Government Gazette*, in the same manner as Industrial Council agreements were extended.

26 Trade Union Advisory Co-ordinating Council.

27 For a discussion of the issue, see J Maree, Introduction to J Maree (ed.), *The Independent Trade Unions, 1974–1984*, Ravan Press, 1987, pp. viii–ix. The alternative term 'democratic' was to my mind even more unsatisfactory, as all trade unions purported to be 'democratic'.

28 Annual Report, 1978. 'Whatever the Commission decides about African and multiracial unions,' the report also states, 'we cannot suppose either the government or employers would like to see strong trade unions which are able to make strong demands for better living standards for the people.'

29 NUMAROSA stood for the National Union of Motor and Rubber Workers of South Africa. The initial meeting of trade unions, which the Union could not attend, was held in Johannesburg in March 1977. The meeting with the NUMAROSA representative in Cape Town was in April or May of that year. See minutes, MCM, 24 April and 22 May 1977.

30 NUMAROSA was also rumoured to have been urged to form a federation by the international trade secretariat (ITS) to which it was affiliated. The international trade union movement, during the Cold War, was split into two camps: the International Confederation of Free Trade Unions (ICFTU), to which the ITSs were affiliated, and the World Federation of Trade Unions (WFTU), representing primarily trade unions in the so-called socialist bloc. The influence of the relevant ITS (the International Metalworkers' Federation) on NUMAROSA is

acknowledged in Steven Friedman's account of the formation of FOSATU. See Friedman, *Building Tomorrow Today*, pp. 180–4.

31 For an account of Mary Moodley's role in the struggle, see Luckhardt and Wall, *Organize ... or Starve*, pp. 324–5, who describe her as living in the coloured area of Wattville. I assume the coloured area of Wattville was at some stage renamed Actonville.

32 TWIU had been, after FCWU, the most important affiliate of SACTU. According to Luckhardt and Wall's history, Mateman was acting general secretary of SACTU at one point. See Luckhardt and Wall, *Organize ... or Starve*, p. 412.

33 There had in fact been an unsuccessful attempt to organise the workers by the Sweet Food and Allied Workers' Union under Leonard Sikhakhane. See minutes, MCM, 23 July 2008. This union later split, with Sikhakhane forming the Food and Beverage Workers' Union.

34 Oscar's report on this visit to Johannesburg is contained in the minutes of the MCM of 17 September 1978.

35 In the eleven months since the 1977 conference, for instance, I recorded over 100 visits to factories and branches all over the Western Cape, as well as in Johannesburg. See Annual Report, 1978.

36 'The lesson of the past year has been that workers are more than ready to join unions,' my report to the conference commenced. 'We can see this from the attendance at our meetings. We can see this from the new factories which have joined us and the old branches that have come to life ... it is not the workers' lack of interest that can explain the weak state of the union today. It is the leaders who have failed to take unions to the workers.' FCWU Annual Report, 1978.

Chapter 10: The N1 at Richmond

1 Minutes, ANC, 26 and 17 September 1978. There was no proposal for AFCWU to contribute to my salary, and in fact it never did so.

2 I operated on the assumption that the Special Branch would get the minutes, but they were less likely to pick up something like this if it was not explicit. There is, however, a record of a resolution deciding what his salary was to be. MCM, 29 October 1978.

3 Amy Biehl was an American postgraduate student working at the Community Law Centre at the University of the Western Cape, when she was killed by a mob led by PAC activists in Gugulethu in 1993.

4 Minutes, MCM, 29 October 1978.

5 Minutes, ANC, 24–25 September 1977.

6 *Rapport Ekstra*, 15 October 1978. The Afrikaans original of the passage quoted is as follows: 'Die geveg van die "Big Shots" in die Arbeidersparty namens die swartes, of plurals, sal, as dit gewen word, nie die vryheid van die Kleurlinge verseker nie maar wel die ondergang van die Kleurling arbeiders ... Ek beweer dat Mnr Curry nou namens die Arbeidersparty op subtiele wyse die produktiwiteit van ons fabrieke aan vreemde etniese groepe wil oorhandig.'

7 MCM, 29 October 1978.

8 I was not personally present at the only other meeting where workers adopted a menacing attitude toward the Union, at Langeberg East London in 1985.

9 Minutes, MCM, 26 November 1978.

10 Minutes, MCM, 28 January 1979.

11 Minutes, MCM, 26 November 1978.

12 Minutes, MCM, 26 November 1978.
13 Minutes, MCM, 17 December 1978.
14 Minutes, MCM, 28 January 1979.

Chapter 11: Durban road
1 Amelia Mahlangu is an official of FAWU now based in Bloemfontein.
2 Rainbow Chicken Farms was then a private company. Although already a major player in the poultry industry, it had never been organised by a trade union, either in Worcester or Hammarsdale, Natal, where it had first been established. None of the other companies operating in the poultry industry had been organised either.
3 The meeting took place a few weeks after Juffrou H's dismissal, on 20 November 1978. See minutes, MCM, 26 November 1978.
4 Community organisations operating as proxies for political organisations were not yet visible.
5 MCM, 17 December 1978.
6 At a meeting with Jones management, we accused them of taking sides, and pointed out they were obliged to continue to deduct until such time as workers had formally resigned. See minutes, MCM, 28 January 1979.
7 Minutes, MCM, 25 February 1979.
8 It was impossible to prove for more reasons than this. For example, the fact that it was an offence meant victimisation had to be proved beyond reasonable doubt. The divisional inspector of the Department of Labour told me during one of our meetings that there had never been a successful prosecution of an employer for victimisation (and it clearly did not worry him that this was so).
9 The date was 1 March 1979. There is a detailed account of the strike in the FCWU circular letter, 12 March 1979.
10 FCWU circular letter, 12 March 1979.
11 Worcester was something of a stronghold of the white working class in the Western Cape. It was for many years one of the few constituencies in which the Herstigde Nasionale Party, a right-wing breakaway from the Nats, put up a candidate for parliamentary elections.
12 The term 'hotnot' is more or less the equivalent of 'kaffir.' A rough translation of what was said is: 'And you Hottentots who are still sitting – why don't you change and get out.' See FCWU circular letter, 12 March 1979.
13 The management at Jones instructed its time office, a breakaway stronghold, to conduct a survey to ascertain whether the workers the workers wanted to pay over the stop order cheques it was holding, and whether they were members of the Union. Although Jones did eventually pay over the stop order cheques to the Union, the survey showed the majority favoured a breakaway. Minutes, MCM, 25 February 1979.
14 The bail was paid out of the trust fund which the attorney Hymie Bernadt operated, known as the Trade Union Trust. The trust had been established at about the time Miss Ray was banned, and the funds included the surpluses the Union had generated from rentals for the Paarl building.
15 Joni Mitchell, 'My Old Man', on the album 'Blue'.
16 Amy was a white intellectual whom some had favoured to replace Becky Lan as general secretary when she was banned.
17 According to a roneod information document issued in my name, entitled 'A

Chronology of the Fatti's & Moni's Dispute', 10 August 1979.
18 In the chronology I refer to the power of attorney as a petition, in order to simplify the narrative somewhat. However, this was not accurate. See 'A Chronology', 10 August 1979, and minutes, MCM, 23 April 1979.
19 In a context where workers have the power to bring production to a standstill, it seems to me, resort to self-help forecloses resort to the law. By the same token, resort to law forecloses self-help, which in this instance was the only option that had any prospect of success. See J Theron, 'Trade unions and the law: Victimisation and self-help remedies', *Law, Democracy and Development*, 1 (1997), pp. 11–38.
20 To institute legal proceedings, as a first step it would have been necessary to lay a complaint of victimisation with the Department of Labour. Then, it would be necessary to await its outcome. In the unlikely event that the Department recommended prosecution, the case would be referred to a public prosecutor, who would have to prove beyond reasonable doubt all elements of the alleged offence, including the intention to victimise.
21 Minutes, MCM, 23 April 1979.
22 Minutes, MCM, 23 April 1979.
23 Minutes, MCM, 23 April 1979.
24 Minutes, MCM, 23 April 1979.
25 Report of the Commission of Inquiry into Labour Legislation (Part 1), Department of Labour and Mines, 1979.
26 The Convention on Freedom of Association and Protection of the Right to Organise (no. 87 of 1948) is regarded as one of the ILO's fundamental conventions. Article 2 requires 'workers ... without distinction whatsoever, shall have the right to establish and, subject only to the rules of the organization concerned, to join organizations of their own choosing without previous authorization'.
27 Wiehahn Report, Part 1, para 3.153.2.
28 See Wiehahn Report, part 1, chapter 4. It is clear from the analysis in this report that the Commission saw the establishment of a court of law with jurisdiction in labour matters as conforming to global trends.
29 'The principal element', according to the report, 'was the economic development of the country during the last three decades.' This had resulted in an increased demand for labour, particularly skilled labour. As a result, 'increasing numbers of unskilled and semi-skilled workers, particularly Blacks [i.e. Africans] had to be trained and utilized to perform higher-level skilled jobs.' Wiehahn Report, Part 1, Chapter 1, para. 1.2.
30 See Wiehahn Report, Part 1, para. 3.32.
31 According to Friedman, there were a total number of four 'companies' (by which it appears he means factories) at which independent trade unions were recognised in 1979, and all were foreign-owned. See Friedman, *Building Tomorrow Today*, p. 147. This provides some indication of the weakness of the movement. The formation of FOSATU in April 1979 was too recent to have had any bearing on the report, and in any event did not lead to any perceptible upsurge in organisation.
32 TUCSA's general secretary was Arthur Grobbelaar.

Chapter 12: Modderdam Road
1 Minutes, MCM, 29 April 1979 and undated circular letter, circa May 1979.

2   See minutes, MCM, 29 April 1979. The MCM suggested the Union place a photograph of the committee in the newspapers, so that everyone could see the Union was not run by one person.
3   A list of organisations that supported the strike is contained in 'A Chronology', 10 August 1979.
4   Annie Mentoor had been a child in Wolseley at the time of the Red Robin strike, and remembered the Union as having been 'a good thing'. At some point she moved to Macassar, where Deepfreezing and Preserving Company was located. Deepfreezing, as it was called, was the subsidiary of a UK company. This did not imply its management had a more enlightened attitude towards the Union. After becoming chairperson of the Union at the factory, Mentoor was laid off at the end of the season, along with others. The Union then employed her as branch secretary of Somerset West.
5   Persons of Indian or Asian origin constituted a separate group in terms of apartheid policy, for all purposes other than labour legislation, although persons of Malay origin did not fall in this category, presumably because they had been in the country so long it was not possible to distinguish them from other coloureds.
6   The strike was at Ford's Struandale plant in Port Elizabeth, and is discussed in more detail in chapter 14.
7   Section 1, Industrial Conciliation Amendment Act (94 of 1979).
8   It should be noted that 'mixing' between white and coloured had also been allowed in terms of the 1956 Act if the numbers of one group were too small to form an effective separate trade union. See section 3, Act 94 of 1979, amending section 4(6) of Act 28 of 1956.
9   The Minister could also impose 'such conditions as he deems expedient' to such declaration. See section 3 of Act 94 of 1979, amending section 4(6) of Act 28 of 1956.
10 'Segregated unions and conditions of registration for African unions unacceptable', the Union said in telegrams to the Minister of Labour when the amendments were being debated in Parliament. 'We ask for full freedom of association for all workers.' The telegram was sent on behalf of the management committees of both FCWU and AFCWU, and also sent to the Leader of the Opposition. Minutes, MCM, 17 June 1979.
11 Government proposed to fine registered trade unions R500 for each migrant or 'commuter' who was a member, or with whom a trade union had a 'relationship.'
12 This was also because the whites-only trade unions had been an important support base for the Nats, particularly in the north of the country. See Friedman, *Building Tomorrow Today*, pp. 158–63. It is also a further indication of how weak the emergent trade unions were that government was more concerned about the response of the whites-only trade unions.
13 A Special Branch officer from Vredendal had been seen in town that day, in the company of a local policeman. This policeman was in the vicinity of the hall shortly before the incident occurred. I laid a charge at the Lambert's Bay police station, but nothing came of it, of course. See minutes, MCM, 29 April 1979. The puzzling thing was that the kombi had also been spray-painted with red paint. It was this that alerted us to the fact that the vehicle had been tampered with. Otherwise we might have driven away, shredding the tyres. That would have been an inconvenience, rather than a disaster, so perhaps it was intended as a warning, whether directed at the locals or the Union itself.

14 Their full name was the Boland Inmaakwerkers Vereeniging. *Vereeniging* is Afrikaans for 'association', which was at that time the title preferred by organisations of white-collar workers and public servants, to distinguish themselves from trade unions representing industrial workers.
15 Minutes, special meeting, management committee, 30 June 1979.
16 Minutes, MCM, 17 June 1979.
17 Minutes, MCM, 17 June 1979.
18 Minutes, MCM, 17 June 1979.
19 There is no record of who was present at the meeting.
20 Minutes, MCM, 17 June 1979.
21 'A Chronology', 10 August 1979.
22 This conception of the relationship between the Union and the community, as it was then conceived, was later to influence my perspective on the relationship between unions and politics.
23 'A Chronology', 10 August 1979.
24 This was in terms of section 10(1)(b) of the Bantu (Urban Areas) Consolidation Act of 1945.
25 'A Chronology', 10 August 1979; Annual Report, 1979.
26 'A Chronology', 10 August 1979.
27 Athalie was teaching herself isiXhosa and had by this time acquired some fluency in the language.
28 On 14 July 1979. 'A Chronology', 10 August 1979.
29 Inkatha also donated money to the strike.
30 Personal letter, 30 September 1979.
31 Annual Report, 1979.
32 Minutes, AFCWU, ANC, 25 and 26 August 1979.
33 Minutes, FCWU, ANC, 25 and 26 August 1979.
34 Report of the national organiser, AFCWU ANC, 25 and 26 August 1979.
35 Minutes, FCWU, 23 September 1979.

Chapter 13: The road from Namaacha
1 Some years afterwards I was told by Ian Farlam that he had relied on a case decided by my father, which was shortly afterwards reversed by the Appeal Court in Bloemfontein. The Appeal Court's restrictive interpretation of 'the law' put a serious damper on anyone wishing to gain access to workers residing on private property, such as farm workers. Farlam himself was later to become a judge of the post-1994 successor of the Appeal Court, the Supreme Court of Appeal, and later presided over the Marikana Commission of Enquiry, as discussed in chapter 25.
2 *Argus*, 31 July 1979: 'Cheers in court as union man acquitted', cited in FCWU circular letter, 1 August 1979. 'A courtroom packed with members of the [FCWU] erupted into cheers and clapping yesterday when the Secretary of the union was acquitted on two charges.'
3 Henning Hintze, according to Ray's autobiography, was director of the German Volunteer Service in Zambia. See Simons, *All My Life and All My Strength*, p. 325.
4 This was in Doornfontein, Johannesburg. Indres Naidoo died of natural causes in 2016. An obituary describes him as being an early recruit to Umkhonto we Sizwe (MK). See *Sunday Times*, 10 January 2016. Jeanette Curtis was murdered in Angola in 1984 by means of a parcel bomb intended for her husband, Marius Schoon.

5   Ray makes cryptic mention of her meeting in Maputo with 'Jan Theron of the Food and Canning Workers' Union' in her autobiography. Simons, *All My Life and All My Strength*, p. 329.
6   Simons, *All My Life and All My Strength*, pp. 166–7.
7   Oscar had even organised the white women working in Laaiplek, and joked how their boss would tell them to go to their 'kaffirgod'. Oscar also told me that the bosses of Marine Products in Laaiplek had dismissed him. The workers went on strike and 'reinstated' him, as he put it. Accordingly he regarded them as his employer. Shortly after he was reinstated, he resigned from Marine Products to become an official of the Union.
8   In her autobiography Ray records her disagreement with Oscar over the conduct of an unnamed union official who had impregnated a woman member. Oscar apparently maintained this was a 'private matter'. See Simons, *All My Life and All My Strength*, pp. 229–30. Probably the unnamed official was Oscar himself, and the product of this relationship was the daughter with whom he stayed in East London, as described in chapter 16.
9   It is quite surprising that the debate about registering FCWU as a trade union for 'coloureds' is not discussed at all in Ray's autobiography.
10 The term 'united front' became more commonly accepted than 'common front', which was the term we actually used. See minutes, MCM, 22 July 1979. The date of the meeting must have been 28 June (rather than July, as recorded) 1979.
11 SACTU's attitude toward FOSATU is discussed in Luckhardt and Wall's history of SACTU. See Luckhardt and Wall, *Organize ... or Starve*, pp. 462–4.
12 See Annual Report, 1979 and minutes, ANC, 25 and 26 August 1979. Under a section headed FOSATU, the report states as follows: 'Some of the unions [affiliated to FOSATU] ... are organizing the same kind of factories we organize. A situation where different unions organize workers doing the same job is not in the interests of workers and of unity. Our Union must therefore decide: is it in our Union's interests to move closer and perhaps affiliate?'
13 According to Luckhardt and Wall, 'the determination of the FCWU and the AFCWU in the western Cape also testifies to the strength of SACTU'. Luckhardt and Wall, *Organize ... or Starve*, pp. 468–9.
14 From about this point on, every funeral of someone who had some association with the struggle, whether direct or not, became a political event, including the funeral of Oscar's eldest son, Karl, who was tragically murdered shortly after his return to the Union.
15 The fact that CTMWA retained the title 'association' was in my opinion indicative of whose interests were really paramount within the organisation: its leadership was dominated by workers in administrative positions in the municipality, and possibly did not see themselves as a trade union at all.
16 M Kagan of the National Union of Distributive Workers wrote to the Union to express his shame and disgust at TUCSA's defeat of the resolution. See minutes, MCM, 23 September 1979.
17 Minutes, MCM, 23 September 1979.
18 See the discussion of the Rikhoto case in Richard Abel, *Politics by Other Means: Law in the Struggle against Apartheid, 1980–1994*, Routledge, 1995, pp. 53–65.
19 Years later I met by chance with a representative of the organisation who expressed surprise that we had any doubts that the money was intended as strike pay.
20 See minutes, MCM, 23 November 1979.

21 See minutes, MCM, 23 November 1979.
22 This was when drafting a joint press statement at the conclusion of the mediation.
23 'Foreign states' did not include the Transkei or other nominally independent homelands.
24 Registration as a racially separate trade union meant adopting a constitution that defined its membership in terms of race, as FCWU had been compelled to do before my time, and specifying the race of its paid-up membership in each industry and each magisterial district where a trade union was organised.
25 Memorandum of Agreement between United Macaroni Factories Limited and FCWU and AFCWU, dated 8 November 1979.
26 Para. 12, Memorandum of Agreement between United Macaroni Factories Limited and FCWU and AFCWU, dated 8 November 1979.
27 'Strike at Fatti's and Moni's is over', *Cape Times*, 9 November 1979.
28 The editorial was a week later. See 'Lessons of a strike', *Cape Times*, 16 November 1979.
29 SACOS is the acronym for the South African Council on Sport, an umbrella body for a variety of organisations representing different sporting codes, which stood for non-racial sport.
30 The Western Province General Workers' Union was formed in 1978, although the Advice Bureau continued to operate.
31 Statement on the settlement reached between Fatti's & Moni's and the FCWU and AFCWU, 12 November 1979.
32 Joint statement, recorded in minutes, MCM, 25 November 1979. The statement also went on to reject the system of provisional registration, as a means of promoting politically pliant sweet-heart unions.
33 The full statement was as follows: 'Our AFCWU has decided not to apply for registration on the present basis. The Act clearly does not permit us to form one mixed union. Even if the Minister were able, which we do not believe he is in terms of the Industrial Conciliation Amendment Act, to form one mixed union, this would still mean we would have racially separate branches and executives, and the control of the union would vest with one racial group. The workers are not prepared to accept a registration in 1979 which is inferior to the registration that was granted to the FCWU in 1942.' See minutes, MCM, 25 November 1979.

Chapter 14: The road to the crèche
1 See, for example, Friedman who, relying on an interview with Alec Erwin, records FOSATU's position as being that 'it ... rejected the system, but the issue was one of tactics, not principle'. The implication is that the position of the Union and WPGWU was one of refusal to register on principle. See Friedman, *Building Tomorrow Today*, p. 169. Subsequently an article which appeared in the *South African Labour Bulletin* characterised trade unions like WPGWU and AFCWU as 'boycotters', which considerably undermined the value of their argument about the relation of law to the state. See Bob Fine, Francine de Clerq and Duncan Innes, 'Trade unions and the state: The question of legality', *SA Labour Bulletin*, 7, 1 and 2 (September 1981), pp. 39–68; and in Maree (ed.), *The Independent Trade Unions 1974–1984*, pp. 191–207.
2 The 'parallel' union was the United Automobile Workers (UAW). It is not clear whether FOSATU contemplated registering UAW as a separate trade union for Africans while its mother body remained registered for coloureds only. It

seems Kally Forrest's history of NUMSA, of which NUMAROSA was one of the predecessors, is silent on this issue. See K Forrest, *Metal That Will Not Bend*, Wits University Press, 2011.
3. The reason it was a high-profile strike was primarily that Ford was a global corporation at a time when calls for global corporations to disinvest from South Africa were mounting.
4. See G Kraak, *Breaking the Chains*, Pluto Press, 1993, p. 159; and Friedman, *Building Tomorrow Today*, pp. 189–91. As recorded by Friedman, the president of FOSATU, in a blunder of spectacular proportions, was asked by management to translate a request that workers return to work on its behalf, and did so. See also FCWU Annual Report, 1980.
5. See minutes, MCM, 27 January and 24 February 1980.
6. Our openness to the community was, to some extent, true to tradition. In Ray's time the Union took up issues regarding workers' housing, and a relic of this tradition was for branches to submit resolutions to conference demanding crèches and the like in their community. A letter from the general secretary would then be sent to the local authority, which might or might not be acknowledged, and there the matter would rest.
7. The case of two such workers who were evicted despite the Union instructing their attorney to defend their cases is recorded in the minutes of the MCM of 23 September 1979.
8. This was the position taken when the Union applied for a Conciliation Board to negotiate improved wages and conditions of work. The Department of Labour was initially persuaded to accept Growers' position, but under the threat of legal action eventually approved a Conciliation Board.
9. Bill Andrews is referred to as 'the "grand old man" of the South African Labour Movement' in *New Life*. He is regarded as one of the founders of the trade union movement in South Africa, and was involved in the formation of the Industrial and Commercial Workers' Union in 1919. See Close, *New Life*, p. 3 and www.sahistory.org.za/william-h-bill-andrews.
10. The Island refers to Robben Island prison. I have not attempted to verify this.
11. This limitation was well before the establishment of the Industrial Court, and with it the notion that a dismissal could constitute an unfair labour practice. It was also before the adoption by the ILO of the Termination of Employment Convention (158 of 1982).
12. Minutes, ANC, 25 and 26 August 1979.
13. One of the arguments I advance in this book is that the extension of labour rights is not achieved through litigation but through organisation. It is not an argument that rules out litigation, but is critical of the reliance on litigation strategies.
14. Minutes, MCM, 23 September 1979 and 25 November 1979.
15. Minutes, MCM, 16 December 1979.
16. Minutes, MCM, 16 December 1979.
17. See chapter 17.
18. Minutes, MCM, 27 January 1980.
19. It is noteworthy that the Department of Labour had a representative as an observer at these negotiations, which were led by Oscar. See minutes, MCM, 27 January 1980.
20. There was one payment of strike pay during the strike, at R15 per worker. The

Endnotes

number of workers paid was 352, which was about half the number that were supposed to be on strike. See minutes, MCM, 16 December 1979 and 27 January 1980.

21 Negotiations took place at a Conciliation Board. The Union was informed the Conciliation Board had been approved in December 1979, some five months after it had been applied for. It met on 15 January 1980. See minutes, MCM, 27 January 1980.

22 Minutes, MCM, 27 January 1980.

23 The wages of some workers were as low as R9 and R10 per week, for women and men respectively, although the position was complicated by the fact that many women workers earned on a piecework system. The minima agreed upon for women and men were R23.92 and R26.22 respectively. See minutes, MCM, 24 February 1980.

24 This is what had happened with the General Workers' Union in the Red Meat strike.

25 Minutes, MCM, 24 February 1980.

26 Charges of assault were subsequently laid against the police, with the assistance of the Union's lawyers. See minutes, MCM, 24 February 1980.

27 Because of the numbers involved, the Union was not able to pay strike pay. However, households that were entirely dependent on an income from Growers were identified, and a 'relief' payment was made to them. See minutes, NEC, 23 March 1980. The farmers must have instructed the management to settle because they were losing too much money, and the entire board of directors was present when they finally did, at a meeting brokered by local church leaders.

28 Lizzie was initially employed as an organiser with the object of working to build Cape Town branch. See minutes, MCM, 16 December 1979.

29 The descriptions 'Eastern Province' and 'Western Province' had no legal status at the time. Both were part of the Cape Province.

30 Minutes, MCM, 25 November 1979.

31 The management of Moberg's later reversed this moratorium on employing African seasonal workers. See minutes, MCM, 25 September 1979.

32 To this end, conference had adopted a resolution urging workers of Moberg's and Jones to form one branch. Minutes, ANC, 25 and 26 September 1979.

33 Minutes, MCM, 16 December 1979.

34 Minutes, MCM, 25 November 1979. Oom Joe was also elected as part of the delegation to the meeting at Langeberg's head office but had declined to come.

35 The application was published in the *Government Gazette* on 4 January 1980. In its application, as I have explained, a trade union had to specify the race of the workers it represented: in this case, coloured workers in the canning industry in the magisterial district of Paarl.

36 See minutes, MCM, 16 December 1979. The Union objected – to no effect – on the grounds that the company was deducting subs for an unregistered union, while it refused to do so for AFCWU.

37 For some reason, Lizzie Phike did not dispute her retrenchment and I only heard about it after the event. For Lizzie's own account of her retrenchment, see S Gunn and R Visser (eds.), *Labour Pains for the Nation*, Human Rights Media Centre, 2007, p. 61.

38 Minutes, MCM, 23 September 1979.

39 Minutes, MCM, 27 January 1980.

40 Minutes, MCM, 24 February 1980.
41 This could either have been at the same management meeting as Mrs R's outburst or an earlier one There is no indication in the minutes that Oom Joe was present on 27 January 1980, but I recall him asking the question and refusing to clarify.
42 Minutes, MCM, 24 February 1980.
43 Minutes, MCM, 24 February 1980.
44 My handwritten notes, meeting of 25 February 1980.
45 It was never disputed that the letter was typed in the factory manager's office.
46 In the minutes of the MCM of 16 December 1979, I reported that the 'financial situation at head office was bad'.
47 Minutes, Special MCM, 15 March 1980.
48 Minutes, MCM, 27 April 1980.
49 So keen at first had the workers been to be represented at the national executive committee that they had decided to collect the money to fly three delegates to Cape Town. See minutes, NEC, 23 March 1980. In the end, they travelled by road.
50 The record as regards the events of the strike is somewhat confused. There is a minute recording a work stoppage beginning on Friday, 25April (NEC, 23 March 1980) whereas a circular letter records a full-blown strike on Monday, 25 April 1980. It appears the full-blown strike must have commenced on Monday, 28 April 1980. See undated circular letter, headed 'Kromco Strike, Grabouw' and commencing 'Dear friends'.
51 Circular letter, circa 30 April 1980.
52 Circular letter, circa 30 April 1980.
53 Circular letter, circa 30 April 1980.
54 Minutes, MCM, 25 May 1980.
55 I later learned from one of the Elfco managers that there had been a meeting of the bosses of the area at the time of the Kromco strike, to review their labour policies. This meeting apparently included all the large employers in the valley, i.e. farmers and operations that had no links to the pack-stores.
56 Kromco like Growers was a registered co-operative.
57 Minutes, MCM, 25 May 1980.
58 See minutes, MCM, 24 August 1980.
59 There were about 200 of them. See minutes, MCM, 25 May 1980. The minutes do not reflect how marginal the increase was for most workers.

Chapter 15: The Graaff-Reinet road
1 By the time of the 1981 annual national conference, I was able to report that 'in all cases the Union negotiated directly with the employers, without the involvement of the Department of Manpower and without using their channels. As a result we were able to decide for ourselves how we wished to be represented at these negotiations. Naturally we always fought to have as many workers as possible as delegates to the negotiations.' See Annual Report, 1981.
2 Minutes, MCM, 24 February 1980.
3 Minutes, MCM, 27 April 1980.
4 Minutes, MCM, 25 May 1980.
5 Letter, 11 June 1980, referring to events about three weeks earlier.
6 See Friedman, *Building Tomorrow Today*, pp. 206–8, 232.
7 Minutes, MCM, 22 June 1980.

8   This was on 16 June 1980. See minutes, MCM, 22 June 1980. See also Friedman's account of the period. According to him, there were two strikes at SAAWU plants prior to the Western Province Preserving and Langeberg strikes, namely at SATV and National Convertor Industries. My recollection is that these ended in defeats for SAAWU. However, Friedman's chronology is not always clear. See Friedman, *Building Tomorrow Today*, pp. 217–23.
9   It appears the minutes are incorrect, in so far as they suggest the strike took place a day later, i.e. on 19 June, and that a settlement was reached on 20 June 1980, whereas it must have been reached on 19 June.
10 Minutes, MCM, 22 June 1980. Norushe was detained in terms of section 22 of the General Law Amendment Act.
11 Welile would surely have known that the management, for whom English was also a second language, had no idea what he was talking about.
12 Minutes, MCM, 22 June 1980.
13 The statement confirmed that Langeberg management had 'agreed with the Executive Committee of the AFCWU East London Branch and officials of the Union's Head Office that we accept the principle of freedom of association and that the management of LKB East London is therefore willing to deal with the representatives of the Union. Furthermore it has been agreed that negotiations will continue till full and permanent harmony is arrived at.' The phrase 'deal with' was probably adopted because Langeberg was not prepared at that point to say 'recognise' in the absence of a formal agreement.
14 According to Friedman, Oscar taught Gqweta his unionism. See Friedman, *Building Tomorrow Today*, p. 218.
15 Zodwa, SAAWU's administrator, was also Norushe's girlfriend.
16 The SAAWU officials were detained in June 1981, and while they were in detention I negotiated with KSM, the flour-mill that SAAWU had organised. See MCM, 28 June 1981.
17 The meeting was on 19 May 1980. See minutes, MCM, 25 May 1980. One of the unexplained amounts paid to Mrs R was for R500. This was more than my monthly salary at that time of R398.34.
18 A letter to this effect was received from the Industrial Registrar dated 29 May 1980. See minutes, MCM, 22 June 1980.
19 In terms of the 1979 amendments to the Industrial Conciliation Act, the period of provisional registration was for one year but could be renewed.
20 The Union had proposed to the Canners' Association that the pineapple canneries be covered by a new agreement, and also Table Top in George. On 22 April 1980 a formal meeting was held between FCWU and AFCWU and the Canners' Association to discuss this proposal. See MCM, 25 May 1980.
21 The negotiations took place on 5 August 1980. See minutes, MCM, 24 August 1980. This did prove to be the last time we negotiated at a Conciliation Board.
22 Minutes, MCM, 22 June 1980. The Union's delegation consisted of Oscar, Liz and Lizzie as well as the writer.
23 Minutes, special general meeting of Dal Josaphat branch, 30 June 1980. On 4 August 1980 a general meeting was convened of workers at Jones and Moberg's, where it was formally decided to form one branch of FCWU, with Aunt Sabbagh as the chairperson and Liz as the secretary. Lizzie Phike was elected secretary of the AFCWU branch. See minutes, MCM, 24 August 1980.
24 At the time, workers in greater Cape Town were almost wholly reliant on

buses and trains to get to work. Later that decade, with the helping hand of the PW Botha government, minibus taxis started to become an established mode of transport. Golden Arrow still operates a bus service at the time of writing. Ironically, it is now owned by a trade union investment company.

Chapter 16: The road to Kidd's Beach

1. The 1980 Annual Report speaks of the 'new hope' that the outcome of the elections in Zimbabwe had generated. See Annual Report, FCWU, 1980.
2. By contrast, the alternatives forms of organisation, such as a system of worker committees in factories, depended on an external agency in order to be effective. This had been the model of the Advice Bureau.
3. This was the long-standing branch secretary of Ceres branch.
4. The provisional registration was renewed in August 1981. See minutes, MCM, 23 August 1981.
5. See minutes, MCM, 25 January 1981. The takeover involved the entire group, including H Jones in Paarl and Industria, and Brink Brothers in Montagu.
6. Soon after the Union had recruited a majority at Collondale Cannery, and elected a committee, the bosses 'retrenched' five of them, including the chairperson. When the workers stopped work and asked to speak to the management, they were all dismissed. See minutes, ANC, 23 and 24 September 1980.
7. Botha was reported as telling this meeting that the government would not tolerate the emergence of an unregistered union movement. See *Daily Dispatch*, 9 and 10 October 1980, cited in minutes, MCM, 26 October 1980. The reference to unregistered unions was primarily to SAAWU and AFCWU, although the Western Province General Workers' Union also began organising stevedores in the East London docks at about this time.
8. Lennox Sebe was the Chief Minister of the Ciskei homeland and, after it was made 'independent', became its first President. Lieutenant General Charles Sebe was the head of Ciskei's Intelligence Service.
9. This was also reflected in the wage determinations published in terms of the Wage Act, which usually differentiated between the different urban areas, such as Port Elizabeth, Durban, the Witwatersrand and Cape Town. The wage for East London and its surrounds was generally the lowest.
10. I personally saw a proposal by SAAWU to the International Confederation of Free Trade Unions (ICFTU), which was also a major funder of the FOSATU trade unions.
11. Minutes, MCM, 26 October 1980.
12. This was in November 1981. It emerged shortly before the transition that Special Branch agents were responsible. See www.sahistory.org/dated-event/griffiths-mxenge-murdered.
13. The Union was supposed to have once had a branch in Durban, but it cannot ever have had a significant membership. This was confirmed for me by Leon Levy, who was formerly Transvaal secretary of FCWU.
14. Minutes, MCM, 26 October 1980.
15. It had only been R5 a week, because what they had earned was so much less than what workers in the Western Cape earned. Minutes, MCM, 26 October 1980.
16. Minutes, MCM, 14 December 1980.
17. Minutes, MCM, 23 August 1981.
18. There were various indications prior to the publication of the Industrial

Conciliation Amendment Bill on 27 March 1981 that employers should deal with representative unions, whether registered or not.
19 'This is our day-to-day experience in organising workers in factories. While the majority work for wages which are below the breadline, in fear of losing their jobs, there are the privileged few who earn many times more. Sometimes they are called "permanent staff", sometimes they are paid monthly to distinguish them from the ordinary weekly paid worker. They have pension funds, medical aid, housing benefits, while the ordinary worker ... has none.' Annual Report, 1980.
20 Moni, Terblanche and another manager came to the Union's head office in February 1980 to tell us that they had too many workers at the factory, and to ask what the Union's view would be if they were to retrench our members. We told him in no uncertain terms the Union would regard any retrenchment as a breach of our agreement. Minutes, MCM, 24 February 1980.
21 Minutes, MCM, 24 August 1980.
22 The meeting included representative from both Isando and Bellville, and took place at the Bellville mill. Minute, meeting with Fatti's & Moni's, Bellville, 11 September 1980.
23 Letter, P. Moni to general secretary, FCWU, 6 October 1980.
24 The reason I use the term 'labour relationship' rather than 'employment relationship' is that, as I explain in the concluding chapter, more and more workers nowadays are in a labour relationship with persons who are not their employer.
25 The negotiations took place on 20, 21 and 24 October 1980, ending at 4 a.m. on 25 October 1980. See minutes, MCM, 26 October 1980.
26 Draft recognition agreement, 6 October 1980.
27 According to Steven Friedman, Chloride recognised SAAWU in November 1980. See Friedman, *Building Tomorrow Today*, p. 222. I was informed about the Chloride negotiations by Gqweta, and also perused the Chloride agreement.
28 The use of this phrase suggests that already in 1980 the idea of the 'flexible firm' was in circulation. This idea is regarded as originating in the UK under Margaret Thatcher. Thatcher was elected in 1979 but only embarked on the process of economic restructuring with which her tenure was associated at a later juncture.
29 Minutes, MCM, 26 October 1980.
30 The other significant industry-level negotiation was in the inshore fishing industry. However, it was only by November 1980 that we had been able to pressure the employers' association, now called the South African Inshore Fishing Industry Association, into a proper negotiation meeting, prior to concluding an agreement. See minutes, MCM, 26 October 1980.
31 Minutes, MCM, 14 December 1980.
32 I have no record or recollection as to what they were doing in East London, but I think it was to attend a general meeting of the branch. The fact that they were detained is recorded in the 1981 Annual Report.

Chapter 17: Malta Road
1 Minutes, negotiations, 11 September 1980.
2 The mark-up charged by the millers was known as the 'millers' margin'.
3 As already noted, the inshore fishing industry in its heyday appears to have been the cash cow that certain of these conglomerates had milked. The three conglomerates with interests in fishing were Premier, Tiger and Fedfood, each

of which owned or had a shareholding in Southern Seas Fisheries, the Oceana group and Marine Products respectively.
4. According to Davies, O'Meara and Dlamini, five conglomerates dominated the manufacturing sector as a whole, together with a few multinational companies and parastatals such as SASOL. The five conglomerates were Anglo American Corporation, Barlow Rand, Federale Volksbeleggings, Anglovaal and SA Breweries. Each of them owned or controlled significant subsidiaries in food manufacturing. See R Davies, D O'Meara and S Dlamini, *The Struggle for South Africa: A Reference Guide*, vol. 1, Zed Books, 1985, pp. 57–65. Fatti's & Moni's was the only milling company that was privately owned.
5. Minutes, MCM, 25 January 1981.
6. This was not just true of Premier. At a later juncture I was told by the human resources manager of one of its divisions that the managing director of Tiger Oats had issued an instruction to all branch managers that no manager was to dismiss workers en masse at any of its plants without his express say-so.
7. The only trade union in the milling industry was a tiny union organised in the craft tradition, representing persons who had qualified as millers, a position equivalent to that of an artisan. I had spoken to the secretary of this union during the Fatti's & Moni's strike: he had come from the UK to work as a miller in South Africa, and was sympathetic to the strike. Unlike most other established unions, his union did not seek to determine wages for the ordinary workers.
8. So far as I am aware, the only subsidiaries of Tiger Oats to be organised by emergent unions were Meadow Feeds in Pietermaritzburg, which was organised by SFAWU at about the same time as or later than AFCWU organised their East London plant, and KSM Milling, which is referred to below.
9. 'Weekly paid' as opposed to 'monthly paid' employees were categories Premier management had introduced, to draw a dividing line between workers involved in production and those in the middle layers. Since we had no monthly paid members, we were not in a position to dispute it at the time.
10. SA Milling minute, 7 May 1981; FCWU minute, 23 June 1981.
11. Since Premier had no scheme of their own to put forward, management eventually accepted the Union's grades of work. See FCWU minutes, 23 June 1981.
12. The negotiations were on 23 June 1981 and completed in one day. See FCWU minutes, MCM, 28 June 1981.
13. FCWU minutes, MCM, 29 March 1981.
14. Minutes, MCM, 23 August 1981.
15. The Benrose plant fell within the definition of fruit and vegetable canning industry as defined in the Conciliation Board Agreement. This had (for the last time) been extended to cover Johannesburg, and it turned out the workers were being paid less than the country wages the Union had negotiated in the Western Cape. The bosses were compelled to pay them back pay.
16. The Union had attained a majority membership at I&J Benrose by October 1980. See minutes, MCM, 26 October 1980.
17. It later transpired this was the Food Beverage Workers' Union, which was affiliated to CUSA.
18. For a discussion of I&J Benrose, see minutes, MCM, February 1981.
19. Minutes, MCM, 27 July 1980.
20. Minutes MCM, 26 April 1981.
21. See minutes, MCM, 29 March 1981.

22 This was in July 1981. FCWU minutes, MCM, 26 July 1981
23 The company was originally known as the Imperial Cold Storage Company, and supplied the British army with frozen meat during the South African War. As a reward for this service, Sir De Villiers Graaff's father was granted a baronetcy.
24 The chairman in question was Dr Jan van der Horst, who was regarded as one of the doyens of South Africa business. It is tempting to speculate as to the role of the other major shareholder of Sea Harvest, the Spanish multi-national Pescanova, in the last years of the dictatorship of General Franco.
25 The links were least obvious in the case of the co-operatives. There were two co-operatives in the milling industry that were a significant presence in the industry, namely Sasko, with its head office in Paarl, and Bokomo, with its head office in Malmesbury. After 1994, these two co-operatives later merged and converted to a company (not necessarily in that order), Pioneer Foods. It was presumably because co-operatives are owned by their members that they were not susceptible to takeovers by the other conglomerates.
26 As explained later, ABF later sold its share of Premier to a consortium led by Liberty Life, another of the big finance houses in South Africa.
27 As well as centralising negotiations in pineapple canning, we had recently persuaded the three major deciduous fruit pack-stores to negotiate one wage agreement. Kromco, Elfco and Growers negotiated jointly for the first time in June 1981. Minutes, MCM, 28 June 1986.
28 Tiger and Imperial, who seemed to be slower in appointing HR and IR managers, in fact relied on consultants in the recognition negotiations that took place at this juncture.
29 A recognition agreement with I&J in respect of its fish processing plants was signed on 31 July 1981. See minutes, MCM, 23 August 1981. The recognition agreement with Land Harvest was signed on 27 August 1981. See minutes, MCM, 25 October 1981.
30 Meadow Feeds had the same management as KSM, the flour mill next door, which SAAWU had organised. When SAAWU's officials were detained, I was asked to negotiate simultaneously for both Meadow and KSM. The issue was the election of the committee, which management wanted to take place in the workplace, in terms of the recognition agreement. The Union's position was always that we conducted elections in terms of our constitution, and retained the right to do so at a general meeting, outside working hours. See minutes, MCM, 28 June 1981 and 25 October 1981.
31 We argued for adopting the term 'workers' rather than 'employees' in agreements, because the terms 'employee' and 'employer' were easily confused. Often there would be someone on the management side of the table, for whom English was a second language, who would inadvertently provide ammunition for our argument by referring to 'employers' when he meant 'employees', or vice versa. We were also concerned to avoid the kind of equivalence the terminology implied, and to differentiate 'workers' from 'employees' such as managers and their administrative staff.
32 How these offences were framed was hotly contested. We recognised the offences of dishonesty but 'gross insubordination', as we saw it, was a licence for managerial abuse of power, which it often was.
33 There were a handful of reported cases during 1981 in which the 'unfair labour practice' definition of the ICA was invoked.

34 Our counter-argument was that disputes over discipline would occur regardless of what the dispute settlement procedure said. Limiting its scope would simply make disputes more protracted and difficult to resolve.
35 Dismissal was, at the time and for some years afterwards, regarded as the most common cause of strikes nationwide, after wages.
36 For example, Premier's draft agreement for SA Milling proposed that if a dispute could not be resolved through meetings between the parties, it should be referred to mediation, and if not resolved there 'the dispute issue shall be referred to arbitration'. In this instance, however, it was proposed to constitute a panel of arbitrators: two appointed by the parties and a third 'umpire' appointed by the two arbitrators. See clause 11, Draft Agreement of SA Milling, circa June 1981.
37 The Premier agreement was typical of many in assuming that a trade union was in authority over its members, and was able to instruct workers as to what they could and could not do. Accordingly, the Union was asked 'to accept responsibility for any industrial action undertaken by its members'. See clause 5, SA Milling Draft Agreement, circa June 1981.
38 The notion of a strike as an exercise of power in furtherance of a collective demand could only arise once a right to strike had been established.
39 This phrase was taken from the Union's draft and adopted in the final agreement with Premier. See clause 5, Agreement between SA Milling Co. and AFCWU/FCWU.
40 'Any permanent employee of the company' should be eligible for membership of the union, Premier proposed in their draft agreement for SA milling, excluding 'security, personnel department, all managers/foremen and administrative personnel'. See Draft Agreement of SA Milling, circa June 1981.
41 A 'casual' worker was the equivalent of what in the US is referred to as a 'day labourer'. In terms of the Factories, Machinery and Building Work Act (22 of 1941), a casual could not work more than three days in any week. In 1983 the Act was replaced by the Basic Conditions of Employment Act (3 of 1983). Essentially the same definition of casual employee was retained, and it also applied in most wage determinations.
42 The watchmen were to be given a choice of either accepting jobs with the new company or being absorbed into the mill. See FCWU/AFCWU minute, negotiation of recognition agreement with SA Milling, 25 September 1981.
43 The recognition agreement with SA Milling was signed on 29 October 1981. This conversation thus took place at some point prior to that. The rise of Pick n Pay appears to have more or less coincided with that of retail conglomerates in the UK and the US, such as Walmart.
44 Tariffs imposed on canned goods from South Africa were also a major constraint on the profitability of the fruit and vegetable canning industry. Fresh fruit was the beneficiary of the shift in consumer tastes away from canned foods, and it did not seem that the producers faced the same problems in accessing European markets, notwithstanding calls to boycott South African fruit. The marketing of the fruit which the pack-stores produced for export was done through a government control board, the Deciduous Fruit Board.
45 The fact that Langeberg was a co-operative may have meant it was less vulnerable than Jones. The members of the co-operative were farmers, who would have been more concerned with a market for their produce in the long term, than a short-term return on capital. Co-operatives like Langeberg had benefited

from government support in the past, and at some point someone slipped us a RFF memorandum suggesting that it was because of this support that RFF (which was owned by Anglo American) was no longer able to compete with them. Langeberg was not mentioned by name, but this was clearly whom it was referring to.

46 Minutes, MCM, 26 July 1981.
47 Minutes, MCM, 22 November 1981.
48 This was on 23 April 1982. The Union objected in a letter from the general secretary, FCWU, to the Industrial Registrar, dated 19 May 1982. The quotation is from the reply of the Boland Inmaak Werkers Vereniging, dated 2 June 1982. However, in terms of the legislation of the time the Industrial Registrar had no alternative but to uphold our objection.
49 Minutes, MCM, 30 May 1982. Initially production was to be transferred to Highlands Canning in Grabouw, which was owned by the same group. However, not long after that, Highlands Canning also closed down.
50 Annual Report, 1982.
51 Minutes, MCM, 24 October 1982.
52 Minutes, MCM, 25 January 1981.
53 A general meeting to re-establish Port Elizabeth branch was held on 31 May 1981. See minutes, MCM, 28 June 1981. Some of the workers who joined had previously been members of AFCWU when Langeberg had a factory there.

Chapter 18: Off Bhunga Avenue
1 The decision to proceed with a conference of trade unions was taken at the end of 1980. Minutes, MCM, 14 December 1980.
2 The general secretary of the IUF was Dan Gallin.
3 It was usually a good indication of a conservative approach to the labour relationship where a trade union referred to 'employees' rather than 'workers' in its title.
4 The coloured women, after all, had long had the right to form and join trade unions, but were not able to exercise it.
5 It seems to me the focus on skilled workers and training also explains the change in the Minister's job title, and the title of the Department, from 'Labour' to 'Manpower Utilisation'. Subsequently, of course, both Minister and Department reverted to their old titles.
6 Steven Friedman gives an account of the session attended by the FOSATU unions. See Friedman, *Building Tomorrow Today*, pp. 242–3.
7 This was the second time in my life I was flown to Pretoria at government expense. The first time, as a student, the Students Representative Council had been invited to give evidence to a commission chaired by Louis le Grange, who was now Minister of Police. My two colleagues and I were sent packing, because we had questioned the impartiality of Owen Horwood, then Vice-Chancellor of the University of Natal (as it then was), in our opening statement.
8 A draft of the Bill was leaked to the press sometime in December 1980. According to Friedman, the draft provoked an 'outraged' response from the emergent unions. From the Union's perspective, however, what was noteworthy was the lack of response on the part of the emergent unions. See Friedman, *Building Tomorrow Today*, pp. 243–4 and FCWU minutes, MCM, 14 December 1980.
9 See chapter 13.

10 In the notes I prepared for a meeting of trade unions in February 1981 I expressed our position thus: 'A common stand on registration seems unlikely; however, common action in support of certain principles is essential.' See 'Points for discussion', 8 February 1981.
11 It was alleged that FOSATU or its supporters had tried to prevail on the editors of the *SA Labour Bulletin* not to publish the article. See WPGWU (1979) in Maree (ed.), *The Independent Trade Unions 1974–1984*, pp. 176–207.
12 In June 1980, for example, Neil had written to me about a decision that FOSATU would not assist AFCWU in any way in the Transvaal, which his contacts in MAWU claimed was imposed on the union from above. Whether or not that was the case, I personally did not see we could object to such a decision given that we were not an affiliate.
13 See undated letter, circa December 1980/January 1981 from Neil Aggett to the writer. Neil explained Fanie Botha's intentions with regard to the legislation thus: 'the state desperately wanted to assimilate and control' unions, even if it necessitated 'great compromises' on its part. 'The other repressive option open to the state may be used once they could isolate and divide stubborn unions.'
14 The organisations represented were CUSA, FOSATU, SAAWU, WPGWU and FCWU/AFCWU. See FCWU minutes of the meeting, 8 February 1981. There were no agreed minutes of this meeting, and the FCWU minutes do not record which persons represented the different bodies, because of security concerns. Pendlani and I represented the Union's head office. My recollection is that Joe Foster put the FOSATU position.
15 The other issues included pensions (since workers were up in arms about proposed legislation on the transferability of pensions), influx control (since there was legislation proposed regarding influx control), industrial conflict, encompassing the issue of strikes, strike funds, and the deportation of striking workers. See FCWU minutes of Johannesburg meeting of unions, 8 February 1981.
16 Letter, WPGWU, dated 11 March 1981 and FOSATU, 'Draft position paper for joint union meeting', undated, circa March 1981. FOSATU's position on recognition was quite surprising. Recognition, as FOSATU saw it, was entirely a matter between a trade union and employers. The following statement regarding bargaining was somewhat jarring: 'Employers can make a positive contribution toward industrial peace by bargaining in good faith with representative trade unions ... and by actively preventing the involvement of the paranoiac and totally insensitive police force.'
17 Not surprisingly, being a federation, there were indications that not everyone in FOSATU was open to such a realignment. Shortly after this meeting, Neil wrote to me to express his concern about FOSATU's position. 'Indications up here are that FOSATU is not taking seriously the possibility or even the desirability of a united stand,' Neil wrote to tell me: 'There were rumours of this for some time, with FOSATU officials talking about the fact that they did not want to be "bullied" into taking a stand on registration, but FOSATU should make its own independent position.'. He was probably reacting to elements within FOSATU that supported the initiative reluctantly, and in the hope it would lead to new recruits for FOSATU. Letter, Neil Aggett to the writer, 7 April 1981.
18 The meeting took place on 22 March 1981. See FCWU minutes, 22 March 1981, and FCWU minutes, MCM, 26 April 1981. The same organisations which had been present in Johannesburg were there.

19 Similarly, SAAWU did not want BAWU (Black Allied Workers' Union), an obscure trade union in the Black Consciousness tradition from which they had originally split. The problem of whom to invite was resolved by agreeing to publicise the conference, and letting would-be participants approach the organisers (which never happened).
20 The dispute between MACWUSA and FOSATU, in my opinion, had to do with the social make-up of Port Elizabeth. It was unlike Cape Town, with its established coloured middle class dominating the middle layers in manufacturing and commerce. It was also unlike East London, where the equivalent positions were filled by Africans. Competition for positions between coloured and African communities was intense, since it was also not part of the Coloured Labour Preference Area. The principal advantage apartheid had afforded the coloured community was education and training.
21 The Industrial Conciliation Amendment Bill of 1981 was published in Notice 235 of *Government Gazette* 7521 of 27 March 1981.
22 This fact is generally underemphasised in commentaries on the Bill. See, for example, Friedman, *Building Tomorrow Today*, p. 243, who described this provision as 'removing the bar on mixed unions'. This is inaccurate. 'Mixed unions' is a term that was understood at the time to refer to those unions which were permitted to have white and coloured members.
23 This was that government would never tolerate the emergence of an unregistered trade union movement. See *Daily Dispatch*, 9 and 10 October 1980, cited in minutes, MCM, 26 October 1980.
24 It would, for an example, be an offence to grant 'financial or other material assistance' to workers engaged in an illegal strike. There was a prohibition on unions affiliating to any political party or organisation. The Registrar was even given the power to close down unions. See section 65, Industrial Conciliation Amendment Bill, 1981, and the FCWU memorandum of objections, dated 22 April 1981.
25 Only a registered trade union could have its subs deducted by stop order. However, stop order facilities were not an especially attractive 'carrot' for registered trade unions in the absence of an effective means to compel recalcitrant employers to grant them.
26 BMWU was the Black Municipal Workers' Union, and it organised municipal workers in Johannesburg.
27 The resolution adopted stated, in respect of registration, that trade unions would 'resist and reject the present system of registration insofar as it is designed to control and interfere in the internal affairs of the union.' See minutes, MCM, 23 August 1981.
28 The resolution with respect to the provisions of the Bill of 1981 reads as follows 'We accept that trade unions are public bodies and accordingly we do not object to providing information with respect to our constitution, finances and representativeness. However, we refuse to subject ourselves to control by anybody other than our own members. We therefore resist and reject the present system of registration insofar as it is designed to control and interfere in the internal affairs of the union. The meeting specifically agreed to support each other in defiance of any abuse in the powers of investigation given to the authorities by the Industrial Conciliation Act.' See minutes, MCM, 23 August 1981.

29 The summit rejected the present Industrial Council system, and recommended that unions that were not members of Industrial Councils should not enter them. Unions already participating in Industrial Councils were supposed to 'refer this back to their respective unions for endorsement'. Annual Report, 1981

30 At H Lewis, the workforce was dismissed following a strike over the dismissal of a committee member. At Model Dairy, the workforce was dismissed following a strike over a demand that management negotiate with the Union. Both were newly organised plants, and Model Dairy was the first dairy the Union had organised. On the other hand, strikes at Sea Harvest and at the newly organised Appletizer plant in Grabouw were resolved without compromising the Union's organisation. See MCM, 25 October 1981.

31 At Table Top in George the members of the committee in the quality control department, a Union stronghold at the plant, were 'provoked' into resigning, and the workers went on strike to demand their reinstatement. The bosses responded by dismissing the workers. See minutes, MCM, 25 October 1981.

32 The workers went on strike over a demand for a Christmas bonus, which was not a demand that had been made in the recently concluded wage negotiations. At the price of preserving the jobs of the remainder, the Union had to accept that 26 workers would lose their jobs. See minutes, MCM, 22 November 1981.

33 'Trade unions today have become fashionable,' I stated in my annual report that year; 'because there has been a great upsurge of interest in unions the impression is created that the union movement is a great deal stronger than is actually the case. In fact the unions are themselves weak, and between the unions there is little genuine solidarity.' See Annual Report, 1981.

34 SACTU had been affiliated to the WFTU. The Union had, over many years, received greetings from the likes of the Hungarian Food Workers' Union or the Guyana Agricultural and General Workers' Union at its conferences. These were presumably WFTU affiliates. The messages were dutifully read, although there was no way of knowing if they were from genuine trade unions. The affiliates of the IUF, on the other hand, were organising corporations we might someday organise, like Coca-Cola. The most prudent stance appeared to be to avoid aligning ourselves on either side of the Cold War divide. Accordingly, the MCM decided the Union would maintain links with IUF, but not affiliate. See MCM, 22 June 1980.

35 This was the Clothing Workers' Union (CLOWU).

36 It was never clear to me on what basis FOSATU brought this challenge and I never heard of its outcome while I was in the Union. The most explicit reference to the claim that FOSATU's legal challenge brought about a change in the law is in Michelle Friedman's popular history of FOSATU, which states as follows: 'increasingly, FOSATU's legal advisers began to use the law to challenge existing labour laws and to create new ones. A major victory for FOSATU was its successful Supreme Court action in 1983 whereby government lost the right to register unions on a racial basis.' It is unclear which case this refers to. See M Friedman, *The Future is in the Hands of the Workers: A History of FOSATU*, Mutloatse Arts Heritage Trust, 2011, p. 60. However, government formally abandoned the requirement that trade unions register on a racial basis in November 1981, when the Bill became an Act.

37 Norushe was first brought to court on 24 February 1981. See minutes, MCM, February 1981 and 26 April 1981.

38 The trial commenced on 3 March 1981. See minutes, MCM, February 1981. Oscar was defended by Advocate Ian Farlam, the same advocate who had defended me in Worcester.
39 I had last seen Hennie Ferrus at a public meeting that the Union organised in Paarl following the breakaway. The ostensible object of the meeting was to publicise our victory in the Growers strike, but in fact it was to win back workers at Jones following the breakaway. Ferrus prophesied I would one day be Minister of Labour, which was more indicative of his own aspirations than mine, I think.
40 Minutes, MCM, 26 April 1981.
41 These included the young Zackie Achmat, whom I met for the first time with Virginia. He was also attired in khaki.
42 Johnny Issel died in 2011. At his memorial service, Essa Moosa, who was an attorney at the time, was one of the speakers. He confirmed Johnny's role in organising the funeral, and that this was the first public display of the ANC flag in the 1980s.
43 As indicated previously, this was because it comprised both registered and unregistered unions and had engaged in politics through an alliance. I later learned how sharply divided SACTU had been over the issue of registration, with Oscar weighing in strongly on the side of those that opposed registration. See R Lambert, 'SACTU and the IC Act', *SA Labour Bulletin*, 8, 6 (1983), pp. 25–44.
44 Letter, general secretary, SAAWU to branch secretary, AFCWU Johannesburg, 28 March 1981. The branch forwarded the letter to the head office.
45 The few members elected Peggy Dlamini as our first branch secretary in Durban. See minutes, MCM, 28 June 1981.
46 In the light of subsequent events, it seems I might have underestimated how profoundly threatened the ANC was by the reforms the government was initiating. There is a discussion of SACTU's position on registration after the publication of the Wiehahn Commission report, but no indication what its response to the 1981 Bill was. However, it appears that position did not change. See Luckhardt and Wall, *Organize ... or Starve*, pp. 459–4.
47 I kept no record of these calls, which were numerous, and took place over a period of months. The only mention of them was to 'incidents of harassment', at the MCM of January 1982. See minutes, MCM, 31 January 1981.
48 The meeting was on 1 November 1981. Minutes, MCM, 25 October 1981.
49 Minutes, MCM, 22 November 1981. I did not personally attend the Anti-SAIC meeting.
50 This was consistent with the prohibition the Bill introduced on trade unions affiliating to any political party or organisation. See note 23.
51 Minutes, ANC, 19 and 20 September 1981. For an account of the wage negotiations at Langeberg Boksburg, see minutes MCM, 26 July 1981.
52 The branches were asked to consider the following question: 'whether it is useful for the FCWU to remain registered bearing in mind that there is opposition to registration amongst members.' Minutes, ANC, 19 and 20 September 1981.
53 The Labour Relations Amendment Act (57 of 1981). The Act came into force on 1 November 1981.
54 Minutes, MCM, 22 November 1981.
55 The next to approach us were workers from a bakery belonging to the Bokomo

group in Malmesbury, which was soon to merge with Sasko. At about the same time, the Union had begun organising mills in the Sasko group, beginning with its Rondebosch mill. Sasko has since become Pioneer Foods.
56 Minutes, MCM, 30 May 1982.
57 Minutes, MCM, 27 February 1983.

Chapter 19: The road to the cemetery

1 The Boksburg factory was also producing canned vegetables rather than fruit, for the domestic rather than the export market. It was therefore not subject to the same cost constraints.
2 The last thing I could find that Neil had written was about this strike, which ended on 20 July 1981. 'It was only through the complete unity of action of the LKB workers and consistent organising of the Union that this big victory was won. It shows that if the workers are organised and strong in the factories we can win wages that are higher than the minimums laid down in the CB Agreement.' My contemporaneous note, headed 'Neil's work'.
3 Minutes, MCM, 25 October 1981.
4 Minutes, MCM, 26 July 1981.
5 When, for example, Thozamile Gqweta was detained for the fourth time in little more than a year, David Lewis and I wrote a letter to the press on behalf of our respective unions, to bring attention to his plight. The letter was dated 23 July 1981. See minutes, MCM, 26 July 1981
6 According to Friedman, there were 16 detainees 'connected' with the emergent trade unions. However, some, mainly from the white Left, were only loosely connected, if at all. See Friedman, *Building Tomorrow Today*, p. 278.
7 See, for example, a 1982 speech by Barend du Plessis, who was then Deputy Minister of Information. He told an international conference there was a need to 'isolate' people who misused trade unions for political purposes. If people transgressed the law, he was reported as saying, they were charged in court, but 'the rest' were 'removed for a little while'. Editorial, *Sowetan*, 24 September 1982.
8 Undated letter from Neil Aggett to the writer, circa 1980.
9 Beverley Naidoo, who interviewed me at a stage when she was still planning to write a novel based on Neil's life, seems to think I thought (or advised that) Neil should do his military service. That is not correct. See Beverley Naidoo, *Death of an Idealist: In Search of Neil Aggett*, Jonathan Ball, 2012.
10 The Labour Relations Amendment Bill was published in Government Notice 768 in *Government Gazette* 7824 of 9 October 1981.
11 An unfair labour practice was initially defined as 'any practice which, in the opinion of the Industrial Court, is an unfair labour practice'.
12 This was the notorious building where the police, and specially the Special Branch, had their Johannesburg headquarters.
13 This was one of the things that the MCM resolved we do. See minutes, MCM, 31 January 1982.
14 FCWU leaflet, 7 February 1982: 'Neil Aggett, Tvl Secretary Food and Canning Workers' Union – died in detention', published by Projects Committee and printed by SRC Press, University of Cape Town.
15 Act 83 of 1967. The Act was also notorious for its broad definition of 'terrorism'. It was repealed in 1991.

16 This is also what we told the press and our members. See minutes, special MCM, 8 February 1982.
17 FCWU leaflet, 7 February 1982.
18 The video *Passing the Message*, by Cliff Bestall and Mike Gavshon.
19 Minutes, MCM, 28 February 1982.
20 In Johnny Copelyn's memoir, he makes the startling claim that 'his [Neil's] death had resulted in our union [the National Union of Textile Workers] calling the first national protest of workers across the country'. This is misleading, to say the least. See Johnny Copelyn, *Maverick Insider*, Pan Macmilllan, 2016, p. 119.
21 Friedman, *Building Tomorrow Today*, p. 284.
22 Although I did not mention any organisation by name, FOSATU was obviously sensitive on the issue. Steven Friedman, in his book, says I 'was sharply rebuked by FOSATU which charged that he had used an event which should have cemented worker unity to sow discord'. I am not aware of any such 'rebuke'. See Friedman, *Building Tomorrow Today*, p. 303.
23 Writing not long after the event, when the ANC was still banned, Friedman records unnamed unionists as saying the funeral had been 'hijacked', and that political activists had 'claimed as their own a unionist who had studiously avoided them'. I was interviewed at length by Friedman, but I did not want to be named as an informant in his book, among other reasons because I was concerned that the narrative he was constructing would be dominated by white voices. See Friedman, *Building Tomorrow Today*, p. 284.
24 The speeches were by David Lewis of GWU, Robert Gqweta, Thozamile's brother, on behalf of SAAWU, and Dennis Rubel, a medical doctor, on behalf of his friends. Dennis was later to work for the Union's Kempton Park branch.
25 Undated personal letter to the writer, circa 1980. This, he went on to say, was what had led to the sort of 'opportunism' he perceived as having emerged in FOSATU.
26 I had of course been affirming the Congress tradition, in a discreet fashion, through the choice of guest speakers at conferences. At the last conference Neil attended, it had been Dora Tamana, a Congress stalwart with no organisational link to the Union. See minutes, ANC, 19 and 20 September 1981.
27 I have relied on Beverley Naidoo's detailed account of how Hogan was set up. See Naidoo, *Death of an Idealist*, pp. 180–6.
28 Whitehead distinguished between the full card-carrying member of the ANC, the active supporters and the sympathiser, and thought Neil was something between an active supporter and sympathiser. 'Aggett an ANC sympathiser', *Rand Daily Mail*, 8 October 1982.
29 See chapter 12.
30 In my speech at St George's Cathedral, I was quoted as saying that the 'the security police … are cooking up a show trial in which the democratic trade unions will be charged with furthering the aims of the working class of South Africa. They say Neil was detained to play a star role in that trial. No doubt Neil is dead because that role was not to his liking.' 'Aggett stood for workers unity', *SASPU National*, February/March 1982.
31 'Aggett broken, says detainee', *Sowetan*, 27 October 1982; 'Witness tells of Aggett's last hours', *Rand Daily Mail*, 26 April 1982.
32 Tom Lodge testified at the inquest, as an expert witness, to the effect that Neil would also not have described Marxism as an ideology. *Sowetan*, 11 October 1982.

33 'Controller. He was in charge of Neil Aggett up to his death', *Sunday Times*, 10 October 1982.
34 Whitehead did not apply for amnesty to the Truth and Reconciliation Commission.
35 The quotation is from Bob Dylan's song 'Absolutely Sweet Marie', on his 1966 album *Blonde on Blonde*.
36 Minutes, MCM, 28 February 1982.
37 See undated AFCWU circular letter, circa 8 February 1982, 'To all fraternal organisations'.
38 Minutes, MCM, 28 February 1982.
39 Minutes, Special MCM, 18 March 1982.
40 Joe Mavi of BMWA died in a car accident in June 1982.
41 The basis for finding her guilty of treason was for having helped plan ANC strategy in the field of labour. The judge said she played an important role in a number of strikes and boycotts, and evidently regarded ten years' imprisonment as lenient, given that in wartime the death sentence was often imposed for this crime, and 'the ANC regarded itself as at war with South Africa on all fronts'. 'Clenched fist as Hogan gets 10 years', *Rand Daily Mail*, 22 October 1982.
42 See editorial, *Rand Daily Mail*, 15 November 1982. On the other hand, Hogan was found guilty of treason.
43 Minutes, Special MCM, 8 February 1982 and Friedman, *Building Tomorrow Today*, p. 284.
44 Friedman, *Building Tomorrow Today*, p. 285; Kraak, *Breaking the Chains*, p. 179.
45 Because Beatrice and Mildred were employed by Johannesburg branch before it split into two, separate meetings of the BECs of the two branches were held to discuss the charges against them, and thereafter a 'combined' general meeting in Tembisa. See MCM, 30 May 1982.
46 The greatest tribute that could be paid to Neil, I said in my report to the conference later that year, was that the organisation of workers in the Transvaal continued apace without him, and for a long time without any full-time Union official. See Annual Report, 1982. In fact this was not strictly true, since Maria was already assisting the Union.
47 A special conference was held where branches reported back on the question posed at the previous conference as to whether to register or not. It was resolved to try to merge the registered and unregistered union at the workplace and at branch level, before deciding whether the Union should be registered or not. See minutes, special national conference, 28 and 29 March 1982. However, the resolution to merge at branch level was not implemented at Paarl.
48 Minutes, MCM, 17 April 1982 and 30 May 1982.
49 Minutes, MCM, 30 May 1982.
50 The dominance of TUCSA trade unions in the Western Cape is another reason why it was unrealistic to think FCWU could deregister: as long as there were other registered trade unions that were not in any way challenging the system of registration, it would only have isolated FCWU even further to do so. The only organisation with a coloured membership on the solidarity committee apart from ourselves was now CTMWA, which had overcome whatever political scruples it had about associating with us. However, it still styled itself as an association rather than a trade union, and still represented workers in only one municipality.

51 So far as I am aware, the only significant factory where a FOSATU union had been organised, the motor assembly plant at Blackheath, had by this point closed down.
52 In my report to the MCM following the stoppage and the funeral, I said as follows: 'The work stoppage ... showed us which unions can be relied upon to support us ... We were stood by [sic] Unions with strong organisation: the General Worker' Union, SAAWU in East London, some unions affiliated to FOSATU.' See MCM, 8 February 1982.
53 Joe Foster, 'The workers' struggle: Where does FOSATU stand?', *SA Labour Bulletin*, 7, 8 (July 1982), pp. 67–86 and in Maree (ed.), *The Independent Trade Unions 1974–1984*, pp. 218–38. Although the paper argues for the establishment of a working-class movement, the only attempt it makes to describe what form such a movement might take is a reference to the 'advanced industrial countries' where 'there are a number of different organisations – trade unions, cooperatives, political parties and newspapers – that all see themselves as linked to the working class and furthering its interests'.
54 In my opinion Friedman exaggerates the impact this paper had on the emergent trade unions, which he claims 'alarmed all its [FOSATU's] rivals'. I am not aware the paper attracted mention in any of the formal trade union unity meetings. What debate it generated was in white Left circles. See Friedman, *Building Tomorrow Today*, pp. 285–6.
55 'Search for a workable relationship', *Grassroots*, May 1982 and *SA Labour Bulletin*, July 1982.
56 'All the great and successful popular movements have had as their aim the overthrow of oppressive – most often colonial – regimes. But these movements cannot and have not in themselves been able to deal with the particular and fundamental problems of workers,' said the FOSATU paper. 'It is, therefore, essential that workers must strive to build their own effective organisation even whilst they are part of the wider political struggle.' 'Our viewpoint is that a union should not split the struggle of workers in the factory from struggles outside the workplace, on community and political issues,' said the FCWU article. 'However, we do believe that separate forms of organisation are needed for these struggles. A trade union is not a community or political organisation.'
57 The Union's experience on a committee to oppose the Bill, facetiously called the 'Disorderly Bill Action Committee', was an early indication of the political polarisation taking place in the community.
58 'There has emerged into our political debate an empty and misleading political category called "the community". All communities are composed of different interest groups ... Under the surface of unity community politics is partisan and divided. FOSATU cannot possibly ally itself to all groups that are contesting this arena. Neither can it ally itself with particular groups. Both parties will destroy the unity of its own worker organisation.' Foster, 'The workers' struggle: Where does FOSATU stand?'
59 'One reason [for this apathy] is that the working class is made to feel inadequate in almost every way – whether it be at school, church or home,', the FCWU article stated. 'There are people who do not want to recognise there are class divisions among the oppressed. For us there can be no lasting alliance between workers' organisations and community organisations unless this division is understood.'

60 *Work in Progress*, 23 June 1982, p. 26.
61 Minutes, MCM, 27 June 1982. The minutes record that this decision was originally taken at a special meeting, but I have no record of this special meeting.
62 I established that the song is 'Sobashiya abazali ekhaya' after reading Gcobani Bobo and Elvis Jack's novel, *The Rise of the Dagger* (2015).
63 Agreed press statement issued after meeting, attached to the writer's own minute.
64 Notable absentees were CUSA and CCAWUSA. However, there were several community trade unions present that had not been at Langa, as well as CTMWA.
65 The five principles were that trade unions had to be 'non-racial', say no to registration, not participate in Industrial Councils, there could be no separation with politics, and no separation with community problems.
66 FCWU minutes kept by the writer. I did not record the reference to SACTU in the minutes, however.
67 Other community trade unions expressed similar sentiments. For example, a speaker from Orange Vaal General Workers' Union, one of the new 'community unions', said, 'retrenchments and price increases do not affect non-registered unions any differently from registered unions'. Writer's own minutes.
68 FCWU minute, trade unions meeting, 3 and 4 July 1982.
69 We saw this as an expression of what FOSATU called 'disciplined unity'. It looks more like discipline than unity, was David Lewis's quip at the Wilgespruit meeting. 'Why can't each union be represented on its own?' See writer's minutes, Wilgespruit trade union meeting, 24 April 1982.
70 Anthony Butler's biography of Ramaphosa provides a credible account of how NUM was established and why Anglo decided to grant it access to its mines before it had organised the workers. Ramaphosa, according to Butler, presented himself to Anglo's Bobby Godsell as 'a person of reason and moderation who recognised that worker organisation should benefit both bosses and union'. An interesting aspect of his account is FOSATU's unsuccessful appeal to SACTU to block this initiative. See Anthony Butler, *Cyril Ramaphosa*, Jacana Media, 2007, pp. 136–7.
71 Jeremy Baskin in his history of COSATU confirms there was such a meeting. It appears he was there himself, although it is not clear in what capacity. See Baskin, *Striking Back*, pp. 37–8.
72 FCWU minute, trade unions meeting, 3 and 4 July 1982. The other 'principle' was that the policy of a new federation should be binding on its affiliates. This would of course be a form of 'democratic centralism'. Ironically, this might be said to be an approach the Party and FOSATU had in common.
73 There is a cryptic note in Baskin's book that suggests that SACTU and the 'community unions' were in alliance at this point of time, but that SACTU changed its position in 1984. See Baskin, *Striking Back*, p. 39.
74 Baskin states that one of the objectives of the meeting was to counter the 'opportunistic manoeuvres' of FOSATU.
75 Given how close Neil Aggett was to individual members of seven, it is tempting to speculate how he would have responded to this intervention. No one can say, of course, how someone might have responded had they lived, and least of all in Neil's case, when his death was in some respects a catalyst for the process that was now unfolding. What we can say with certainty is that he was always scrupulously loyal to what the Union had decided, which was that there had to be an open and inclusive process.

76 FCWU minute, trade unions meeting, 3 and 4 July 1982.
77 FCWU minute, trade unions meeting, 3 and 4 July 1982.
78 CUSA wanted to oust me as chair the next day on account of my remarks, but did not persist. I afterwards speculated that Norushe might have played a positive role had he been present, but he had been detained in Port Elizabeth only two weeks before, while visiting the branch, and was released a matter of days after the meeting. The curious thing was that he was not even questioned while inside. See minutes, MCM, 27 June and 25 July 1982. Norushe later wrote to head office on behalf of the branch to state that 'we must not allow the black racists to delay us in this venture. We assert that it is not the colour of a man that's going to determine the pace of our struggle but the commitment and action by all progressive minded people'. See minutes, MCM, 30 January 1983.
79 The Union tried to make a different point. Although 'worker control' was a principle everyone apart from the seven seemed to have in common, there was no agreement about what it meant.
80 The GWU position was stated at the end of meeting. FCWU minute, trade unions meeting, 3 and 4 July 1982.
81 See minutes, ANC, 25 and 26 September 1982. The conference also decided that a new federation should be 'open to all unions representative of workers, irrespective of membership and policies, and that the policies of the new federation be decided democratically by the unions that form it'.
82 Minutes, ANC, 25 and 26 September 1982.
83 My father presided in the marathon trial of 24 PAC members in Victoria West. He made highly critical comments about a Security Police informant in acquitting the accused and was not appointed again to preside over a political trial. See Muriel Horrell, *Survey of Race Relations*, SAIRR, 1969, p. 65.
84 The finding was handed down on 20 and 21 December 1982. See undated circular letter, circa December 1982/January 1983.

Chapter 20: Congella Road
1 'The enemy is not stupid,' was my response to Leila. 'It knows only too well the divisions between us.'
2 Liz Abrahams evidently played a key role in this regard. According to Patrick Ricketts, who spoke at a memorial service for Liz in 2009, there were more MK recruits from Paarl than anywhere else in the Western Cape. Ricketts himself joined MK, and subsequently became a colonel in the SANDF. Although he did not, understandably, refer to their race, it seems most of these recruits were in fact coloured.
3 The Union did in fact hold a reasonably successful meeting in Cape Town, but the Johannesburg meeting nearly ended in disaster when it transpired that the local leadership of AFCWU were unable to explain why FCWU was registered. See minutes, meeting of trade unions and community organisations called by FCWU/AFCWU, 20 October 1982; minutes, MCM, 24 October 1982 and minutes, meeting called by AFCWU, 26 January 1983. The tense climate in East London and Port Elizabeth put paid to the possibility of engaging with community organisations there.
4 Minutes, MCM, 19 December 1982.
5 After protracted negotiations Langeberg took the dismissed workers back, in dribs and drabs, but not the committee, who management claimed had

instigated the strike. A dispute in this regard was referred to mediation. Some of the committee were eventually re-employed.

6 Minutes, MCM, 31 January 1982. In the minutes I ascribe the problems of organising in the Durban area to 'the activities of other unions which have made a mess of organising the workers. As a result the workers have no confidence in unions.'
7 Some among them, such as Union Flour Mills, belonging to the Premier group, remained unorganised for the duration of my tenure as general secretary of FCWU and of FAWU.
8 Minutes, MCM, 30 May 1982.
9 Minutes, MCM, 24 October 1982.
10 Minutes ANC, 25 and 26 September 1982.
11 In his memoir, *Maverick Insider*, Johnny Copelyn makes no mention of the fact that Virginia cut her organisational teeth in another union. The Union supported the NUTW initiative because, as indicated, we regarded TUCSA-affiliated unions that represented textile workers and clothing workers in the Western Cape as being in bed with the bosses.
12 1983 Annual Report.
13 'We reject the government's constitutional proposals. We reject the Labour Party's acceptance of these proposals … We resolve to resist any attempt to mislead people into accepting these proposals. We cannot fight for unity on the factory floor and allow a constitutional dispensation which discriminates between people of different races and excludes the majority.' Minutes, MCM, 30 January 1983.
14 Letter, 20 December 1982 from GWU and minutes, MCM, 30 January 1983.
15 Minutes, NEC, 26 and 27 March 1983, and writer's notes of discussion on trade union unity at that meeting.
16 FCWU minutes of trade union unity meeting, 9 and 10 April 1983.
17 Government had reason to be worried about a united response to its constitutional reforms from the trade unions. Shortly before the second report of the President's Council was published, F.W. de Klerk, a conservative voice in the cabinet of PW Botha, told businessmen that labour relations had been identified as the 'Achilles' heel' of South Africa by its enemies. They should ask themselves whether they had taken the necessary steps to thwart the strategy of making labour a political battleground, he was reported as telling the businessmen. See 'SA's real enemies active in labour, says De Klerk', *Rand Daily Mail*, 4 November 1982.
18 Other community unions argued for reviving regional solidarity committees, as proposed at Langa. See FCWU minutes of trade union unity meeting, 9 and 10 April 1983.
19 Others supported me in asking him to withdraw the statement, including Rev. Marawu, a GWU organiser and Congress stalwart of note. FCWU minutes of trade union unity meeting, 9 and 10 April 1983.
20 This was perhaps clearest in the case of the GWU. Stevedoring represented too narrow a base to sustain the union, but it could not expand into the transport sector without risking a clash with other unions that were organised there, notably FOSATU's T&GWU.
21 FCWU minutes, minutes of trade union unity meeting, 9 and 10 April 1983.
22 The only one of the community unions that stayed out of the feasibility

committee was Orange Vaal GWU, which circulated a document before the Athlone meeting setting out their position. The tone was altogether different from that adopted in Port Elizabeth. 'If unity was a problem for trade unions under yesterday's boom conditions its absence under today's crisis conditions is fast becoming a life or death question for the labour movement ... as unions lose more and more members through retrenchments and victimisations and, for those still lucky enough to have a job, wages mean less and less as prices continue to go up.' See OVGWU proposals, document dated February 1983.

23 See chapter 23. It appears that at least two affiliates of FOSATU, Sweet Food and NAAWU, had full-time shop stewards at this stage. According to Mordecai Mabasa, who was employed at Nestlé in Isando, Chris Dlamini was already a full-time shop steward at Kellogg's when he joined Sweet Food in 1981. It appears from Copelyn's memoir that NAAWU had recognition agreements providing for full-time shop stewards and union offices inside certain motor assembly plants at the time FOSATU was formed. He appears to regard this as a mark of their organisational sophistication. See Copelyn, *Maverick Insider*, pp. 164–5.

24 FCWU minutes, feasibility committee meeting of 1 and 2 July 1983.

25 'Within South Africa a tradition was followed of narrowly defined industrial unions', FOSATU argued in a discussion document. 'FOSATU did not want to follow this tradition because it was clearly doomed to failure in the larger, more industrialised South African economy of the 1970s.' Undated FOSATU discussion document, circa 1983.

26 This was despite the policy of the Barlow Rand group being one that was supposed to respect the autonomy of plant management. Initially the issue was about grades of work. See minutes, MCM, 24 October 1982.

27 An indication of this was that NUTW was only now trying to establish itself in the Western Cape, even though this was where most textiles and clothing factories were located.

28 This emerged after the formation of COSATU. I do not know how widespread this practice was.

29 CUSA joined the feasibility committee after the Athlone meeting.

30 FCWU minutes, feasibility committee meeting of 1 and 2 July 1983. There were also semi-official minutes kept by David Lewis (they were never approved).

31 David's comment on this issue of foreign funding was, 'We have received donations from outside the country, to our great embarrassment, and know that almost all the unions present have done so. At the same time it has created problems.' See FCWU minutes, feasibility committee meeting of 1 and 2 July 1983.

32 The only other union that did not rely on donor funding was CTMWA, which like us, was only financed by members' subscriptions, but was still organised in only one municipality where it had a closed shop. It could not therefore be a strong ally.

33 Sisa Njikelana of SAAWU proposed a notion of 'organised members' instead of paid-up membership because of the practical difficulties of collecting subs by hand. See FCWU minutes, feasibility committee meeting of 1 and 2 July 1983. However, all the Union's branches outside the Western Cape relied on hand collections. Even within the Western Cape, hand collections accounted for about half the Union's income.

34 Letter, general secretary of FOSATU, 19 October 1983.
35 The first reference to these meetings is in the MCM minutes of 27 February 1983. The reference to main-line organisations is in the MCM minutes of 24 April 1983.
36 This explained why there was no organisational memory of the Union in a place like Durban. I also never heard a positive recollection of SACTU from anyone who was involved in the Union at that time, or about any campaign in which it was involved. This suggested to me that the involvement of the Union had been at a leadership level. See Friedman, *Building Tomorrow Today*, pp. 31–3; Baskin, *Striking Back*, pp. 15–16; and R Lambert, 'Political unionism in South Africa: The South African Congress of Trade Unions', DPhil thesis, University of the Witwatersrand, 1988. According to Lambert, 'virtually the entire leadership of SACTU were recruited into Umkhonto we Sizwe. They were trained to work "simultaneously" in the unions and the new military wing of the Alliance' (p. 452). He regards the fact that in doing so 'they unintentionally destroyed the organisation that was becoming effective' as an 'unintended consequence' (pp. 462–3).
37 Even in the 'golden fifties' there had been a question as to what risks union leadership should take in engaging in political struggles. Ray Alexander records in her autobiography, for example, that both Oscar and she had wanted to 'defy' as part of the Defiance Campaign launched by the ANC in 1952, and that after a debate lasting the whole day, the Union had refused permission for them to participate. See Simons, *All My Life and All My Strength*, p. 229.
38 These events are not dealt with in Seekings's history of the UDF. See J Seekings, *The UDF: A History of the United Democratic Front in SA, 1983–1991*, David Philip, 2000, pp. 49–66.
39 Minutes, MCM, 24 April 1983.
40 Although there were initiatives in other parts of the country to establish a united front, it does not seem there had been any approach to FOSATU or any of the other unions involved in the unity talks anywhere else, apart from SAAWU and GAWU. See Seekings, *The UDF*, pp. 61–4. Seekings caricatures the reasons certain trade unions did not affiliate to the UDF at p. 63, where he states that 'the unionists were committed to socialist goals ... but should preserve themselves as unblemished vessels of the socialist project'.
41 Minutes, MCM, 29 May 1983.
42 For a contrary view about the state of SAAWU, see Andrew Roux, 'SAAWU consolidates', *SA Labour Bulletin*, 10, 2 (1984).
43 Even in the proposed trade union federation, the establishment of area committees was a major issue, because of fears local structures would be subverted or hijacked. FOSATU was a protagonist of area committees. FCWU and others were opposed, although we later modified our position. See minutes, feasibility committee meeting of 1 and 2 July 1983, and FCWU minutes, feasibility committee meeting of 1 and 2 July 1983.
44 FCWU minutes, meeting of organisations against new laws, 5 June 1983.
45 FCWU minutes, meeting of organisations against new laws, 5 June 1983.
46 The UDF was dissolved in March 1991. See Seekings, *The UDF*, pp. 280–3.
47 We did in fact have a cordial and constructive meeting, and agreed to support the UDF's call for a boycott of the elections for the community councils that government was setting up. The meeting took place on 9 October 1983 in Benoni. See minutes, MCM, 30 October 1983.

48 The Epol plants were at Isando, Roodepoort, Springs, Pretoria West, Cape Town and East London. See my minutes of national negotiations with Epol, 1983.
49 Minutes, MCM, 29 January 1984.
50 There was evidence to this effect at the Truth and Reconciliation Commission.
51 Minutes, MCM, 26 February 1984.

Chapter 21: The road to Pretoria West

1 The workers of Mdantsane took the trains to get to work, and activists used the trains to mobilise against the Ciskei government. In fact, one of the reasons workers took the trains was that the Ciskei government had taken over the company that operated the bus service and profited from it. When bus fares were increased, the Ciskei government used their army and police to try to quell a boycott. This nicely illustrated the class character of the government's constitutional dispensation. See 1983 Annual Report.
2 This was in terms of the Black Local Authorities Act, 102 of 1982.
3 It became obvious that elections were imminent when the Labour Party decided against holding a referendum of coloured voters. They no doubt feared that they would lose a referendum. See minutes, MCM, 29 January 1984.
4 Minutes, MCM, 26 February 1984.
5 The minutes of April 1984 record a visit to head office at which we were informed the UDF was 'now not able' to hold the meetings, and our dissatisfaction about this. The matter was discussed again in May. See minutes, MCM, 29 April and 27 May 1984.
6 Noko was so popular with the coloured members that they wanted to vote for him in the election of a president for AFCWU at Wilgespruit. Paarl AFCWU objected, and it was necessary to have a vote as to whether they should be allowed to vote, before they ultimately did do so. See minutes, ANC, 20 and 21 August 1983.
7 This belief was of course unfounded. AFCWU and GWU were not registered, and were party to the decision that SAAWU and GAWU complained about. This is discussed more fully below.
8 Labour Relations Amendment Act, 2 of 1983. The other important amendment adopted at the same time provided that a 'labour broker' would be the employer of workers it 'procured or provided to a client'. See also MSM Brassey and H Cheadle, 'Labour Relations Amendment Act 2 of 1983', *Industrial Law Journal*, 4, 1 (1983), pp. 34–7.
9 The first reported case involving SAAWU was a dispute in terms of section 43 that was referred at the end of 1984. See SAAWU *v* Faiarte Ceramics (Pty) Ltd, *Industrial Law Journal*, 6 (1985), 262.
10 I see this trend, variously described as juridification or judicialisation, as integral to the neo-liberal project.
11 In the case of shift millers at SA Milling (who were in effect artisans), a dispute in this regard went to the highest level in the Premier group. See minutes, MCM, 30 October 1983. In the case of Robertsons, which belonged to an American multinational, we were told that the fact that the head of their laboratory had become a member represented a red line for their parent company. We decided, for pragmatic reasons, to employ Inthiran Moodley (the worker concerned) as an official instead.
12 Minutes, MCM, 29 May 1983.

13 The factory belonged to the Fedfood group. We were already recognised at mills and bakeries belonging to this group, and at Marine Products, Oscar's old factory, so it could hardly refuse to negotiate with us. See minutes, MCM, 25 September 1983.
14 Most of the workers were earning R27 a month. Wages were increased to R20 per week.
15 Minutes, MCM, 29 January 1984.
16 Gumede and Others *v* Richdens (Pty) Ltd t/a Richdens Foodliner (1984) 5 *Industrial Law Journal* 84.
17 I would argue that the court's focus on procedural fairness at the time was calculated to deflect a consideration of the real reason for retrenchments, where it called into question the unfettered form of capitalism increasingly propagated by the neo-liberal project. Nowadays, however, the Labour Appeal Court has no scruples in affirming the right of an employer to retrench workers in order to maximise its profits. See chapter 25.
18 This was because in food manufacturing, as in a number of industries, trade unions had not succeeded in uniting workers across state boundaries. Australia was also a country in which the turn to the law had been radical. Collective bargaining at that time took the form of an arbitration process.
19 This is to reiterate the argument made in chapter 11, namely that in hard cases an organisation has to elect whether to mobilise its own resources or to rely on the law. It is hard to see how you can have the best of both worlds in these circumstances. See also J Theron, 'Trade unions and the law: Victimisation and self-help remedies', *Law Democracy and Development*, University of the Western Cape, 1 (1997).
20 This was in terms of section 43 of the LRA of 1956. The amendments came into effect on 1 September 1982. See Brassey and Cheadle, 'Labour Relations Amendment Act 51 of 1982', pp. 31–3.
21 See Shezi and Others *v* Consolidated Frame Cotton Corporation, *Industrial Law Journal*, 5 (1984), pp 1–15, a case involving NUTW. In arriving at this interpretation, the court expressly rejected the approach taken in two UK cases which the employer had relied on. One states that 'a man is not reinstated in his employment when he is just put back on the payroll'. Another states as follows: 'The natural and primary meaning of "to reinstate" … is to replace him in the position from which he was dismissed, and so restore the status quo ante the dismissal.' See Jackson *v* Fisher's Foils [1944] 1 All ER 421 and William Dixon *v* Patterson [1943] SC (J) 78.
22 In contrast to this analysis, Friedman does not have any critique of the role of the Industrial Court and lawyers acting for the unions. For example, he states, 'the biggest victory came in July, 1981 when the Supreme Court ordered an East Rand employer, Stag Packings, to reinstate seven fired NUTW members' (Friedman, *Building Tomorrow Today*, p. 318). In fact, the workers were dismissed in July 1981. The significant Supreme Court ruling was in June 1982. It upheld an appeal against an *in limine* point to the effect that it was not possible for a court to order reinstatement in a case of victimisation. Thereafter the matter would have gone back to court, unless it was settled. See National Union of Textile Workers and Others *v* Stag Packings Pty Ltd and Another 1982(4) SA 151 and Friedman, *Building Tomorrow Today*, pp. 314–26. As indicated, reinstatement by the Industrial Court was not reinstatement in the job.

23 At the stage that FCWU merged with SFAWU and other unions, the bills of lawyers acting for SFAWU were being submitted directly to the funders for payment. Payment was, as I understand it, from a fund controlled by the ICFTU, so there is no reason to suppose that the procedure would have been any different in the case of other FOSATU affiliates or other unions.

24 I have no record, and can no longer remember, what precisely the two workers expelled were alleged to have done that provoked the workers. However, the alternative course of action open to the workers would have been to demand that the two workers concerned be disciplined, as the Fatti's & Moni's workers did in Skilpad's case. Some might argue the Union could also have demanded a closed shop, in which all workers were required to become union members. Personally, I was opposed to the closed shop, as already noted.

25 Minutes, MCM, 29 July 1984.

26 Blue Continent employed less than 100 workers, and it did not fit squarely into any of the industries in which we were organised. It also did not fit squarely into the industries covered by Barlow Rand's food manufacturing subsidiaries, Tiger and Imperial, which is presumably why it was owned by Barlow Rand itself. The workers at Blue Continent wanted wage negotiations, but management were holding out for the negotiation of a recognition agreement. See minutes, MCM, 29 January 1984.

27 Four workers were reinstated. The fifth had in the meantime been employed at Sasko's Pretoria mill, which the Union had also organised. See minutes, MCM, 27 January 1985.

28 Minutes, MCM, 26 February 1984. Ngoma had at this point already been removed as chairperson, and appealed to head office. I declined to intervene, on the basis that the branch was entitled to remove its chairperson. The correspondence between Ngoma and me is in the FCWU archive.

29 Minutes, negotiations with KSM Queenstown, 10 February 1984. There was also a debate in the *SA Labour Bulletin* on the strike between R de Villiers and Ian Macun. See I Macun, 'Correspondence: A/FCWU in Queenstown', *SA Labour Bulletin*, 10, 3 (1984).

30 One of the protagonists of situational ethics was a prominent Anglican bishop in the UK, John Robinson. See www.wikipedia.org/wiki/Situational_ethics.

31 Circular letter, 16 January 1985 and minutes, MCM, 27 January 1985.

32 The Union members concerned claimed to have known nothing about the Labour Party's involvement in the meeting. What infuriated them further was that the Action Committee had itself done nothing to address the housing problem. Minutes, MCM, 29 April 1984.

33 Minutes, MCM, 30 September 1984.

34 Minutes, MCM, 9 December 1984.

35 In contrast to Epol's Cape Town operation, Meadow had a state-of-the-art mill outside Paarl that employed very few workers, apart from the drivers.

36 Undated FCWU training document headed 'Epol Cape Town', circa 1985. Epol's Cape Town plant had played a central role in organising other mills and bakeries in Cape Town.

37 See Brassey and Cheadle, 'Labour Relations Amendment Act 2 of 1983', and chapter 25.

38 The Advice Bureau that the General Workers' Union retained as a separate entity might have served as a model for the growing number of workers who could not

easily be accommodated in an industrial trade union. However, in the post-1994 period the Advice Bureau was converted into a trade union, and not very long after that it collapsed altogether.
39 This attack can be regarded as commencing in August 1981 in the US, when President Reagan crushed a strike by air-traffic controllers by dismissing over 11,000 workers, and banning them for life from being employed by the federal government. At about the same time a document disclosing plans for massive closures of mines in the UK came to light. Tensions about these plans, and the decision to close three pits, triggered the 'Great Strike' of 1984–5, which ended after 16 months with the employers and Thatcher victorious. It does not seem coincidental that these two initiatives were launched at the same time. As Andrew Glyn has persuasively argued, there was a real fear that organised labour threatened the stability of the capitalist system in the 1970s. See A Glyn, *Capitalism Unleashed: Finance, Globalization and Welfare*, Oxford University Press, 2006, pp. 1–19.
40 We held meetings with GWU, CCAWUSA, FOSATU and also with CUSA.
41 For accounts of the other meetings, see minutes, MCM, 26 February 1984.
42 FCWU minutes, meeting of 4 February 1984.
43 'It may be asked who can judge who is ready and who is not,' I said. 'To us, a union which has not decided to form a federation is not ready. If a federation is not prepared to disband and to organise in industrial unions, it is not ready. A general union that is not prepared to restrict its area of organising, and which has not taken steps in the last year to do so, is also not ready.' Minutes, third feasibility committee meeting, 3 and 4 March 1984.
44 'Our hope is that the federation of unions will soon be established ... We also hope it will drive toward the organisation of all workers, and set a standard of organisation that others will follow. For workers in food factories, we hope it will mean the establishment of one union for all the workers throughout South Africa.' See FCWU Annual Report, 1984.
45 At a national level, this would be a congress (an alternative title for conference). The position on worker leadership was of course an explicit rejection of the SAAWU approach in which a paid official could be (and was) president of the trade union.
46 Minutes, ANC, 24–26 August 1984.
47 I was part of the subcommittee set up by the feasibility committee to look at this question. FOSATU, CCAWUSA, GWU and CUSA all admitted to receiving money from ICFTU at this meeting. See minutes, fourth feasibility committee meeting, 28 and 29 March 1984.
48 Memorandum on the finances of a new trade union federation for the feasibility committee, undated FCWU document, circa April 1984.

Chapter 22: Hospital way
1 The manufacturing sector in the UK is reported to have shrunk by two-thirds. See *The Guardian*, 16 November 2011: 'Why doesn't Britain make things anymore', by Aditya Chakrabortty.
2 Conservatively counted, FCWU membership as at 31 December 1984 was 11,065 and AFCWU membership was 12,187.
3 Oscar arrived with John Pendlani, who was billed as guest speaker, but ceded this role to Oscar. See minutes, ANC, 24–26 August 1984. During the debate

about the UDF, Oscar asked a question of clarification on the Union's stance from the floor and appeared completely satisfied with my reply.
4 See minutes, MCM, 26 April 1981. She had replaced the charismatic K, who was the last person from Johnny M's time still in a leadership position in the Union. The fact that K was dishonest, like so many of her predecessors, had taken long to emerge because she seemed such a straight arrow. After K was dismissed, the Vakbond made common cause with her, in an attempt to initiate a breakaway in Ceres. See minutes, MCM, 17 April 1982.
5 All AFCWU monies were hand-collected, as a result of the bar on stop orders for unregistered trade unions. See minutes, ANC, 24–26 August 1984.
6 Thus, workers earning less than R50 a week would pay 40 cents, those earning R50 or more but less than R70 would pay 50 cents, and those earning above R70 would pay 60 cents. Minutes, ANC, 24–26 August 1984.
7 In milling and baking, for example. See 1984 Annual Report.
8 My salary at the time was R500 per month, and was increased to R550. See minutes, MCM, 30 September 1984.
9 The Labour Monitoring Group, 'The November 1984 stayaway', *SA Labour Bulletin*, 9, 6 (1985), pp. 74–100.
10 See Friedman, *Building Tomorrow Today*, pp. 446–5; Baskin, *Striking Back*, pp. 43–5.
11 Our plan to establish a branch in Vereeniging had been set back following the closure of the Epol plant there. This resulted in the loss of 250 jobs.
12 However, some of the organisations that participated would have been UDF affiliates. See Seekings, *The UDF*, pp. 124–9.
13 These included Chris Dlamini of FOSATU, Moss Mayekiso of MAWU, and Piroshaw Camay of CUSA. The Union, along with others, formally declared a dispute with SASOL as part of a strategy to force it to reinstate the dismissed workers. See minutes, MCM, 9 December 1984.
14 It seems to have been the first occasion that the three major organisations representing employers in industry, including the Afrikaanse Handelsinstituut, had jointly and publicly taken a stand on a political issue. See Friedman, *Building Tomorrow Today*, pp. 450–1.
15 Minutes, NEC meeting, 30–31 March 1985, held at St Philomena's Home, Sydenham.
16 Minutes, NEC, 30 and 31 March 1985.
17 Minutes, MCM, 28 April 1985.
18 I used the two-day stay-away on the Rand as an example of how unions could engage in politics in the 1985 Annual Report. 'What we can see more clearly in the recession is that we cannot fight for higher wages for workers in a factory or industry while doing nothing about the situation of the working class as a whole ... But riots are not going to pressure the government to make changes that will benefit the working class. This will only happen when the working class can put pressure on the government in an organised way.'
19 This was an innovation credited to Matthew Goniwe, in Cradock. See Seekings, *The UDF*, pp. 143–4.
20 Pendlani, of course, no longer had any position in the Union, but was a respected 'elder statesman'.
21 Circular letter dated 12 December 1984 from John Nkadimeng, general secretary, SACTU. For a contrary view of this letter, see Baskin, who describes the letter as

'carefully worded'. Baskin, *Striking Back*, p. 45.
22 Letter, J Foster, 25 April 1985 and reply from general secretary, FCWU, dated 25 May 1985. FOSATU said that the SACTU letter was one of the reasons for its proposal at the meeting of the feasibility committee that took place on 7 June 1975. The other reason was that CUSA had requested a postponement of the Ipelegeng meeting, although this would seem irrelevant.
23 'A blunder of the first order' was how I described FOSATU's action in my report to the Union's conference. See 1985 Annual Report.
24 See FCWU minute of meeting of 7 June 1985, which took place in CUSA's offices. Baskin's account of this critical turning point in the unity talks is incorrect. He states that there had been a decision within FOSATU in early 1985 to 're-open' the talks, and implies that other organisations in the feasibility committee were approached about this shortly thereafter, and assented to this approach. See Baskin, *Striking Back*, pp. 45–6. However, no such decision was formally communicated to other organisations in the feasibility committee, and none of the other unions assented to it, as was clear from the dismay they expressed at the meeting.
25 Baskin records that Dlamini said, in an interview, that he had joined the SACP in the 1980s, but that he would not reveal when. See Baskin, *Striking Back*, p. 62, and Devan Pillay, 'The Communist Party and the trade unions', *SA Labour Bulletin*, 15, 3 (1990).
26 As well as Dlamini, John Gomomo of MAWU became a member of the Party, although it is not known when. See Pillay, 'The Communist Party and the trade unions'.
27 Minutes, inter-union meeting, Soweto, 8 and 9 June 1985. The official minutes were kept by David Lewis, but only the minutes of the first day were circulated. The FCWU minutes are of both days.
28 I do not agree with Baskin's account of the meeting, and specifically his suggestion that the surprise of the meeting was the position of CUSA affiliates. It was surely a foreseeable consequence of FOSATU inviting AZACTU unions in the manner it did that CUSA and AZACTU might make common cause. See Baskin, *Striking Back*, pp. 47–8.
29 The meeting at CCAWUSA's offices took place on 21 June 1985. The unions represented were FCWU, CTMWA, NUM, GWU, CCAWUSA, NUTW, NAAWU, MAWU, PWAWU and SFAWU.
30 NACTU was formed in 1986. See Baskin, *Striking Back*, pp. 157–9.
31 Maxwell Xulu, who regarded himself as a socialist, was branded as a spy in 1991, based on reports from ANC intelligence, and ousted from the leadership. Evidently Xulu himself never admitted to the charge and there are reasons to doubt its veracity. See Forrest, *Metal That Will Not Bend*, pp. 459–62.
32 The president of COSATU was Elijah Barayi, from NUM. According to Jay Naidoo, the speech was written by Marcel Golding. Naidoo distances himself from its contents in his autobiography. Jay Naidoo, *Fighting for Justice: A Lifetime of Political and Social Activism*, Pan Macmillan, 2010, pp. 100–1.
33 Inkatha and the Inkatha Women's Movement also came out in support of the Fatti's & Moni's strike.
34 Another example of this zigzag was Jay Naidoo's going public about his meeting with the ANC before any affiliates knew about it. For Naidoo's account of the meeting, see Naidoo, *Fighting for Justice*, p. 104. Baskin claims that the meeting

was a 'chance encounter'. See Baskin, *Striking Back*, pp. 73–5.
35 In Ramaphosa's speech to the inaugural congress, he was quoted as saying that 'a giant' had arisen.
36 Although I had no part in the process of approving the credentials of unions, I was subsequently to learn that the membership figure of some unions had been grossly inflated. I had hoped the establishment of a new federation would bring an end to this tendency.
37 See minutes, NEC, 1 and 2 February 1986.
38 Jay Naidoo has described this trip as being 'to build COSATU's financial sustainability'. See Naidoo, *Fighting for Justice*, pp. 107–9.
39 See circular letter, 19 November 1985; minutes, Head Office Controlling Committee, 23 November 1985; and circular letter, 6 January 1986. The branch official was John James. The other person associated with the Union who was detained together with Lizzie was Maggie Wilson, who was employed by the medical benefit fund. Lizzie's own account is in Shirley Gunn and Rachel Visser (eds.), *Labour Pains for the Nation*, Human Rights Media Centre, 2007, pp. 66–70.

Chapter 23: Albert Road
1 In theory the MCM only governed the Union between NEC meetings, but because there were only two NEC meetings a year, one of which coincided with the annual national conference, it was in practice highly influential. The structure that eventually replaced the MCM was called the Head Office Controlling Committee.
2 The reasons for the constitutional changes are detailed in the 1985 Annual Report. As pointed out in the report, the introduction of a regional structure had been under discussion for a few years.
3 The SFAWU AGM was on 9 February 1985. See minutes, FCWU NEC, 30 and 31 March 1985.
4 See chapter 25.
5 Minutes, NEC, 25–26 October 1985
6 First we catapulted a worker from a fish factory on the West Coast into the position of assistant general secretary. When that did not work out, we employed Eddison Stephen, a worker from Sasko's Rondebosch mill.
7 Lizzie Phike became a national organiser for the fruit and vegetable sector when Liz Abrahams became branch secretary of Paarl branch after the merger between FCWU and AFCWU. The national organiser for the fish-processing sector was Appolis Madikiza, who was a worker in Hout Bay before becoming an official of Cape Town branch. Mandla Gxanyana was responsible for milling and baking.
8 Among others, Debbie Budlender was employed as a researcher, primarily to assist the national organisers, and was later drawn increasingly into the position of head office administrator. Denis Rubel, the medical doctor who spoke at Neil Aggett's funeral, was employed by the medical benefit fund in Paarl, where it had established a clinic. On moving to Johannesburg he was recruited by Kempton Park to work for it. The medical doctors that succeeded Denis at the Paarl clinic included Liz Thomson and Leslie London.
9 African Products in Meyerton, in the Vaal, was one of very few food-manufacturing plants that employed white workers in numbers, probably because of the 'high-tech' nature of the process (converting maize to starch and

sweeteners). During tea break on a visit to the plant, I was reminded of the kind of white men our members had to deal with daily. A group of us, officials and workers, were standing outside the management offices in the weak winter sun, chatting animatedly, when a phalanx of white men in blue overalls walked past, like an icy blast, determinedly refusing to acknowledge our existence.

10 The PW Botha government was supporting UNITA, with the backing of the US. When an offensive by the MPLA, supported by Cuba and the Soviet Union, looked like it might succeed, PW Botha sent in more troops. This, incidentally, allowed the white Left to assume a new role, in opposing conscription.

11 The CTMWA owned the building where it was based. The only other COSATU affiliate that I know of that might have owned a building at this point was NAAWU.

12 COSATU leadership pleaded for affiliates to move in, to ease the burden of paying the rent. I did not think this was a good idea, as it would make the building a target, as it in fact became.

13 The decision was taken at COSATU's central executive committee. See FAWU, 1986 Annual Report. A strong expression of the disquiet in FCWU ranks at COSATU's lack of regard for the importance of financial self-sufficiency is found in the minutes of the NEC in February 1986. See minutes, FCWU NEC, 1–2 February 1986, Wilgespruit.

14 Minutes, joint meeting of the NECs of SFAWU and FCWU, 2 and 3 February 1986.

15 This is a problem for all membership-based organisations. The usual safeguard is a quorum. However, it is very difficult to determine a quorum that is both an adequate safeguard and not so onerous that general meetings become impractical.

16 See 1985 FCWU Annual Report. The strike took place at about the same time at which the bakery employers were dismantling the regional Industrial Council for bakeries in Cape Town. Not long after the Durban strike, they dissolved the Durban Industrial Council as well.

17 SA Breweries coined the term 'temporary sole supplier' to counter the accusation that it was a monopoly.

18 FCWU's attempts to provide education remained sporadic. However, I cannot say that when FAWU later appointed its own education secretary it made a perceptible difference to the provision of education.

19 The modern-day equivalent of Standard Three is Grade Five. The reason she could not go further was that the nearest school to which she had access only went up to this grade, a common scenario on the *platteland*.

20 This is what the General Workers' Union and T&GWU did in the transport sector. These unions made no attempt to incorporate the members of the unions that organised generally, and it was years later that a merger with the SA Railway and Harbours Union (SARWHU) took place. SARWHU was one of the affiliates of SACTU that had not survived the years of repression. It was 'relaunched' in mysterious circumstances, and was somehow represented at the inaugural congress of COSATU. See Baskin, *Striking Back*, pp. 47, 147.

21 The two factories organised by RAWU were DairyBelle and Dairymaid. Both belonged to the Imperial group, and the combined membership was about 800 workers. There is a list of organised factories in the 1986 Annual Report, after the merger to form FAWU. However, it is not reliable, as even at the stage at

which that report was compiled there had been no process to verify the paid-up membership at the branches. See 1986 Annual Report.

22 The inaugural conference was held on 31 May and 1 June 1986. Amos Masondo, who was later to become mayor of Johannesburg, was the chairperson. He had some position in GAWU but was regarded as neutral, because he was not seeking office in the merged union.

23 The decision was taken at the central executive committee of 7–9 February 1986, and the meeting took place in Lusaka on 5–6 March 1986. See Baskin, *Striking Back*, p. 94.

24 TUCSA was dissolved at a special conference held in December 1986. See Baskin, *Striking Back*, p. 159.

25 See Baskin, *Striking Back*, pp. 304–7. Baskin argues that COSATU learnt to be more pragmatic towards historically conservative trade unions following talks regarding the admission of the Garment and Allied Workers' Union in 1988, although it is not clear on his account whether GAWU actually became a member in its own right prior to or upon merging with ACTWUSA the following year. Another significant ex-TUCSA affiliate that became part of COSATU as a result of a merger was MICWU (the Motor Industry Combined Workers' Union).

26 The phrase 'manager of discontent' to describe the role of a trade unionist was coined by the American sociologist C. Wright Mills.

27 I refer here to the constitutions of both the Union and COSATU. In terms of the Union's constitution, as already explained, an office-bearer such as the president had to be a member, i.e. a worker. In terms of COSATU's constitution, an office-bearer had not only to be a worker but also an elected leader in the workplace where he or she worked.

28 See minutes, FAWU NEC, 16 and 17 August 1986. As regards the issue of full-time shop stewards, the NEC agreed to refer the matter to branches and regions for further discussion, but in fact nothing came of this resolution, for reasons that will become apparent.

29 Interview, former member of Sweet Food and FAWU, 14 July 2016.

30 So far as I am aware, my 1990 article in *SA Labour Bulletin* contains the first and only suggestion that the introduction of full-time shop stewards was problematic prior to the Marikana massacre. See Theron, 'Workers' control and democracy: The case of FAWU' *SA Labour Bulletin*, 15,3 (1990), 39–64.

31 According to Baskin (*Striking Back*, p. 60), Barayi was a 'personnel assistant' at Blyvooruitzicht mine. This was a subsidiary of Anglo American.

32 See minutes, NEC, FAWU, August 1986. Clover was the tradename for an agricultural co-operative, National Co-operative Dairies Ltd (NCD). It subsequently converted from being a co-operative to a company. It is now controlled by Hosken Consolidated Investments Ltd, which describes itself as a 'black empowerment investment company', with a board dominated by former trade unionists from SACTWU. See www.hci.co.za.

33 The United Workers' Union of SA (UWUSA) was formally launched in Durban on 1 May 1986. It is debatable whether it ever functioned as a trade union.

34 The sleep-in was a strategy to prevent the bosses bringing in scabs and dismissing workers, and a number of trade unions adopted this strategy during this period. My comments are directed at the strikes in which FAWU was involved, of which I have personal knowledge. There were three in the 1986 report. See FAWU Annual Report, 1986, and Baskin, *Striking Back*, pp. 83–5.

35 See P Benjamin and H Cheadle, 'Proposed amendments to the Labour Relations Act', *SA Labour Bulletin*, 13, 1 (1987). The proposals were published in the form of a draft Bill at the end of 1986, and in slightly revised form in September 1987.
36 Labour lawyers were arguing for laws that encouraged strikes that were 'functional to collective bargaining'.
37 The strike at BTR Sarmcol was in May 1985, and the Industrial Court case, which the union lost, was only the beginning of a legal war of attrition. It is not clear who won the war. Although the union eventually won the case, this was more than six years after the workers had been dismissed. I also do not know how many, if any, of the dismissed workers got their jobs back. It seems to me the eventual victory of the union had much to do with the fact that the country was in a political transition, and the desire of the courts to reposition themselves in the new order that was emerging. See BTR Industries SA Pty (Ltd) and Others *v* Metal and Allied Workers Union and another 1992 (3) SA 673 (A); 1992 (3) ILJ 803 (A). See also Abel, *Politics by Other Means*, pp. 125–72.
38 It was in fact only after the Union had begun to make some headway in organising other Clover plants that the bosses became more amenable to discussing a settlement there. The matter was still in abeyance when I left the Union.
39 The Union was later hit with a massive legal bill from this attorney's firm for work undertaken for Sweet Food prior to the merger. It included professional fees for tasks as mundane as applying for unemployment insurance for dismissed workers.
40 The membership at OK Bazaars was the core of CCAWUSA. OK Bazaars paid much lower wages than Pick n Pay, which was already dominant in the retail sector, but also sold its wares to the black working class. That made it an easy target for the union, but also meant that OK had tighter margins.
41 This was also a consequence of the forced pace at which CCAWUSA and two much smaller unions in the retail sector merged. The reasons for the split were portrayed as political at the time. For Baskin's account, see *Striking Back*, pp. 202–11.
42 This is notwithstanding attempts to put a positive gloss on it. See for example I Obery, S Singh, and D Niddrie, 'Disciplined mine strike a test of strength', *Work in Progress*, 49 (1987).
43 NUM had demanded a wage increase of 30 per cent. At the end of the strike, some 50,000 workers lost their jobs. Others, having been dismissed, were re-employed at the wages the bosses were offering when they went out. On the strike generally, see Baskin, *Striking Back*, pp. 224–39.
44 In 1989 NUM formalised the situation in which its president was also employed by it. See Baskin, *Striking Back*, p. 460. According to Baskin, the decision was hotly debated. Presumably it entailed NUM amending its constitution. Baskin does not say when Motlatsi became a full time employee. Motlatsi was 'elevated' to the board of Anglo's gold mining subsidiary in 1998, and at the time of writing was vice-chairperson of AngloGold.
45 At the first ordinary conference after the merger, I wrote in the 1986 Annual Report about members 'imitating their bosses by treating officials in the same oppressive way as their bosses treat them'. See 1986 FAWU Annual Report on 'the position of officials'. The first annual national conference of FAWU was held a matter of months after the inaugural conference, at my insistence, supposedly to consolidate the merger.

46 Minutes, ANC, FAWU, 10 and 11 October 1986.
47 Minutes, FAWU NECs of 24–26 April 1987 and 5–7 June 1987.
48 At the NEC meeting of 5–7 June 1987 there were little over 50 persons present, including officials. This was significantly less than routinely attended MCM meetings prior to the merger.
49 This 'gang of four' had been involved in worker organisation in the early 1970s and joined SACTU in exile. The reason for their expulsion was that they had questioned SACTU's subservience to the Party. The four were Rob Petersen, Paula Ensor, David Hemson and Martin Legassick. Rob Petersen was in the same law-class year as me at university and had been my friend.
50 Copelyn was still general secretary of NUTW at the time.
51 For a critical view of the negotiations over the amendments to the LRA, see Martin Jansen, 'Weakness of the Anti-LRA campaign', *SA Labour Bulletin*, 14, 5 (1989) and the response by Geoff Schreiner, 'On strengthening the Anti-LRA campaign: A response to Jansen', *SA Labour Bulletin*, 14, 7 (1990). One of the controversial amendments replaced the open-ended definition of an unfair labour practice with a code. In my book, an acceptable code would have simplified the law and reduced the scope for litigation. It would also have dovetailed with a system in which disputes were resolved by a tripartite tribunal rather than a court of law. In contrast to the hullabaloo about the code, there was hardly a murmur about the amendment establishing a Labour Appeal Court (LAC) presided over by a judge. The LAC would hear appeals against decisions of the Industrial Court.
52 See minutes, FAWU, Head Office Controlling Committee, 24–25 October 1986. The matter was discussed again in 1987. See minutes, FAWU HOCC, 26–28 March 1987 and 25–26 April 1987.
53 The process of negotiating a national recognition agreement was initiated by FCWU in early 1986 and concluded in early 1987, although I do not have the date on which it was concluded. See minutes, FCWU Head Office Controlling Committee, 8 March 1986.
54 An HR manager we dealt with defended promoting union leaders on the basis that the unions could be expected to identify the most capable leaders, and these were inevitably the best candidates for promotion. It was as though this was a service that unions performed for employers, in exchange for the privilege of recognition.
55 Possibly this form of empowerment would not have emerged if alternative initiatives by worker organisations to generate employment for their members had been more successful. NUM tried to do so by establishing co-operatives for those who had lost their jobs during the great strike. NUMSA did the same for dismissed Sarmcol workers. But it does not seem these were carefully considered initiatives, and the odds were stacked against them from the outset. See chapter 25 and Michelle Adato, 'Democratic process, mediated models and the reconstitution of meaning in democratic organisations: Trade union cooperatives in South Africa', DPhil dissertation, Cornell University, 1996.
56 See minutes, FAWU NEC, 24–26 April 1987.
57 The factionalism in FAWU was already a talking point in COSATU. 'There are camps in our union', is how I put it in the 1986 Annual Report.
58 FAWU Annual Report, 1987. What transpired at the 1987 conference is also discussed in the FAWU Annual Report of 1988.

59 Minutes, FAWU ANC, 18–20 September 1987. 118 delegates voted to adopt the report and 61 to refer it to the NEC.

Chapter 24: The Transkei road

1. Xashimba was an electrician's assistant at Spekenam, and was fluent in Afrikaans and isiXhosa.
2. The issue was a clause in the recognition agreement which entitled union officials to be present in wage negotiations. This was something which we expected the bosses to accept as a matter of course.
3. Vleissentraal was the dominant group in the red meat industry. The red meat industry was, of course, notorious for its hard-line stance against GWU, leading to the red meat strike in 1980.
4. I remember, for example, that a reference to leadership in a workers' organisation being easily corrupted in the *Grassroots* article years before had provoked puzzlement amongst unionists I spoke to.
5. Food and Allied Workers Union *v* Spekenam Supreme (2), *Industrial Law Reports*, 9, 4 (1988), pp. 628–39.
6. Food and Allied Workers Union *v* Spekenam Supreme (1), *Industrial Law Reports*, 9, 4 (1988), pp. 627–8.
7. See minutes, FAWU NEC, 28–29 April 1988.
8. Minutes, NEC, 6 May 1988.
9. Statement by the national office-bearers regarding the special national conference, 28 May 1988.
10. Sydney Mufamadi conveyed the invitation from SACTU for a delegation representing both sides to travel to Lusaka.
11. The telex was sent from the telex machine of Dairymaid. Minutes, HOCC, 17 June 1988.
12. 21 branches voted to reverse the NEC's decision, and to reinstate the REC decision to dismiss Richard, and 3 voted against. The position of 4 branches was unclear. See minutes, special national conference, 25–26 June 1988.
13. Minutes, ANC, 23–25 September 1988.
14. That was probably also why the NACTU leadership developed cold feet, and declined to participate in the summit. For an uncritical account, see Baskin, *Striking Back*, p. 319.
15. The criticism that Nduzulwana expressed was directed at the fact that the summit would only include affiliates of NACTU and COSATU, and that this was a decision imposed by the leadership undemocratically. See minutes, special NEC meeting, 20 March 1988 and Report on special NEC meeting on 28 March 1988 compiled by Elliot Nduzulwana. On the workers' summit generally, see also Baskin, *Striking Back*, pp. 316–43.
16. Letter dictated to Lesley London, dated 13 June 1989.
17. COSATU had been represented at two high-level meetings about negotiations while I had been in the Transkei. The first was with community organisations in the MDM to arrive at a common position 'in the event of negotiations materialising' regarding a new constitutional dispensation, such as had led to Namibia's independence. See 'Report of meeting between COSATU and community based organisations', undated, 1989. At the second, with the ANC NEC in June 1989, it transpired that negotiations had already started.
18. A telex from NUMSA head office was circulated within the Union, recording that

there was a global consensus that there had to be negotiations, which included the US and the UK, the USSR, as well as the front-line states. See telex from NUMSA head office to Eastern Cape and Border regions, dated 13 June 1989.

19 Annual Report, FAWU, 1989.
20 The box was headed 'The spring of 1919: The birth of the Communist International', and quoted what Lenin had to say about the historical significance of the Third International: '[it has] ... discarded its opportunist, social-chauvinist, bourgeois and petty bourgeois dross, and has *begun to implement* the dictatorship of the proletariat.'
21 As indicated in chapter 3, Frank Marquard was the first president of FCWU.
22 The issue of affiliation was debated at the 1986 conference, where it was decided to refer the matter to the branches for further discussion, and the NEC was mandated to take a decision. See minutes, annual national conference of 10 and 11 October 1986, and 1987 Annual Report.
23 There is a fairly cryptic record of the reasons for the decision taken at the NEC meeting of 4 August 1989. However, the 1989 Annual Report deals with these reasons more fully, including the fact that NACTU's FBWU was affiliated to the IUF. The same report goes on to record how a 'unity meeting' with FBWU in April 1989 had failed.
24 These are among the core policies of what we now know as neo-liberalism. 'Privatisation, deregulation and decentralisation' were identified as strategies that were being adopted to counter the emergent union movement, in the meeting between COSATU and other organisations about building the MDM.
25 FAWU Annual Report, 1989, p. 22.
26 One of the appendices to the 1989 Annual Report was a document headed 'IUF' containing, among others, undated documents by SACTU on the international trade secretariats (ITSs) and the importance of taking a non-aligned position.
27 It can be argued that taking over worker organisations or otherwise controlling them is consistent with the role that the Party actually plays in its alliance with the ANC, namely to secure its Left flank against dissidents.
28 There were proposals to amend the constitution made by the head office and other proposals made by branches. My understanding is that all these amendments were adopted, although this is not clear from the minutes in my possession. See minutes, national conference, 12–15 September 1989.
29 The letter had not reached Nduzulwana, and been forwarded to the head office. It was first tabled at the NEC. See minutes, NEC, 4 August 1989.
30 Minutes, ANC, 12–15 September 1989.
31 For example, an undated document entitled 'Cape Town branch report', which was written in defence of the national leadership of the Union, states that the source of problems in the branch was 'their [the soon-to-be-ousted branch leadership's] political beliefs or practice'.
32 Political resolution, annexed to 'Statement by the "Campaign for Democracy" in FAWU over the problem in FAWU Cape Town branch', submitted under cover of a letter dated 30 April 1990 to all affiliates of COSATU. See also founding affidavit by Miles Hartford dated 20 March 1990, in proceedings between himself and five others and the Union and five others filed under case no. 3247/90 in the Supreme Court, Cape Town.
33 A letter from the branch secretary objecting to the validity of the general meeting was sent to the national office-bearers beforehand. The minutes of the NEC

meeting that approved the outcome of this meeting make no reference to this and other objections to the validity of the meeting. See minutes, NEC, 23–25 February 1990.
34 The grounds on which Miles was dismissed seem to have been patently spurious. The main ground appears to have been that, as one of three signatories to the bank account, he had signed a number of blank cheques in advance prior to his going on leave. This was fairly standard practice. I did it myself.
35 Mtiya was also the well-regarded chairperson of the Western Cape region of COSATU.
36 In this position, Ngcuka would notoriously decide that although there was a prima facie case of corruption against Jacob Zuma, he would not be prosecuted.
37 'Statement by the "Campaign for Democracy" in FAWU over the problem in FAWU Cape Town branch', submitted under cover of a letter dated 30 April 1990 to all affiliates of COSATU. There is an extract from minutes of AGM of Cape Town branch of 4 February 1990 annexed to this statement, showing that 71 votes were cast in the vote for the chairperson of the branch.
38 The documents I refer to were headed 'Cape Town branch report', undated, and 'Cape Town branch under attack: From who?', dated 5 March 1990.
39 Nduzulwana is referred to as 'that evil man'. In order to make a relatively inconsequential point, the same document says about Miles: 'Comrades, this is something Mr [surname] must know very well because he is an educated person from the University but he was deliberately trying to mislead workers. He was hoping that we as workers will be ignorant enough to fall for his trick.' See 'Cape Town branch under attack: From who?'
40 Pamphlet issued by FAWU headed 'FAWU taken to court', circa May 1990. In fact, the court proceedings had been withdrawn some time before the five workers were expelled.
41 Letter, general secretary, FAWU, dated 16 May 1990.
42 A different version of this meeting is presented in the biannual report at the 1993 national conference, which states that the article appeared before the meeting, and the meeting was arranged in response to the article. This is incorrect.
43 This was R650,000 for the building and R120,000 for renovations, i.e. a total of R770,000. Completely different figures are presented in the biannual report to the 1993 national conference. In the copy that was passed on to me, it states that the 'Premier Group sold the hostel to the union at R56,000. The union paid to Premier a sum of R _____.' The blank space is filled in in ink as being R80,000. Presumably delegates were informed verbally that this was the amount.
44 See annexures to the 1989 Annual Report and minutes, annual national conference, 12–15 September 2012. In my article in the *Labour Bulletin* I estimated the building was worth between R3 and R5 million. The minutes of the conference record 'unanimous support for the building of the headquarters of the union and also the contribution of R3 per member', yet I doubt whether there was ever a serious intention to collect such a contribution, or that they did so.
45 The Union apparently told workers that I had got the facts wrong. I do not know which facts I was alleged to have got wrong, and the Union did not ever dispute what I said in print or raise it with me in person.

Chapter 25: The desolate market
1 I am referring specifically to the three decades following the Second World War.

See T Picketty, *Capitalism in the 21st Century*, Belknap Press, 2013, pp. 96–9.
2 The Accord was named after the building where the Department of Labour has its head office. An essential element of the Accord was labour's acceptance of one of the controversial 1998 amendments: that a Labour Appeal Court headed by a Supreme Court judge would preside over the settlement of disputes.
3 The two of them soon fell out, however, leading to yet another split. Kawa was killed in a motor car accident not long afterwards.
4 Power-sharing implied no fundamental change to the existing order. See Department of Political Education discussion paper, May 1991. I assume this refers to the Department of Political Education of the ANC or the Party.
5 Friday Mabikwe died in 2012.
6 See www.rhodesfruitgroup.com. The canning operation of RFF is now taking place in Limpopo and Swaziland and, since 2010, also at Tulbagh.
7 At some point Langeberg also sold the East London cannery, which now operates as Summerpride Foods. Collondale Cannery is also still operating. The last time I heard, both factories were unorganised.
8 I wrote up a case study about the project George was involved in and the endeavour to turn him into an entrepreneur. See J Theron, 'From workers to entrepreneurs', Occasional Paper, Labour Law Unit, University of Cape Town, 1995. Ironically, his job as a translator was with the CCMA, which decided after some years that it would be cheaper to engage translators as contractors rather than to employ them.
9 Evidence of ZP Kondile, Amnesty hearing, 12 October 1998, Truth and Reconciliation Commission. See http://www.sabctrc.saha.org.za/documents/amntrans/johannesburg/.
10 Labour Relations Amendment Act, 2 of 1983. See MSM Brassey and H Cheadle, 'Labour Relations Amendment Act 2 of 1983', *Industrial Law Journal*, 4, 1 (1983), pp. 34–7.
11 There are various definitions of standard employment, but all refer to someone who is employed on a full-time basis and for an indefinite period, i.e. permanently. Non-standard employment includes part-time employment, and a major contributor to the increase of part-time employment at this time was the extension of shopping hours.
12 It could be argued that the Union was now too big to have its conference every year, yet no more workers attended these conferences than had before.
13 Own notes, Visit to Ceres, 6 August 1990.
14 Although farm workers (with hardly any exceptions) were unorganised, farms in Ceres and elsewhere were shedding jobs at a no less rapid pace than in manufacturing.
15 In the case of branches that rented their own offices and employed their own officials, the percentage retained by the branch could be larger, and in practice the head office did not enforce payment of an affiliation fee.
16 Own notes, Visit to Ceres, 6 August 1990.
17 Susan Goliath was one of the persons with whom I had stayed in regular contact. I formally interviewed her on 23 May 2012.
18 General secretary's report to the FAWU national conference, 5–9 July 1993.
19 This is according to press reports. See *Mail & Guardian*, 16–30 July 1993: 'Tensions and divisions mar FAWU Congress'.
20 I cannot see that Joe Slovo's belated attempt to distance the Party from Stalinism

made the slightest difference to the way its members behaved in practice. See J Slovo, 'Has socialism failed?', 1990.
21 General secretary's report to the FAWU national conference, 5–9 July 1993. See also the 1991 biannual report, in which members are told that 'it is important to keep our union as it is ... following Marxist-Leninist teachings'. The report also lists which leadership figures in FAWU occupied positions in Party structures.
22 Interview, 23 May 2012.
23 Among the front-line states, Julius Nyerere of Tanzania is a prominent example.
24 See Songezo Zibi, 'Unembargoed: SA is crying out for unwavering, ethical leaders', *Business Day*, 27 July 2015.
25 The National Economic Development and Labour Council was established in terms of Act 35 of 1994.
26 Workers in the industry were at this point represented by SACTWU, formed as a result of a merger between GAWU and ACTWUSA. According to Johnny Copelyn, Manuel called upon the COSATU contingent in Parliament (of which Copelyn was part) to condemn a protest by SACTWU against his decision. The COSATU contingent backed Manuel and it seemed the protest fizzled out. See Copelyn, *Maverick Insider*, 2016, pp. 323–4.
27 See chapter 17.
28 I believe the strike resulted in a disastrous defeat for the Union. There is an account of the strike in Baskin, *Striking Back*, p. 417.
29 This was in 1999, the same year in which Old Mutual, having demutualised, listed on the London Stock Exchange. The minister responsible for the SAB decision was Alec Erwin, as Minister of Trade and Industry.
30 See FAWU and others *v* SA Breweries Ltd, 2004. *Industrial Law Journal*, 25, p. 1979. The case was brought on behalf of the retrenched workers at Newlands brewery. Although the Union won the case in the Labour Court, the workers did not go back to the brewery. Instead they accepted packages by way of a settlement. Presumably if they had not done so, SAB would have appealed the judgment.
31 See Copelyn, *Maverick Insider*, p. 317. On a list of the richest persons in the country, Johnny Copelyn was number 34, above Graham Mackay, the MD of SAB who has since died, but below Marcel Golding, another trade unionist turned capitalist. See *Sunday Times*, 7 December 2014: 'The Rich List'.
32 I am not able to develop the argument for worker co-operatives adequately here. See J Theron with M Visser, '"Remember me when your ship comes in": Co-operatives and the need to shift from a wage culture', *Development and Labour Monograph 1 of 2009*, University of Cape Town, 2009.
33 Report of an investigation commissioned by the NEC of FAWU, by Clive Thompson, 27 November 1996. It is not clear from this report whether the general secretary had acted in his trade union capacity or in his capacity as a member of the board of the investment company.
34 The board is composed of representatives of business, organised labour and government, along tripartite lines.
35 There is a version of the events that took place described in the Industrial Court case of SACCAWU and others *v* Irvin and Johnson Ltd, NH 11\2\4298 (CT), 31 December 1997.
36 The CCMA did not have the same latitude to make rules as the Industrial Court had had. Although there was still an 'unfair labour practice' in terms of the new LRA, it had a restricted meaning.

37 Ramaphosa was number 161 on a list of the richest men in the country. See *Sunday Times*, 7 December 2014: 'The Rich List'.
38 I was told this by a senior Premier manager who clearly had personal knowledge of this decision, and obviously did not agree with it. This was some time after the group had been taken over, as explained below. Probably he does not want to be named.
39 *Cape Times*, 16 October 1997, cited in J Theron, 'Terms of empowerment', Occasional Paper, Institute of Development and Labour Law, University of Cape Town, 1998.
40 When Premier was in fact taken over, its new bosses would complain about paying 10 per cent more than their major competitors, while smaller mills were paying a quarter of the Premier wage. See General Food Industries Ltd *v* FAWU (2004) 25 ILJ 1260(LAC) at 1266.
41 Section 198, Act 66 of 1995. What is essential about this provision is that the labour broker becomes the employer of workers he or she provides to a client, as opposed to merely being an intermediary.
42 For a more detailed account, see my paper, 'Terms of empowerment', *SA Labour Bulletin*, 23, 1 (1999), pp. 6–14. John is given the pseudonym Mathew Booi in these pieces.
43 The beauty of this – from the bakery management's perspective – was that they could keep them on a 'temporary' basis indefinitely, since the LRA did not define what was 'temporary' about the service. However, amendments to the LRA adopted in 2014 now limit the period for which workers can be employed by a labour broker.
44 According to Ann Crotty, the rump of the milling and baking business was sold 'for not much more than the cost of one of Premier's mega milling facilities'. See *Cape Times* Business Report, 28 October 1999.
45 Secretary's report, 1999 biannual conference; see also General Food Industries Ltd *v* FAWU (2004) 25 ILJ 1260 (LAC).
46 *Sunday Independent*, 8 March 1998. 'COSATU chiefs step in to patch up bitter feud in divided food workers' union', by William Gumede. Both this and a subsequent report claim that Ray Alexander gave her blessing to the formation of the new union.
47 *Business Day*, 19 January 1999. 'Union boss faces disciplinary hearing over funds', by Themba Hlengani.
48 Writer's own notes, 29 January 1999. I subsequently heard that this group went ahead to form a separate trade union and told workers I was backing their initiative.
49 There are no statistical data that reflect this, because the Quarterly Labour Force Survey of Statistics SA is based on household surveys, and collects data per sector. Employment in services such as cleaning and 'temporary employment services' are almost certainly captured as employment in services, even though these services would not exist but for the primary or secondary sector (manufacturing, in this instance) that generates them.
50 Lawyers have terms like the phrase 'cascading outsourcing' to describe this phenomenon. It should be noted that in the case of Atwell's and the mill, I have not been able to establish whether Staffgro's wages were lower than Workforce's.
51 The Minister of Labour is empowered in terms of the Basic Conditions of Employment Act of 1997 to publish sectoral determinations for a sector,

setting minimum wages and conditions of work. The fact that the first sectoral determination was for this sector, and it was described as 'contract cleaning', is testimony to the lobbying by the employers' association concerned.
52 The fact that there was a minimum wage set by the sectoral determination did not preclude workers from demanding a higher wage.
53 I was told so by someone who had formerly been a senior official of the Union, at Lizzie Phike's funeral.
54 Secretary's report, 1999 biannual conference. The officials dismissed in the period of two years covered by the report included Mandla, Ernest Buthelezi, the assistant general secretary, and Wagiet. In Mandla's case proceedings were still pending. Both Buthelezi and Wagiet were largely successful in cases against the Union, and secured substantial payouts.
55 Ray Alexander's speech to the FAWU national conference, 26–30 July 1999.
56 Both Mandla and Dereck lost their unfair dismissal applications.
57 See chapter 2.
58 Organisational rights can only be exercised in the workplace. In terms of section 213 of the LRA, the workplace is defined as the place(s) where the employees of the employer work.
59 See Employment Equity Act (55 of 1998). For my fuller analysis of its failure to address workplace inequity, see J Theron, '*Plus ça change*: Reinventing inequality in the post-apartheid workplace' in O Dupper and C Garbers (eds.), *Equality in the Workplace*, Juta, 2009.
60 See SA Municipal Workers Union *v* Ekurhuleni Municipality, Labour Court (J1120/11), dated 31 May 2012.
61 See Report of the September Commission on the Future of the Unions to COSATU, 1997 (the 'September Commission'), pp. 5–6. 'Democratisation of the workplace' is one of the purposes of the LRA. See section 1, LRA of 1995.
62 See September Commission, pp. 1–2.
63 'Labour regulation' is a somewhat amorphous category, encompassing both legislation and policies. Amendments to labour legislation tabled in 2000 were widely perceived to be aimed at rolling back provisions that employers saw as onerous. They were adopted in a watered-down form in 2002, by way of amendments to the LRA and BCEA. In 2005, a further major overhaul of labour legislation was mooted by the Deputy Finance Minister, Jabu Moleketi, reputedly at the behest of President Mbeki. See *Business Times*, 22 May 2005: 'Economic overhaul underway'.
64 Research work that my colleagues and I did for the Department of Labour showed that small businesses did indeed face real difficulties in complying with aspects of the new labour laws. Yet there was no evidence to suggest that small business created jobs.
65 This was the adoption of the 'presumption as to who is an employee' in terms of section 200A of the LRA of 1995.
66 See SA Transport and Allied Workers Union and Congress of South African Trade Unions *v* Jacqueline Garvas and Others, Constitutional Court Case 112/11, 13 December 2012.
67 Undated letter, faxed to author on 10 April 2007.
68 At the time of writing, the SA Democratic Teachers' Union was implicated in a scandal involving the 'sale' of teaching posts. See the Ministerial Task Team report on the 'jobs for cash' investigation, 20 May 2016, www.dbe.gov.za.

69 This was at a workshop on labour broking convened by the research group at the university that I co ordinated at the time.
70 Interviews were conducted as part of my research work for the university with workers of LKB and LJP (the labour broker in question) in November 2007.
71 'Why SA's chief executives are coining it', *Mail & Guardian* Business, 19–25 June 2015.
72 This has been documented in Greg Marinovich's book on the Marikana massacre. See G Marinovich, *Murder at Small Koppie: The Real Story of the Marikana Massacre*, Penguin, 2016, pp. 65–75.
73 Clauses 5.8 to 5.15 of the recognition agreement detail these and other provisions regarding full-time shop stewards. See Recognition Agreement between Lonmin Platinum comprising Eastern Platinum Ltd and Western Platinum Ltd and the National Union of Mineworkers, 2011.
74 In 2014, according to COSATU's own data, more than half the members of its trade union affiliates earned more than R5,000 per month, whereas half of all workers in employment earned less than R3,500 per month. See pamphlet, COSATU's Vulnerable Worker Task Team, 'Join the drive to organise all vulnerable workers', 2014. This income differential would be even larger if the social wage were taken into account (e.g. medical aid and retirement provisions).
75 In Patrick Craven's account, the extent of the reflection appears to have been a resolution adopted at COSATU's eleventh Congress calling for an end to 'the growing social distance between members and leaders'. See Patrick Craven, *The Battle for COSATU: An Insider's View*, Bookstorm, 2016, p. 30. Craven has documented the events from 2012 to 2015 leading to the split in the federation.
76 For the relationship between neo-liberalism and climate change denialism, see Philip Mirowski, *Never Let a Serious Crisis Go to Waste*, Verso, 2013, pp. 334–42. See also Melanie Klein, *This Changes Everything*, Penguin, 2014, pp. 120–60.

Index

Abrahams, Liz 8, 74, 82–4, 177, 195, 221, 226, 297, 362
   banning order for 74
   death of 362–3
   general secretary of FCWU 74, 220–1
   role in establishment of UWO 259
Ackerman, Raymond 286
Adams, Annie 44
Adams, Hester 47, 51, 106, 341
African Food and Canning Workers Union (AFCWU) 43, 47, 49, 73, 77–8, 83, 88–9, 92, 96, 103, 112, 119, 123–5, 163, 165, 204–5, 208, 210–11, 229–30, 233, 238, 243–4, 252, 261, 279, 290
   as affiliate of SACTU, 73, 153
   Cape Town branch of 89–90, 225–6
   East London branch of 286
   formation of 50
   funding of 290–1
   Johannesburg branch 92–3, 100, 105, 226
   Johannesburg Conference (1978) 97–8
   Mbekweni AGM 113
   members of 44–5, 50, 76, 85–6, 88–9, 98, 100–1, 108, 110, 146, 160, 195, 289, 309
   Paarl branch of 113, 147
   Port Elizabeth branch of 226
   recognition issues 160
   leadership role in East London 175–6
   unification with FCWU 293–5
African National Congress (ANC) 6, 9, 24, 27, 136, 151, 154, 160, 235–6, 244, 247, 267, 270, 274, 288–9, 296, 298, 300, 327, 330, 335, 348, 364
   as banned organisation 24, 73, 112
   ideology of 25, 307
   members of 9–10, 112, 327, 329, 347
   Polokwane Conference 360
   unbanning of (1990) 6
   Women's League 362
   Youth League 362
Afrikaans (language) 15–16, 45, 48, 63, 70, 72, 133–4, 248, 271
   press 37, 102
   teaching of 36
   translation 143
Afrikaner establishment 20, 35, 40, 50, 55–6
   Afrikaans businesses 176
Aggett, Neil 114, 176, 188, 226, 229, 239–41, 259, 317
   detention and death of 242–9, 258
Alexander, Ray 8, 30, 38, 46, 65, 148–53, 315, 330, 355–6
   background of 28
   banning of 69
   founder of FCWU 28, 37, 153–4
Amon, Aletta 45
Andrews, Bill 169
Andrews, Ursula 121
Anglican Church 213, 284
Angola 305, 313
   Independence of (1975) 23
apartheid 49, 71, 220, 253, 301, 340
   housing provision 49
   urban planning 132
April, Dominee 107, 262
arbitration
   see dispute resolution
Argus 128

Index

Ashton Canning 54, 55, 57, 342
  workers of 47, 55
Associated British Foods 217
Attwell's Bakery 238, 306, 353, 356
Australia 279
  Melbourne 365
autonomy (independence) 26–7, 201
  general 19
  of branches 51, 344
  of TUs 10, 219
Azanian Confederation of Trade Unions (AZACTU) 299
  role in formation of NACTU 301
Azanian People's Organisation (AZAPO) 284, 296, 301

Bakery Employees' Industrial Union 227–8
  complaints made by 238
  shortcomings of 234
baking industry 109, 227, 234, 286, 306–7, 352
Balfour, Mabel 96
  banning of (1963) 74
  Transvaal Secretary of AFCWU 73, 94
banning order 32, 64, 74, 241, 261
Bantu Affairs Administration Board (BAAB)
  offices of 45, 50
  inspectors of 157
Bantu (Urban Areas) Act
  provisions of 142
bargaining councils 92
Barlow Rand Group 213, 217
Baxter, Susan (Moosa, Munaadiah) 108, 345
Berg River Textiles 45
Bernadt, Hymie 280
Biehl, Amy 99
Biko, Steve 21, 135, 203
  death of 191
Black Consciousness 135, 164–7, 229, 256, 261, 301
  organisations 284
Black Municipal Workers' Union (BMWU) 231, 255
Black Sash 16, 18

Bloch, Jonathan 27
Bloom, Tony 318
  Chairperson of Premier Food 217
Blue Ribbon Bakery 352
Boesak, Allan 138–9
Boland Inmaakwerkers Vakvereeniging (Vakbond) 133–5, 202–3, 225, 326
  application for registration 177–8, 180, 194
  members of 177, 225
  opposition to 153, 176, 237, 326
Booi, Stanford 157
Bophuthatswana Development Corporation
  partnership with Premier Foods 262
Botha, Fanie 153, 228, 231
  Minister of Labour 132
  Minister of Manpower Utilisation 202
  view of unregistered trade unions 206
Botha, P.W. 189, 296
  administration of 289, 293
Botswana 21
BTR Sarmcol 314
Burger 128
Buthelezi, Chief 301

Camay, Piroshaw
  General Secretary of CUSA 230, 267
Canner's Association
  see Fruit and Vegetable Canners Association
Cape Areas Housing Action Committee (CAHAC) 268
Cape Chamber of Commerce 141
Cape Employers' Association 139
Cape Herald 128
Cape Times 13, 128, 162–4, 178
Cape Town Municipal Workers' Association (CTMWA) 97, 299
  offices of 264, 309
  membership of 156
Cape United Democratic Front
  proposals for creation of 269
Carnation Foods 261

427

Carolus, Cheryl 268–70
capitalism 21, 339, 365
Catholic Church 145, 261
Cele, Dereck 355
Ceres Fruit Growers (Growers) 168,173,184–5
  personnel manager of 308
  Strike (1980) 174,185–6, 221
Ceres Fruit Juices 79
  dispute at 86–7
Chemical Worker's Industrial Union
  dispute with SASOL 292–3
Cillié, Judge Piet 84
Cillié Commission 83–4
class 15, 220
  middle class 15, 25, 26–7, 88, 92, 119–20, 131, 167, 262
  working class 23, 26, 251, 254, 292,359, 365–6
    fragmentation of 3–5, 224, 291, 342, 365
Clover Pietermaritzburg 313–14
codes of conduct 319
  Sullivan Code 55–6
Cold War 93
  end of 329
collective agreements
  agreement for fruit and vegetable canning industry (the agreement) 46–7, 49–50, 53, 74–6, 95
  closed-shop 208
  recognition agreements 208, 210–11, 218–19, 232–33
  'cut-off' point 276
collective bargaining
  centralised negotiations, 210
  direct wage negotiations 169, 185
  extension of (see extension of collective agreements)
  impact of 277–8
  recognition negotiations 85, 208–9, 219–20,
    negotiation of job grading 214–5
Collondale Cannery
  workforce of 205
  Strike (1980) 202, 205
  Coloured Preference Area Policy 49–51, 102
  impact of 55, 110

Commercial, Catering and Allied Workers Union of South Africa (CCAWUSA) 351
  members of 230, 241
  offices of 300
  OK Bazaars strike 314
  splitting of 314
Commission for Conciliation, Mediation and Arbitration (CCMA) 351, 354, 356, 360
communism 24, 27
Communist International 328
Community Action Committee 141,145
Community House 325
conciliation board
  see dispute resolution system
Congress Alliance
  members of 297
Congress of South African Trade Unions (COSATU) 10–11, 296, 303–4, 320, 326, 340, 355, 357
  affiliates of 304, 315, 329, 342, 345, 351, 359, 363
  congress of (1997) 315–16, 328
  constitution of 364
  formation of 5
  inaugural conference of 299–300, 306
  Freedom Charter debate 316, 328
  leadership of 8–9, 300–1, 304–5, 309, 313, 320, 333, 361
  negotiations with SACCOLA 6
  role in Laboria Accord 340
Congress of the People (COPE) 362
Conservative Party (UK) 23
Constitutional Court 359
Cooper, David 24
Cooper, Di 316
Copelyn, Johnny 350
  General Secretary of NUTW 262, 317
co-operatives 35, 56, 173, 348
  formation of 350, 354
Council of Unions of South Africa (CUSA) 232, 252, 265–6, 287
  affiliates of 230, 256, 267
  position on non-racialism 300
  role in formation of NACTU 301

Crawford, Athalie 90–1, 110, 112, 119–20, 144, 129, 237, 259, 305
Cronin, Jeremy 8–10, 12
  trial of (1976) 27
Curtis, Jeanette 32
Curtis, Neville 32

Daily Dispatch 178
DairyBelle 282–3
  as subsidiary of Imperial 282
Daniels, Norman 105
Deepfreezing and Preserving (Deepfreezing) 53, 278
  Macassar factory 176
  retrenching of workforce 225
Del Monte Corporation
  subsidiaries of 55
Democratic Alliance (DA) 9
democratic centralism 269–70
Department of Labour 46, 61, 76, 87, 105, 125, 138, 174, 202
  offices of 229
  staff of 79, 118, 123, 257
Department of Manpower Utilisation 174, 240
disciplinary procedures 221
dispute resolution system 74, 79, 156, 221
  compulsory vs. voluntary arbitration 222, 281–4
  in terms of LRA (1956) 280
    conciliation board 86–8, 120, 137
  in terms of recognition agreements 208, 221–3
  in terms of LRA (1995) 351
  mediation 156, 158–60
  strikes and lockouts 222
disputes of right 126, 281–3
dismissal 162, 174
  at will 169
  remedy of reinstatement 161, 280
  retrenchment 175, 177, 207, 225, 277–9, 286, 287, 289, 291, 294, 314, 349, 352–3, 355, 358
  right to hearing 48–9, 221
  status quo remedy 279–80, 321
  victimisation 116, 122, 167
Dlamini, Chris 9, 287, 298–9, 305–7, 311, 319, 323, 333, 345
  Deputy President of COSATU 301
  full-time shop-steward 311–12
  imprisonment of 299
  President, FOSATU and SFAWU 287
Dönges, Eben
  family of 40
Du Toit, Hennie 263, 319
Durban Strikes (1973) 22, 31, 41, 193
Dutch Reformed Mission Church 107, 138

East London Strikes (1980–1) 175–8, 184, 191–2, 202, 233, 239, 260
  negotiations during 192, 194, 211
Elgin Fruit Packers (Elfco) 173, 183–4
  workforce of 185–6
employee 40, 59, 103, 219, 222–4, 277, 279, 292, 349, 358–9
  definition of 31, 133
  right to join TUs 31
employment 4–5, 40, 50–1, 59, 61, 64, 122, 143, 156, 161, 171, 208, 223, 262, 285, 291, 304, 339, 342, 349, 353, 356–7, 359, 365
  non-standard 343
  standard 343
empowerment 5, 264, 315, 319, 349, 354
  black economic 347–8, 350
  first wave of 352
  role of education in 308
Engel, Virginia 90–1, 110, 112, 129, 144, 244, 259, 269
Epic Oil 331
Epol group, 276, 286–7
  black managers at 318–19
  East London 293–4
  Isando 310
  Pretoria West dispute 275–6, 281–3
Erntzen, John
  General Secretary of CTMWA 97, 299
Erwin, Alec
  General Secretary of FOSATU 160
European Economic Community
  donation to COSATU 306

Eveready
  boycott of 156
extension of collective agreements 92, 95
  conciliation board 86–8, 120, 137
  industrial councils 80, 111
externalisation 347, 349–50, 353–4, 356–7, 359

Fassie, Brenda 322
Fatti's & Moni's 83, 109, 122, 144, 157, 183, 189, 209–10, 213–15, 238, 252
  acquired by Tiger Group 266
  AFCWU approaching of (1979) 116
  AFCWU members at 111, 113
  disciplinary procedures used by 221
  Isando mill 145, 187, 207
  support committee 145, 246
  workforce of 92, 121, 139–40, 156, 223, 225, 228, 282, 334
    Strike (1979) 126, 128, 132, 135, 144–5, 162–3, 166, 169, 175, 177–8, 207, 212, 218, 340
    boycott of 130, 137–9, 141–3, 146, 158–9, 163
    re-employment of workers during 144, 158–9, 162
Federation of South African Trade Unions (FOSATU) 155–6, 165, 232, 252–4, 265, 287–8, 298, 300, 313, 345
  affiliates of 167, 187, 228, 233, 246, 266–7, 269, 292, 310
  organising strategies of 266
  pamphlet on Neil Aggett 243
Fedfood Group
  subsidiaries of 278
Ferrus, Hennie 109, 155,
  funeral of 235–6
fishing industry 213
  inshore 61–2, 78, 106
flexibility, 223–5, 358
Farlam, Ian 148
Food and Allied Workers' Union (FAWU) 3, 11–12, 36, 332, 360
  formerly FCWU 9
  inaugural conference of 311
  Leadership Code 11, 323, 331
  members of 3–4, 11, 231, 315
  relationship with SACP 10, 329–30, 343–6
  self-sufficiency of 6
  strike (1956) 35, 90
FAWU Investments (Pty) Ltd 350
  personnel of 355
Food and Canning Workers Union (FCWU) xi, 32, 41, 44, 46–7, 55–6, 59–60, 63, 70–3, 78, 81–2, 89, 95, 103, 120–3, 136, 143, 157, 163, 170, 172, 177–9, 187, 228–30, 252, 259, 288, 293, 306, 308–10, 312, 315, 331, 333–4, 339, 344
  Archive xi
  as FAWU 9
  breakaway (1972) 104–5, 111, 113–14, 133–4, 177
  conference (1977) 78–9
  constitution of 307
  Dal Josaphat (Dal) branch 36–7, 43, 45–6, 181, 193–5
  forfeiture of registered status (1980) 202
  founding of (1941) 28, 135, 153–4
  Grabouw Conference 335
  HOCC of xi
  Johannesburg branch of 83
  MCM of xi, 38, 45
  management committee (MCM) xi, 38, 45, 57, 262
  members of 28–31, 35–7, 43–4, 47–9, 58, 69, 78–9, 84–5, 98, 101–4, 146, 152, 160, 194–7, 211, 289, 320–1
  membership criteria and verification 214–17, 301
  national executive committee (NEC) xi, 38, 293–4, 323, 325, 328, 334
  New Life 29–30, 43
  Paarl branches of 43, 45, 47, 50–1, 70, 101–2, 196, 202
  relationship to SACTU 73, 153, 155
  splits in 50–1
  structure of 51
  translation issues in meetings 143–4
  unification with AFCWU 293–5

Ford Motor Company
  Struandale Plant 167
  Strike 231
Food & Beverage Workers Union (Food Bev) 230
Food Preservers' Union of Australia 279
Foster, Joe 253, 262, 269
  General Secretary of FOSATU 246
freedom of association 126, 132–3, 160, 208, 276
  recognition of 231
Freedom Party 72
fruit and vegetable canning industry 61, 105, 176, 182–3, 195–7, 201–2, 342
  FCWU members working in 37
  industrialisation 64
Fruit and Vegetable Canners Association
  29, 61–2, 79, 239
Fruit and Vegetable Canning Workers Medical Benefit Fund (Union's medical fund) ix ,27, 65

Gants Foods 53
Garment and Allied Workers' Union
  formation of 311
Garment Workers' Union
  see Western Province Garment Workers' Union
General Allied Workers' Union (GAWU) 264–5, 270, 300
  as UDF union 275
  members of 231, 288
  in unity talks 267
General Food Industries (Gen Food) 354
German Democratic Republic (GDR) (East Germany) 150
Germany
  Fall of Berlin Wall (1989) 329, 339
Global Financial Crisis (2007–9) 365
Goliath, Susan 308, 315, 344, 346
Gomas, Johnny
  expulsion from ANC 9
Goncalves, Hester 61
Good Hope Bakery 130–2, 143, 161
  workforce of 141

  defections to 158, 164
Golden Arrow
  fare increases sought by 197
Government Gazette 110
Gqweta, Thozamile 193, 206, 231–3, 241, 250, 264
  President of SAAWU 193
Graaff, Sir De Villiers
  leader of United Party 16
van Graan, Manie 163, 211, 225, 341
  President of FCWU 146
Grabouw strikes 182–6
Grassroots 253–4
Group Areas Act 319
Guardian 30
Gxanyana, Mandla 260–1, 263–4, 272, 294, 327, 334, 345, 355
  General Secretary of FCWU 325

H. Jones and Co. 35, 45–6, 51, 134, 176, 279, 362
  acquired by Langeberg 225
  application of Coloured Preference Area Policy at 50–1
  as subsidiary of Picardi 40
  breakaway 294
  FCWU members working for 37, 49, 103–4
  Strike (1941) 30, 40, 164
  workforce and management of 8, 40–1, 72, 74, 113–15
Hambridge, Maria 251–2
Hartford, Miles 305, 332
Heath, Edward
  proposed industrial legislation under 23
  resignation of (1974) 23
van Heerden, Auret 249
Hendricks, Gertie 121
Hindson, Doug 244, 251
Hogan, Barbara 145, 246
Holomisa, Bantu 325, 347
  expulsion from ANC 347
Hout Bay Action Committee 285
Howa, Hassan
  President of SACOS 163–4

Irvin & Johnson (I&J) 213–15, 218, 306, 351–2

membership verification 216–17
Springs plant 229–30
Woodstock factory 171, 179, 351
Immorality Act 44
Imperial 218, 309
subsidiaries of 217, 282
Independent Mediation Service of South Africa (IMSSA) 283, 345, 351
Industrial Conciliation Act (1956) 31
as in force in 1976 133
as amended in 1979 133
as amended in 1981 (Labour Relations Act of 1956), 238
industrial councils 61, 187, 232
agreements 80, 111, 228
bargaining in 32
establishment of 126
industrial court 223, 242, 278, 281, 283, 313, 322
impact of 278, 280
recognition of right to hearing 221
industrial tribunal 87, 280
influx control 142
Inkatha 145, 296, 343
International Confederation of Free Trade Unions (ICFTU) 256, 267
funding provided by 234
International Labour Organisation (ILO) 126
International Union of Foodworkers (IUF) 227, 329–30
funding of 234
isiXhosa (language) 63, 70, 104, 144, 176, 191–2, 205, 271–2
translation of 110, 143, 145, 220
Issel, Johnny 101–2, 119–20, 134–6, 235

Jungle Oats 306

Kawa, Sithembele 325–6, 340
Kellogg's 311–12
van Kerwel, Jean 121
Kets, Johan 117–18, 282,
Kgosana, Philip 13
Khumalo, Donsi 256–7
Kikine, Sam 241

General Secretary of SAAWU 205, 236
Kilowan, Abdol 41
Kilowan, Nellie 41, 43, 46, 49, 71–3
Komose, Deborah 273
de Klerk, F.W.
unbanning of political organisations (1990) 6
Krom River Apple Co-operative (Kromco) 173, 183
workforce of 184
Strike (1980) 184
KSM Queenstown 284
KWV 35

Laboria Accord 340
labour 124, 175, 339
broker 287, 353, 355, 360–1
dispute 86, 275–6, 281–3, 292–3
labour rights 209, 221
legislation 117, 132, 166
organisation 30
relationship 169, 209, 220, 222, 357
Labour Appeal Court 358
Labour Court 352
Labour Party (South Africa) 102, 109, 119, 155, 254, 262, 289
members of 39, 285
voter registration efforts of 274–5
Labour Party (UK) 23
Labour Relations Act (LRA)(1956) 237
amendment of (1983) 343
amendment of (1984) 275, 279, 287
amendment of (1988) 317
Labour Relations Act (LRA)(1995) 313, 340, 353, 356
provisions of 348
Laing, RD 23–4
Lamprecht, André 282–3, 284
Land Harvest 220
ownership of 217
Langeberg 57, 175, 177, 189, 192, 240
acquired by Tiger Foods
acquisition of
H. Jones and Co 225
Standard Canning 56

application of Coloured Preference
  Area Policy at 50–1
as subsidiary of Standard Canning
  56
Ashton plant 57, 75, 106
Boksburg plant 99–100, 108, 176,
  188, 237, 239
Daljospahat plant (Moberg's) 36,
  45, 48, 51, 75, 82–3, 182–3,
  271, 362
  closure of 225
  FCWU members working at 37, 49,
    104
  funeral of manager's son 259
East London plant
  strikes at 192, 260
    breakaway at 294
  Montagu plant 54
  relations with FCWU 176, 179–
    80, 86, 195
Worcester plant 38
Leatt, Dr James 157
Legal Resources Centre 142–3
Lenin, Vladimir
  What Is to Be Done? 84
Leon, Sonny
  leader of Labour Party (South
    Africa) 39, 254
Lesotho 45
Lewis, David 31, 71–2, 229, 234, 334
liaison committee 41, 121, 170–1, 191
  disbanding of 191–3
  members of 121, 192, 207
  potential use in containment of TUs
    85
Liberty
  as major shareholder of Premier
    Food 318
Liebenberg, Sandy 339
Life 13
London, Lesley 326
Lozah, Elijah 155
Luthuli, Chief Albert
  President of ANC 329

Mabasa, Mordecai 320
Mabikwe, Friday 124–5, 139, 145–6,
  160–1, 164, 252, 265, 332, 334,
  340–1, 366

Machel, Samora 289
MacKenzie, Annie 60–1, 63
Mafeje, Archie
  withdrawal of appointment by
    University of Cape Town 19
Mafikeng, Elizabeth 8
  President of AFCWU 45
Mafumadi, Sydney 10–11, 12, 232,
  257, 264, 320
  Assistant General Secretary of
    COSATU 300
  General Secretary of GAWU 231,
    288
Mahlakahlaka, Bernard 273
  Chairperson of AFCWU at Land
    Harvest 217
Mahlangu, Isaac 334
Malepe, Peter 325, 332
  Vice-President of FCWU 310, 326,
    332
management 40–1, 48–9, 53, 57, 82,
  84, 104, 114, 117–18, 121, 129,
  148–9, 157, 161, 169, 171, 173–6,
  179, 183–5, 188, 207–8, 210, 214–
  20, 234, 239–40, 252, 260, 276,
  282–84, 340, 343, 356, 361
  centralised vs. decentralised 352,
    344
  lower 192, 278, 280
  black management 56, 276, 219–
    20, 318–19
  personnel vs. HR 353–4, 357
  plant 282
Mandela, Nelson 361
  autobiography of 18
Manuel, Trevor 268–70
manufacturing industry 40, 365
  labour-intensive 279
Manyosi, Moffat 46, 182
Marais, Dawid 18
Marie, Bobby 261–2
Marikana Massacre 312
  political impact of 363
Marquard, Frank 329
  President of FCWU 29–30
Marxism 247–8
Mashinini, Emma
  General Secretary of CCAWUSA
    230

Mass Democratic Movement (MDM) 296, 316
Matanzima, Kaiser 155
Mateman, Don
  General Secretary of TWIU 94–5
Mati, Lulamile 294
Mavi, Joe 250
Mazwi, Gatsby 231, 255-56
Mbeki, Thabo 12, 361
  criticisms of Jeremy Cronin 9–10
  sacking of Jacob Zuma 10–11
Médecins Sans Frontières (MSF) 157
mediation
  see dispute resolution
Meloi, Daniel 121
Mentoor, Annie 130, 163, 174
Mercedes-Benz 204
Metal and Allied Workers' Union (MAWU) 262
  affiliate of TUACC 94
  members of 244, 301, 313
Mfengu 124
Mhlaba, Raymond 328
migrants 7, 24, 66, 231, 343, 366
  efforts to prevent joining TUs 159
  Zulu 301
Marcus, Klaas 173–4
milling industry 23, 212–13
mine workers Strike (1987) 9
  as legal strike 314
Mlambo, Joseph 276, 278
Mlokoti, Clarence 219–20
Moberg's
  see Langeberg Daljosaphat
Mogoatlhe, Israel 345
Molatsi, James
  President of NUM 314
Moni, Peter 207–8
  Managing Director of Fatti's & Moni's 121, 145
Moodley, Mary 93–5
  secretary of FCWU Johannesburg branch 83
Moodley, Intharin 272
Moosa, Munaadiah 345
Morobe, Murphy 316
Motor Assembly and Component Workers' Union of South Africa (MACWUSA) 231, 255, 263, 267, 272

GWUSA 264, 267, 272
  establishment of 256–7
  members of 257
Motsoeneng, Sebei 334
Mozambique 149–50
  Independence of (1975) 23
  Maputo 94, 150, 154
Mpemvushe, James 191–2, 285
  detaining of 191
Mpetha, Oscar 8, 50, 63, 77–8, 84, 96–7, 104, 111–13, 124, 146, 152–3, 160, 165, 169, 175, 190–1, 193, 197, 289
  detaining of 197, 202
  family of 346
  General Secretary of AFCWU 73, 98
  targeted by SB 196
Mtiya, MacWellington 332
Mugabe, Robert 257
Muslim News 128
Mvubelo, Lucy 160
Mxenge, Griffiths 205
Mxhanto, Mzamo 142
Mzozayana, Welile 192

Naidoo, Indres 151, 154
Naidoo, Jay 261, 305–6, 307
  General Secretary of COSATU 301
Namibia 59, 328
National Automobile and Allied Workers' Union (NAAWU)
  as affiliate of FOSATU 269
National Council of Trade Unions (NACTU) 326
  formation of 301
  role in creation of Laboria Accord 340
National Democratic Revolution (NDR) 307, 313
  as two-stage theory 25
  objectives of 310
National Development Plan (NDP) 365
National Economic Development and Labour Council (NEDLAC) 356
National Party (Nationalists/Nats) 15–16, 18, 30, 131, 258
  constitutional 'reform' of 289

industrial policies of 31
members of 14, 40, 172
nationalist political project of
  15–16, 18
political rhetoric of 133
Transvaal 239
National Union of Clothing Workers
policies of 160
National Union of Metalworkers of
  South Africa (NUMSA) 360, 363
National Union of Mineworkers
  (NUM) 256, 363
as COSATU affiliate 316
members of 256, 314
National Union of Motor Assembly
  and Rubber Workers of South
  Africa (NUMAROSA) 93
as FOSATU affiliate 167
opposition to recognition of 156
National Union of South African
  Students (NUSAS) 20, 25
congress delegates 21
Wages Commission 22
National Union of Textile Workers
  (NUTW) 262
as affiliate of FOSATU 269
members of 262, 269, 317
nationalism 16
Ndinisa, Luska 8, 104, 134, 146, 342
Nduzulwana, Elliot 325–6, 331, 340
van Neel, Lilian 285
neo-liberalism 10, 355, 358, 365
Nestlé 83
Ngcuka, Bulelani 332
  National Director of Public
    Prosecutions 332
Ngoma, Joshua
  attempted expulsion of 284–5
Njikelana, Sisa 203–4, 231, 241, 250
Nkadimeng, John 297–8, 330
  General Secretary of SACTU 297
Noko, Alfred 342
  President of AFCWU 275
  Vice-President of AFCWU 146,
    179
non-racialism 83, 136, 165–6
opposition to 300
Norushe, Bonasile 182, 189–90,
  235, 273

Oakglen Canning 48
closure of 225
office-bearer 6, 36, 44, 251, 301
distinguished from official 44
national 89–90, 113, 146
official 100–1, 109, 114–15, 119, 124,
  134–6, 164, 166–7, 172, 179, 184,
  193, 202–3
branch 89, 344
paid 44, 47, 51–2, 58, 72, 74–6, 84,
  86, 108, 113, 170, 188, 193–4,
  262, 265, 267–8, 290, 303, 305,
  308, 311–12, 314–15, 326, 345
salaries of 303, 347
Old Mutual 217, 318
as major shareholder of Tiger
  Group 318
Olivier, Jan 117–18
organisational rights 356
development of 348
disputes 351
under LRA (1995) 348
Orderly Settlement of Black Persons
  Bill (Disorderly Bill)
opposition to 268
outsourcing
see externalisation

Pan Africanist Congress (PAC) 98–9
activists of 109
anti-pass campaign of 13
militia associated with 50
unbanning of (1990) 6, 328
Patel, Leila 259
Pendlani, John 50, 76–7, 91, 104–5,
  163, 217, 238, 257, 275, 294, 297,
  326–7
  President of AFCWU 50, 78
Picardi 40
  H. Jones and Co. as subsidiary of
    40
  H. Jones & Co
  Brink Brothers (Montagu plant) 54
Pici, John 4–6, 227–8, 238, 341, 343,
  353, 355, 359–60
background of 4–5
death of 366
empowerment of, 235–60
Pick n Pay 130, 224, 286, 314

sales pitch of 224–5
Phike, Lizzie 8, 104, 113, 146, 175–6, 193, 294, 305, 310, 345
   detention of 302
   funeral of, 362
Pickard, Jan
   background of 40
Pieterse, Nosey 340
Pityana, Sipho 257
populism 284–5, 301, 310
Port Elizabeth Black Civic Organisation (PEBCO) 167
Poto, Victor 155
Poqo
   Paarl march (1962) 50
Portugal
   Carnation Revolution (1974) 23
   Lisbon 24
   military of 23
poultry industry 108, 115–16
   halaal certification 118
Premier Food 213, 215, 277, 280, 318, 353–4
   dispute resolution system used by 222
      hostel of 6–7, 318, 334–5
      management philosophy of 214–15, 218
   divisions of 215, 227, 332
   FCWU representation 252
   ownership of 217, 250, 318, 352
   retrenchment in 286, 352
      partnership with Bophuthatswana Development Corporation 262
President's Council 254, 257, 268
   Reports of 262
Progressive Party 16
Protestantism 75
Rainbow Chickens 108–9, 276
   acquisition of Epol 286
   outcome of charges by 140, 144, 148
   policy toward TUs (1979) 116
   strike at 115–19, 128, 144, 221

Ramaphosa, Cyril 257, 266–7, 298–9, 352, 363
   General Secretary of NUM 256

Rand, Barlow 281–2, 304
Rand Daily Mail 178
Rapport
   Ekstra (supplement) 102–3
Rasool, Ebrahim 12
   Premier of Western Cape 8
Ray Alexander Union Centre 36
Reagan, Ronald 10
   electoral victory of (1980) 223
Real Africa
   shares held by 350
recognition
   of TUs 85, 93, 206–9
   agreements (*see* collective agreements)
Red Robin
   Strike 60–1
registration 105, 133, 153, 177–80, 194, 202–4, 206–7, 228–30, 232, 244, 255, 293
   certificate of 37
   opposition to 156, 165–6, 237–8, 275, 293–4
reinstatement
   see dismissal
Rembrandt 36
Retail and Allied Workers' Union (RAWU) 309, 322–4, 341
   members of 323
revolutionary change
   vs. reform 24
Rhodes, Cecil 48
Rhodes Fruit Farms (RFF) 47–8, 56, 106, 177, 342
   closure of 225
   FCWU members working at 37, 146
Riotous Assemblies Act 118, 148
River, Berg 36–7, 58, 76
River, Hex 99
River, Olifants 58
River, Salt 84
River, Touws 175
Roberts, Alan 11
Robertson Spices 261
Rogers, Ellen 'Lossie' 8–9, 63–4, 106

SA Breweries (SAB) 307, 311, 318, 349, 352
   externalisation in 349

subsidiaries of 311, 348
SA Dried Fruit (SAD) 87, 108–9, 210
   Worcester Factory 108
SA Inshore Fishing Industry Association (Inshore Association)
   setting of wage scale 61–2
SA Institute of Race Relations
   offices of 26
SA Transport and Allied Workers' Union (SATAWU) 359–60
   acceptance of contract cleaning 359–60
Saaiman, Spasie 121, 125, 161
SASOL
   dispute with Chemical Worker's Industrial Union 292–3
Sea Harvest 63, 169, 171–2
   workforce of 8, 106
   shareholders of 217
Second World War (1939–45) 14
Sepedi (language) 188
September, Reggie 25, 346
September Commission 363
Shangaan (language) 188
Sharpeville Massacre (1960) 38–9
   media coverage of 13
Shoprite
   acquisition of OK Bazaars 314
Sigcau, Stella 347
Simba-Quix 278
Sisulu, Walter 327
Small, Adam 28
Smuts, General 15
Smythe, Annie 93
Sobukwe, Robert 190
   launch of PAC's anti-pass campaign 13
socialism 24–5, 30, 310, 364–5
Solidarity (Polish trade union) 234
Solomon, Polly 134, 182, 341
South Africa 5, 13–15, 21–3, 148, 224, 259, 311
   Bophuthatswana 262
   Cape Colony 191
   Cape Province 112
   Cape Town 4, 9, 13, 25, 35, 39, 43, 64, 70, 83, 90–2, 95, 98, 109, 111–12, 140, 149, 156, 164, 176–7, 189, 204, 225–6, 228, 232, 262, 286, 299, 305, 310, 322, 328, 331, 341
   Cape Flats 216, 268, 366
   Grand Parade 27, 43
   Gugulethu (township) 131, 145
   Newlands 307
   Nyanga (township) 65, 84, 99, 109, 131, 142, 157
   Ciskei 142–3, 157–8, 203–4, 211, 273, 279
   Delmas 296
   Durban 22, 72, 193, 205, 236, 243, 249, 260–2, 293–4, 301, 319–20, 341
   Eastern Cape 15, 36, 112, 179, 213, 220, 324, 327, 329, 333, 340
   East London 175–9, 182, 189, 191, 203–4, 211, 229, 231, 236, 270, 279, 284, 294–5, 326
   Molteno 278
   Port Elizabeth 93, 156, 167, 190, 217, 220–1, 231, 255, 260, 262–3, 272, 297, 299
   Eastern Province 175
   first democratic elections (1994) 5, 346
   Free State 100
   Bloemfontein 293
   government of 149, 169
   Gugulethu 7, 341
   Johannesburg 43, 72, 83, 93, 95–6, 120, 124, 144–6, 181, 187–91, 214, 226, 230–1, 237, 240, 243, 254, 300
   Karoo
      Willowmore 36
   Laaiplek 58
   Mobeni 205–6
   Natal 118, 150, 311, 323, 333
   North West Province
      Klerksdorp 289, 341
      Potchefstroom 289
   Parliament 7, 16, 172, 274
      House of Representatives 274
   Pietermaritzburg 296, 313
   Polokwane (Pietersburg) 318
   Pretoria 16, 93, 229, 252, 257, 275, 341

Sebokeng 292
Sharpeville 292
Somerset West 53, 130
Soweto 292, 298
   Kliptown 264
Transkei 4, 77, 98, 155, 241, 260, 321, 327, 347, 349, 358–9, 366
Transvaal 16, 84, 145, 188, 216, 226, 246, 296, 323
Western Cape 9, 26, 40, 49, 61, 93, 95, 118–19, 129, 167, 188–9, 228, 238–9, 244, 252, 254, 271, 293, 324, 327
   Ashton 54, 57, 75, 274
   Beaufort West 293
   Bellville 110–11, 130, 187, 207
   Caledon 183
   Ceres 55, 168, 174, 290, 344
   Durbanville 110–11
   Genadendal 183
   Gouda 55
   Grabouw 54
   Hopefield 106–7
   Laaiplek 76
   Lambert's Bay 58–60, 62–3, 76
   Mbekweni 45, 49, 95, 113, 302, 362
   Montagu 54, 57
   Paarl 8, 25, 35, 38–40, 47, 49, 53, 58, 70, 105–6, 111, 113, 120, 125, 130, 133, 147, 179, 202, 226–7, 326, 344, 362
   Riviersonderend 183
   Saldanha 106, 120, 168, 185
   Saron 55
   St Helena Bay 76–7, 152
   Stellenbosch 121
   Tulbagh 54, 96, 168, 211, 275
   Vredenburg 106
   Wellington 37, 44–5
   West Coast 58–9, 62, 64, 98, 217
   Wolseley 55, 168, 175
   Worcester 38, 54, 79, 88, 99, 108–9, 118–20, 140, 148, 168, 235, 341, 345
   Zwelethemba 116
South African Allied Workers Union (SAAWU) 23, 206–9, 239–40, 252–3, 260, 267, 270, 300
   as UDF union 275
   Langa Summit 232–3, 235–6, 239, 255
   members of 193, 203–5, 211, 231, 236, 241
   offices of 203, 205
South African Broadcasting Corporation (SABC) 296
South African Commercial, Catering and Allied Workers Union (SACCAWU) 351, 355
South African Communist Party (SACP) 8, 30, 248, 346
   members of 8–9, 30
   relationship with FAWU 10
   unbanning of (1990) 6
South African Congress of Trade Unions (SACTU) 7, 87, 105, 149, 152, 236, 247, 256, 288, 298, 324
   affiliates of 155, 264, 329
   FAWU visit to 330
   founding members of 73
   relationship to emergent unions 154, 297
South African Consultative Committee on Labour Affairs (SACCOLA) 326
   role in creation of Laboria Accord 340
South African Council of Churches 146
South African Council on Sport (SACOS) 163
South African Labour Bulletin 7, 333–4
South African Milling Company (SA Milling) 213–14, 218, 223, 252, 306
   dispute resolution provision 222–3
   Isando mill 215
   recognition negotiations 218–9
   Salt River mill, 213–4, 225
South African Party (SAP) 15–6
South African Preserving Company (SAPCo) 54, 223, 342
   as subsidiary of Del Monte Corporation 55
   personnel of 56, 96, 146
South African Students' Organisation

(SASO)
   establishment of 20–1
   members of 21, 135
South West African People's Organization (SWAPO) 328
Southern African Clothing and Textile Workers' Union (SACTWU) 348
Soviet Union (USSR) 28, 30, 234, 248, 329
   collapse of (1991) 339, 343
Soweto Committee of Ten 145
Soweto Uprising (1976) 31, 64–5, 190
   fatalities during 38–9
   state responses during 65
Special Branch (SB) 24–5, 39, 65, 89, 94–5, 133, 136, 148, 179–80, 237, 239, 241, 243, 250, 272, 302
   personnel of 77
      agent provocateurs 128
      recruitment efforts 144
      questioning conducted by 189
      targeting of individuals by 196
Spekenam 35, 83, 91
   dispute at 321–23, 330, 341, 361
   leadership during 330, 332, 342
   first contact with 90
Stalin, Josef 30
Standard Canning
   acquired by Langeberg 56
   subsidiaries of 56
stay-away 43, 46, 64, 292, 296
stop order facilities 177
   introduction of 46–7
Sunday Times 236, 243
Supreme Court 84, 142, 157, 266, 313
   local divisions of 258
Swaziland 148–9, 154
Sweet Food & Allied Workers' Union (Sweet Food) 230, 287, 293, 303, 306–7, 312
   AGM 306, 308
   membership profile 308
   organisation of 307

Table Top 107, 123, 216–17, 233, 365
Temba Bakery 262
Terrorism Act (1967) 32
   opposition to 18
   Section 6 242

textile industry 53, 80, 90, 348
Textile Workers' Industrial Union (TWIU) 105
   as affiliate of TUCSA 268
   members of 94
Thatcher, Margaret 289
   electoral victory of (1979) 131, 171
Theron, Ernest 332
   President of FAWU 345
Thornton, Amy 120
Tiger Group 213, 218, 352
   acquisition of Fatti's & Moni's 266
   attitude of 285
   Meadow Feeds 214
      Oceana 217, 350
   shareholders of 318
Tiger Oats 342
Tilney, Corder 203
Tiro, Abram
   NUSAS representative for Turfloop 21
Trade Union Advisory Coordinating Council (TUACC) 92–3, 95, 99
Trade Union Council of South Africa (TUCSA) 31, 65, 109, 114, 153, 227, 250, 340, 343
   affiliates of 90, 94, 156, 252, 268
   dissolution of (1986) 311, 313
trade union representative 85, 87, 119, 219
   shop steward 38, 46, 78, 96, 113, 121, 191, 216, 333, 341, 343, 351, 361
      full-time shop-steward 266, 310–12, 314–15, 326, 335, 345, 357, 363–4
trade unions (TUs) xi, 3, 7, 12, 31–2, 41, 73, 87, 92–3, 95, 126–7, 132, 151–2, 201, 218–19, 246, 253, 268, 312, 317, 342, 348, 360, 366
   and pension/provident funds 317–18
   apartheid in 133, 228–9
   autonomy of 10, 219
   community 253
   craft TUs 31, 127
   critiques of 146, 201–2
   decentralised vs. centralised model of organisation 51, 71, 89, 260,

263, 265, 273, 303–4, 344, 352
  emergent 85, 93, 127, 132, 135, 153, 156, 178, 214, 231–2, 234, 243, 246
  federations 89
  head office controlling committee (HOCC) xi
  industrial vs. general TUs 264–5
  investment companies 348, 350, 352, 355, 364
  non-racial 234–5, 237
  parallel TUs 30–1, 43, 65
  self-sufficiency of 6, 364–5
  unregistered 31, 165, 206, 234
trade union unity
  Athlone meeting 264–5
  Ipelegeng meeting 298–300
  Johannesburg meeting (1979) 160, 164, 166
  Johannesburg meeting (1981) 229–30
  Langa summit 231–3
  Port Elizabeth meeting 255–7
  Wilgespruit meeting 253–4
Truth and Reconciliation Commission (TRC) 347
Tutu, Desmond 156
  Secretary General of South African Council of Churches 146

unfair dismissal 223
  disputes 351, 354
  right to a hearing 48–9
Umkhonto we Sizwe (MK) 259, 289
Union of South Africa (1910–61)
  Constitution of 17
United Democratic Front (UDF) 273, 284–5, 287, 302, 322
  affiliates of 285, 298–300
  as banned organisation 296
  formation of xi, 9, 269–71
  members of 274, 316
  Million Signature Campaign 274–5
  repression of 296
United Kingdom (UK) 22–3, 131
  London 15, 22–4, 27, 151, 158
  mine workers strike 289
United Nations (UN) 126
United Party

  members of 16
United States of America (USA) 54, 169, 339
  Central Intelligence Agency (CIA) 324
United Women's Organisation (UWO) 270
  establishment of 259
  members of 268
United Workers' Union of South Africa (UWUSA)
  formation of 301, 313
Unity Movement
  ideology of 69
University of Cape Town
  Libraries
    FCWU Archive (Special Collections, UCT) xi–xii
  students of 19
  Students' Representative Council (SRC) 19–21
University of Limpopo
  as Turfloop 21
University of the Western Cape 28
University of the Witwatersrand 145

vanguard party 95, 233–4, 330
Vavi, Zwelinzima 9–10, 12
  suspension of 364
Verwoerd, Dr Hendrik 49
  administration of 14
  assassination of (1966) 18
  leader of National Party 14
victimisation
  see dismissal
Victoria, Queen 306
Vorster, B.J. 254
  Minister of Justice 14, 18

wages 75, 80, 87–8, 108, 208, 291–2, 307, 342
  differentials 172
  farm 173
  minimum 75, 83, 210
  piecework 61–2
  reduction of 74
  women's 81
Wages Commission 80
Western Cape Trader's Association

(WCTA) 129, 132, 139–40
  members of 129, 131, 139
Western Province African Chamber of Commerce 132
Western Province Garment Workers' Union 90, 167, 311
Western Province General Workers Union (WPGWU) 165, 229, 233
  formation of 164
  offices of 237
Western Province Preserving (WPP) 175, 278, 286
  closure of 293
Strike (1980) 190–1
Western Province Workers' Advice Bureau 31–2, 64, 230, 348
  as General Workers' Union (GWU) 230, 262–3, 269, 295, 316–17, 334
  formation of WPGWU 164
  organisation models promoted by 32
Whitehead, Lt Steven
  interrogator of Neil Aggett 246–7
Wiehahn Commission (Commission of Enquiry into Labour Legislation) 85, 229, 254, 275, 281
  Report of 87, 93, 106, 108, 124, 126, 132–3, 154
Wilgespruit Fellowship Centre 254
Wilson-Rowntree 260
  boycott 270
Wolthers, Wolly 214–15, 218, 250
women 91–2, 106
  as unskilled workers 40–1
  representation in Vakbond 134
  wages of 81
worker
  contract workers (migrant workers) 109, 141
  driver 276
  fisher 350
  meaning of 22
  owner-driver 349
  permanent 224
  seasonal workers 362
  semi-skilled 40–1
  temporary 224
  unskilled 40–1

watchman/security guard 7, 19, 48–9, 84, 98
worker control 257, 345
  black 257
worker education
  programmes of 308
worker organisations
  see advice offices, trade unions, worker co-operative
workerism 10, 270, 310, 312, 316
workplace 4, 31, 41, 47, 49, 51, 72, 75, 84–6, 109, 116, 159, 161, 168–9, 187, 192, 195, 202, 207–9, 215, 218–19, 221–3, 234–5, 239, 252–3, 277, 284, 290, 293, 304, 311–14, 335, 353, 356–7, 360
  definition of 37
  forums 79, 303
  hierarchy 40, 105, 115, 122, 263, 355
World Federation of Trade Unions (WFTU) 329
Wrighton, Peter 318–19
Wynand, Magrieta 170, 180

Xashimba, George 321, 323, 332, 359
  chairperson of Spekenam 342
Xhosa (ethnic group) 206
Xulu, Maxwell
  Treasurer of COSATU 301

Young Christian Workers 261
Yon, Wilma 28, 38,
  banning of 64–5

Zambia 149
  Lusaka 324
Zimbabwe 257, 298
  first democratic elections (1980) 189, 201
  War of Liberation (1964–79) 201
Zimbabwe African National Union – Patriotic Front (ZANU-PF)
  founding of (1987) 189
Zini, Government 255
Zulus (ethnic group) 301
Zuma, Jacob 154, 323, 361, 363–4
  sacked as Deputy President 10–11